BASICS OF BIOTECHNOLOGY

BASICS OF BIOTECHNOLOGY

By

Dr. A. JAYAKUMARAN NAIR

M.Sc., Ph.D.

Department of Biotechnology

University of Kerala, Thiruvananthapuram

UNIVERSITY SCIENCE PRESS

(An Imprint of Laxmi Publications Pvt. Ltd.)

BANGALORE • CHENNAI • COCHIN • GUWAHATI • HYDERABAD
JALANDHAR • KOLKATA • LUCKNOW • MUMBAI • RANCHI
NEW DELHI • BOSTON, USA

Published by :
UNIVERSITY SCIENCE PRESS
(*An Imprint of Laxmi Publications Pvt. Ltd.*)
113, Golden House, Daryaganj,
New Delhi-110002

Phone : 011-43 53 25 00
Fax : 011-43 53 25 28

www.laxmipublications.com
info@laxmipublications.com

First Edition : 2004
Reprint : 2008

OFFICES

India

✆ **Bangalore**	080-26 61 15 61
✆ **Chennai**	044-24 34 47 26
✆ **Cochin**	0484-239 70 04
✆ **Guwahati**	0361-254 36 69, 251 38 81
✆ **Hyderabad**	040-24 65 23 33
✆ **Jalandhar**	0181-222 12 72
✆ **Kolkata**	033-22 27 37 73, 22 27 52 47
✆ **Lucknow**	0522-220 95 78
✆ **Mumbai**	022-24 91 54 15, 24 92 78 69
✆ **Ranchi**	0651-221 47 64

USA

Boston
11, Leavitt Street, Hingham,
MA 02043, USA

UBB-9237-195-BASICS OF BIOTECHNOLOGY
C—15569/08/03

Typeset at : Goswami Printers, Delhi.
Printed at : Repro Knowledgecast Limited, Thane

PREFACE

Biotechnology as a fast growing applied science has already impinged everyone's day-to-day life. It is one of the major technologies of the twenty-first century like IT. Its multidisciplinary activities has made it everybody's science. By this time biotechnology has given its impact in almost all areas of public life such as Public Health, Pharmaceuticals, Food and Agriculture, Industry, Bioenergetics Energy sector and information technology. Now it is very clear that biotechnology is not only a key technology for the 21st century, but also the science of the future. It has potential to ensure food security, dramatically reduce hunger and malnutrition, and reduce rural poverty particularly in the developing countries like India. Considering its commercial potential and its possible impact on the economy, government of India has taken a number of measures to build up trained human resource in biotechnology and promote research and development and its commercial aspects in the country. The introduction of biotechnology as a subject discipline in undergraduate levels in all universities of the country and in school at senior secondary level by CBSE are such initiatives to bring the subject to the basic level and make the students aware of the subject, its techniques, its economical potentials and its socio-political and ethical considerations.

The multidisciplinary activities of the biotechnology include recombinant DNA techniques, gene cloning, genetics and genomics, cell biology, application of microbiology, bioprocess and downstream processing, basic biochemistry and proteomics, cell culture, hybridoma technology bioinformatics and information technology. **'Basics of Biotechnology'** is prepared mainly for those students who pursue biotechnology as a subject for undergraduate courses including B.Tech. and B.Sc of all Indian universities. It follows the syllabus prescribed by UGC and the major universities that run this course. It covers all the essential fundamental aspects of modern biology such as biochemistry, molecular biology, genetics, cell biology, biotechnology as an industry and its socio-economical and ethical aspects in a very comprehensive manner. Reading this text book will give the learners an idea about the relationship between the subject and health, nutrition, agriculture, environment, industry etc. and will understand different application of biotechnology in everyday life. Basics of Biotechnology will be very helpful in developing interest in the beginners to study biotechnology as a discipline and pursue higher studies and research in biotechnology and other professional courses in future.

The book consists of three units and each unit is divided into some chapters. Unit I is divided into four chapters, Unit II consists of two chapters and the Unit III is divided into three chapters. Each of the chapters has been discussed exhaustively as per the requirements of the syllabus and hope that the book will be very useful for undergraduate students and teachers to get basic concepts and understand the techniques of Biotechnology. Suggestions for improvements in the style and content of the textbook, corrections of errors if any will be greatly appreciated from the readers and will be incorporated in future editions.

I am extending my sincere gratitude to Mr. Saurabh Gupta, Director, Laxmi Publications Pvt. Ltd. for inviting me to write the Basics of Biotechnology for the College and engineering students of Indian Universities and IITs. My heartfelt thanks for the encouragement given by Dr. Babu Gopalakrishnan and Rajeev Gopalakrishnan and K.P.Gopalakrishnan Nair throughout my efforts. I am also grateful to Mrs. Chitra Gopalakrishnan and Mr. Kunal for persuading me to take the plunge without hesitation. And of course my wife and my daughter for their prayers and playfulness that turn my house into a home. Finally, I want to thank Laxmi Publications Pvt. Ltd. for bringing out the textbook in a very attractive manner.

A. Jayakumaran Nair
jekks@yahoo.com

ABOUT THE AUTHOR

Dr. A. Jayakumaran Nair presently working in the Department of Biotechnology, University of Kerala is an experienced researcher and teacher in Biotechnology. His areas of expertise include isolation and characterization of novel types of genes and proteins, development of therapeutic antibodies and hormones through plant genetic transformations, screening of novel types of bacterial cultures, tissue and cell cultures and computational biology. Dr. Nair was formerly a Senior Research Fellow, Research Associate and Fellow Scientist of the Council of Scientific and Industrial Research (CSIR), Government of India. He was a Research Associate at Indian Agriculture Research Institute (IARI), New Delhi and Centre for Biochemical Research (CSIR), New Delhi. He also worked as Research Associate and Scientist at the Centre for Biotechnology, Jawaharlal Nehru University, New Delhi. He has received 'Young Scientist Award' (1995) at the Kerala Science Congress and 'Young Scientist Award' (1995) from the Department of Science & Technology, Government of India. Dr. Nair has authored many research articles and books and has contributed to numerous national and international seminars and conferences. Dr. Nair holds a Ph.D. in Biochemistry from Regional Research Laboratory (CSIR), Kerala and a M.Sc. in Plant Sciences from University of Kerala.

Select Projects include :

- *Molecular cloning and characterization of snRNA gene U6 from filamentous fungi—Trichoderma reesei and Aspergillus niger.(Centre for Biotechnology, Jawaharlal Nehru University, New Delhi).*
- *Isolation, characterization and expression of CGTase gene from a new isolate of bacillus species. (Centre for Biotechnology, Jawaharlal Nehru University, New Delhi).*

- *Synthesis, cloning and over expression of Human Epithelial Growth Factor (EGF) gene in E. coli.(Centre for Biochemical Technology, New Delhi)*
- *Genetic Engineering of Lathyrus sativus for obtaining Ox-DAPRO neurotoxin free cultivar. (Indian Agricultural Research Institute, New Delhi).*
- *RFLP analysis of Oil Palm Genotypes—Dura, Pissifera and Teneera (RRL, CSIR, Trivandrum).*

SYLLABUS

KURUKSHETRA AND M.D. UNIVERSITIES
BTT-102-E

BASICS OF BIOTECHNOLOGY
(COMMON FOR ALL BRANCHES)

L T P/D
3 1

Theory : 100 Marks
Sessional : 50 Marks
Total Marks : 150 Marks
Time : 3 Hrs.

Unit-I

1. **Cell Structure and Function**
   Prokaryotes and Eukaryotes : Cell Wall, Membrances, Nucleus, Mitochondria, Chloroplast, Ribosome, Vacuoles, Bacteria and Viruses : A brief descriptions.
2. **Biomolecules**
   A Brief account of Structure of Carbohydrates, Lipids, Proteins.
3. **Cell Division**
   Mitosis and Meiosis.
4. **Genes**
   Classical—Brief Idea about Mendel's laws and Chromosomes, Nature of Genetic material, DNA and RNA, DNA replication.

Unit-II

5. **Gene Expression**
   Central dogma, genetic code, molecular mechanism on mutations, regulation of gene expression, housekeeping genes, differentiation and development mutations and their molecular basis.

6. **Genetic Engineering**

An introduction to genetic engineering : Cloning (vectors, enzymes) ; cDNA and genomic libraries, Transgenics, DNA fingerprinting, Genomics.

Unit-III

7. **Development of Biotechnology**

Nature and Scope of Biotechnology.

8. **Applications of Biotechnology**

Bioprocess and fermentation technology. Cell culture. Enzyme technology, Biological fuel generation, Single cell protein, Sewage treatment, Environmental Biotechnology, Biotechnology and medicine, Biotechnology in agriculture and forestry industry, Food and Beverage, Technology Production of Biological inventions, Safety in Biotechnology.

CONTENTS

Dedicated To My Father

Late Sri M. Ananthakrishnan Nair

and

Mother M. Kamalamma

UNIT 1

1 CELL STRUCTURE AND FUNCTION

1.1 CELL—THE BASIC UNIT OF LIFE

Cell is the basic unit of life in all living forms from smallest bacterium to the most complex animal. According to the cell theory all living things are composed of one or more cells. One-celled organisms are called **unicellular** organisms and those with more than one cell in their body are called **multicellular** organisms. Virus particle does not have any cells and therefore is termed as **acellular**. No matter which type of cell we are considering, all cells have certain features in common: **cell membrane, nucleic acids, cytoplasm, and ribosomes.** Cells are small 'sacks' composed mostly of water. The 'sacks' are made from a phospholipid bilayer. This limiting outer membrane, the plasma membrane separates the outer environment from the contents of the cell. It is the communication link between the cell and its environment. The plasma membrane is semi-permeable, allowing some things to pass in or out of the cell and blocking others. Microscopes make it possible to magnify small objects such as cells in order to see the details of their structure. Both light and electron microscopes were used to study cells. Study of cells with a microscope is called cytology. There are some fundamental activities, which are common for most of the cell types from bacteria to the nerve cells in human. The study of these basic cellular processes is called cell biology.

Cells are 90% fluid (cytoplasm), which consists of free amino acids, proteins, glucose, and numerous other molecules. The cell environment (*i.e.* the contents of the cytoplasm, and the nucleus, as well as, the way the DNA is packed) affects the gene expression/regulations, and thus are very important parts of inheritance.

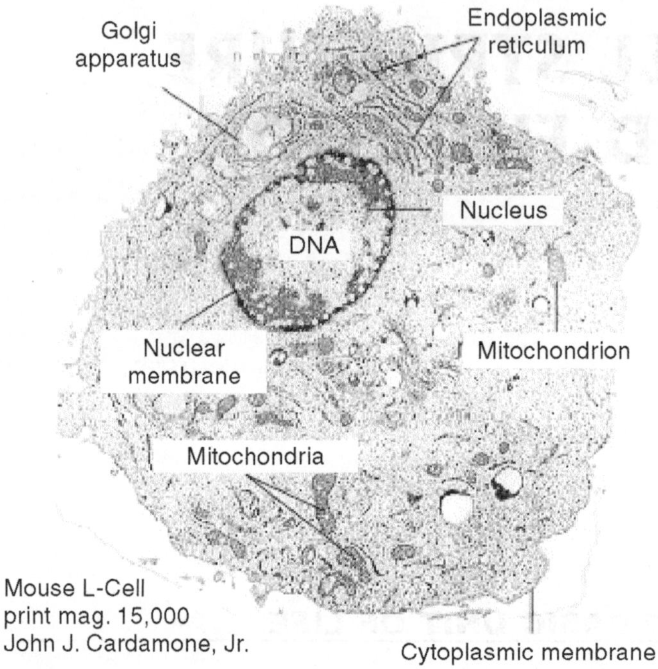

Figure 1.1 Structure of a typical eukaryotic (animal) cell.

1.2 PROKARYOTES AND EUKARYOTES

Based on the microscopic and biochemical differences living cells are basically classified into two major groups—**Prokaryotic cells** or **Prokaryotes** and **Eukaryotic cells** or **Eukaryotes.**

Characteristics of Prokaryotic Cells

Prokaryotes mainly include bacteria and blue green algae (cyanobacteria). Simply stated, prokaryotes are molecules surrounded by a membrane and cell wall. Prokaryotic cells lack characteristic eukaryotic subcellular membrane enclosed 'organelles', but may contain membrane systems inside a cell wall as an extension or infoldings of the cell membrane. The nucleus is not well organized and is with out any membrane. It consists of the genetic material or the chromosome embedded in the middle of the cytoplasm. The term prokaryote indicates this primitive nature of the nucleus. Prokaryotic cells may have photosynthetic pigments, such as is found in cyanobacteria "blue green algae". Some prokaryotic cells have external whip-like flagella for locomotion or hair like pili for adhesion. Prokaryotic cells come in multiple shapes: cocci (round), bacilli (rods), and spirilla or spirochetes (helical cells). All prokaryotes are unicellular organisms and eukaryotes include both unicellular and multicellular organisms.

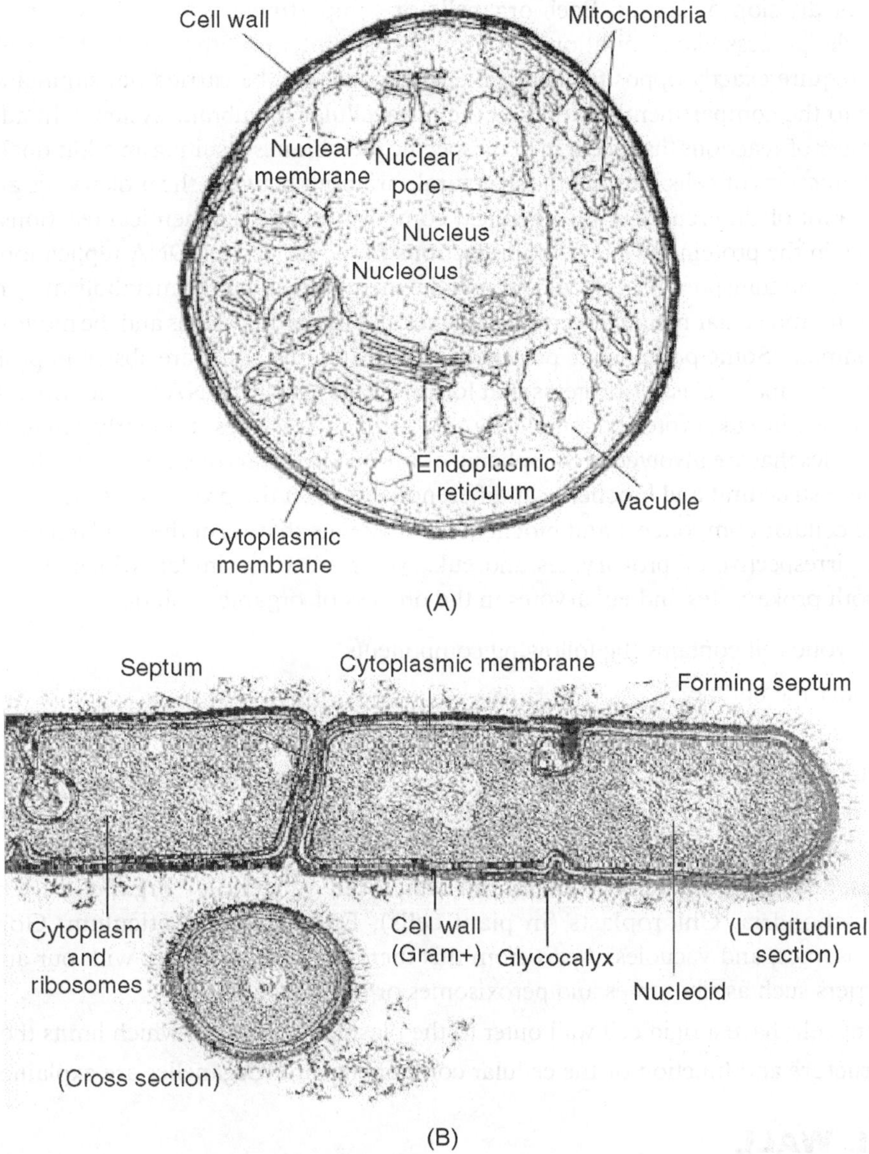

Figure 1.2 Electron micrograph of eukaryotic (A) and prokaryotic (B) cell—a comparison.

Characteristics of Eukaryotic cells

The eukaryotic cells are characterized by the presence of well-defined intracellular membrane systems. They have a well-developed membrane separating the genetic materials from the cytoplasm resulting in the formation of a true nucleus. This membrane system is also responsible for the formation of intracellular compartments and structures. The subcellular compartments established by the intracellular membrane system are known as the organelles, which are responsible for a unique degree of subcellular functional specialization or division of labour. Each organelle or compartment is specialized for carrying out a specific metabolic process like respiration, photosynthesis, protein synthesis etc. Different biochemical reactions that require exactly opposite chemical environment can be carried out simultaneously in the same cell, due to the compartmentalization by the intracellular membrane systems. In addition to this there are a number of reactions that occur in or on specific membranes resulting an additional environment for the distinct functions of cells. Further the structural variations between the prokaryotic and eukaryotic cells, there are a lot of differences in the chemical composition and biochemical reactions. There are a lot of variations in the protein nucleic acid interactions in the process of DNA replication, translation, transcription etc. and transport of molecules across the membranes, energy metabolism, gene expression and its regulations, molecular nature of certain enzymes and its modifications and the mode of interaction with its environment. Some proteins are present in eukaryotes and some are absent in prokaryotes. For example, histones, a major class of proteins that form complexes with DNA for the conversion of DNA in to chromosomes in eukaryotes is completely absent in prokaryotes. Similarly, small nuclear RNA (snRNA) molecules that are involved in the splicing of mRNA in eukaryotes are totally absent in bacteria. Keeping all these structural and biochemical differences between the prokaryotes and eukaryotes there are some basic cellular components and biochemical reactions that form the fundamental character of all living cells, irrespective of prokaryotes and eukaryotes are very similar, which shows a common ancestor for both prokaryotes and eukaryotes in the process of organic evolution.

A typical eukaryotic cell contains the following components :

1. Plasma membrane.
2. Nucleus.
3. Cytoplasm (semifluid).
4. Cytoskeleton-microfilaments and microtubules that suspend organelles, give shape, and allow motion.
5. Presence of characteristic membrane enclosed subcellular organelles, which include mitochondria, Chloroplasts (in plant cells), Endoplasmic Reticulum, Golgi complex, Lysosomes, and vacuoles. Apart from this there are some particles with our any membrane barriers such as ribosomes and peroxisomes or microbodies.
6. Plant cells have a rigid cell wall outer to the plasma membrane, which limits the cytoplasm.

The structure and function of the cellular components and Organelles are explained below.

1.3 CELL WALL

Cell wall is a characteristic feature of plant cells. It is also present in fungal and bacterial cells. Usually it is a rigid envelop present outer to the cell membranes. Because of this plant cells have a specific shape, which is not seen, in the case of animal cells. The plat cell wall is a polymeric carbohydrate composed of glucose units. Starch is also a carbohydrate composed of glucose units. But they differ in the glycosidic linkage that connects two glucose units in the polymer. In starch and glycogen the glucose units are linked by $\alpha(1 \rightarrow 4)$ linkages where as in the case of cellulose it is $\beta(1 \rightarrow 4)$ linkage. Because of this cellulose is unbranched and straight chain and is suitable for mechanical support. Since they are in straight chain the adjacent cellulose molecules are held together by hydrogen bond, which gives a very high tensile strength for the cellulose fibres. In the case of starch the polymer gets a spiral or helical structure, which is suitable for its function as a storage molecule. Figure 1.2 shows the presence of cell wall in the cells both eukaryotic and prokaryotic. But a rigid cell wall is totally absent in the case of animal cell (Fig. 1.1).

1.4 CELL MEMBRANE

1. Cell membranes are selective barriers that separate individual cells and cellular compartments.
2. Membranes are assemblies of **carbohydrates**, **proteins**, and **lipids** held together by non-covalent forces. They regulate the transport of molecules, control information flow between cells, generate signals to alter cell behavior, contain molecules responsible for cell adhesion in the formation of tissues, and can separate charged molecules for cell signaling and energy generation.
3. Cell membranes are dynamic, constantly being formed and degraded. Membrane vesicles move between cell organelles and the cell surface. Inability to degrade membrane components can lead to **lysosomal storage diseases**.
4. **Lipids** of cell membranes include phospholipids composed of glycerol, fatty acids, phosphate, and a hydrophobic organic derivative such as choline or phosphoinositol. Cholesterol is a lipid component of cell membranes that regulates membrane fluidity and is a part of membrane signaling systems. The lipids of membranes create a hydrophobic barrier between aqueous compartments of a cell. The major structure of the lipid portion of the membrane is a lipid bilayer with hydrophobic cores made up predominately of fatty acid chains, and hydrophilic surfaces.

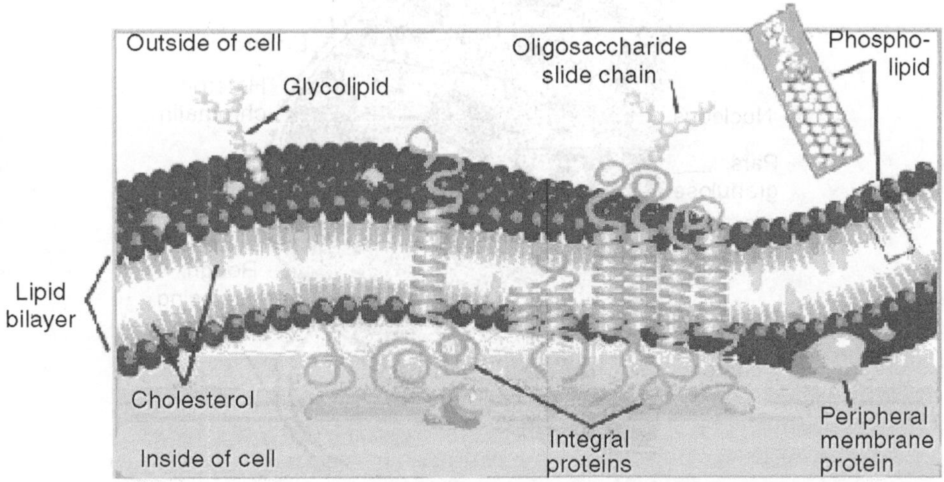

Figure 1.3 Diagrammatic sketch of Fluid Mosaic Model for cell membranes.

5. **Membrane proteins** determine functions of cell membranes, including serving as pumps, gates, receptors, cell adhesion molecules, energy transducers, and enzymes. Peripheral membrane

proteins are associated with the surfaces of membranes while integral membrane proteins are embedded in the membrane and may pass through the lipid bilayer one or more times.

6. **Carbohydrates** covalently linked to proteins (glycoproteins) or lipids (glycolipids) are also a part of cell membranes, and function as adhesion and address loci for cells.
7. The **Fluid Mosaic Model** describes membranes as a fluid lipid bilayer with floating proteins and carbohydrates.
8. **Cell junctions** are a special set of proteins that anchor cells together (desmosomes), occlude water passing between cells (tight junctions), and allow cell-to-cell direct communication (gap junctions).

1.5 CELL ORGANELLES

Cells contain a variety of subcellular particles specialised for carrying out a specific function. Most of them are membrane-enclosed structures. The material that is seen with in the cell enclosed by the cell membrane can be divided into a central concentrated structure known as nucleus and the surrounding semifluid, the cytoplasm. Both cytoplasm and nucleus together forms the protoplasm. The protoplast includes the cell membrane also.

1. Nucleus. The nucleus is the hallmark of eukaryotic cells; the very term eukaryotic means having a 'true nucleus'. It is the largest of the organelles, and is the location of the majority of different

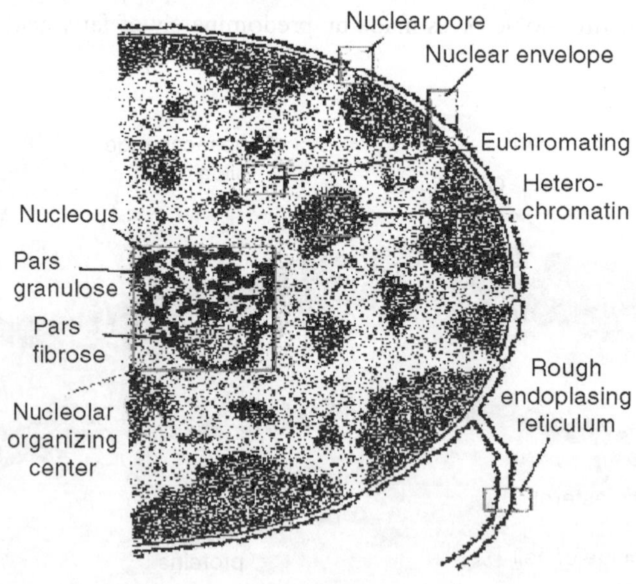

Figure 1.4 Electron micrograph showing the structure of nucleus.

types of nucleic acids. It has got an important role in controlling the shape and features of the cell. Deoxyribonucleic acid, DNA, is the physical carrier of inheritance and with the exception of plastid DNA and mitochondrial DNA (cpDNA and mDNA), all DNA is restricted to the nucleus. Ribonucleic acid (RNA), is formed in the nucleus by coding off of the DNA bases. RNA moves out into the cytoplasm. The nucleolus is an area of the nucleus (usually 2 nucleoli per nucleus) where ribosomes are constructed.

The nucleus is enveloped in a pair of membranes enclosing a lumen that is continuous with that of the endoplasmic reticulum. However, the nuclear envelope is perforated by thousands of nuclear pore complex (**NPC**s) that control the passage of molecules in and out of the nucleus. The nuclear pore is formed by the fusion of outer and inner membrane. The outer membrane is the continuation of Endoplasmic reticulum. The term nucleoplasm is still used to describe the contents of the nucleus.

2. Chromatin. The nucleus contains the chromosomes of the cell. Each chromosome consists of a single molecule of DNA complexed with an equal mass of proteins. Collectively, the DNA of the nucleus with its associated proteins is called **chromatin**. Most of the protein consists of multiple copies of 5 kinds of **histones**. These are basic proteins, bristling with positively charged arginine and lysine residues. (Both Arg and Lys have a free amino group on their R group, which attracts protons (H^+) giving them a positive charge.) Just the choice of amino acids you would make to bind tightly to the negatively charged phosphate groups of DNA. Chromatin also contains small amounts of a wide variety of **nonhistone proteins**. Most of these are transcription factors. (*e.g.*, the steroid receptors) and their association with the DNA is more transient.

During cell division chromatin becomes more condensed and becomes **chromosomes** of specific shapes. During interphase chromatin become dispersed and become very light and cannot be visible very easily (except for special cases like the polytene chromosomes of *Drosophila* and some other flies). However, the density of the chromatin (that is, how tightly it is packed) varies throughout the nucleus :

- Dense regions are called **heterochromatin**
- Less dense regions are called **euchromatin.**

Heterochromatin is found in parts of the chromosome where there are few or no genes, such as centromeres, telomeres, transposons and other 'junk' DNA is densely packed. Those genes present in heterochromatin are generally inactive; that is, not transcribed.

Euchromatin is found in parts of the chromosome that are active in gene transcription; it is loosely packed in loops of 30-nm fibers. These are separated from adjacent heterochromatin by insulators. The loops are often found near the nuclear pore complexes (which makes sense for the gene transcripts to get to the cytosol).

3. The Nucleolus. During the period between cell divisions, when the chromosomes are in their extended state, 1 or more of them (10 in human cells) have loops extending into a spherical mass called the nucleolus. They are the sites of synthesis of three (of the four) kinds of rRNA (ribosomal RNA) molecules (28S, 18S, 5.8S) used in the assembly of the large and small subunits of **ribosomes**. The synthesis of these rRNA takes place in the denser regions of nucleolus known as nucleolar organizer region. (The 5S rRNA molecules are synthesized at other locations in the nucleus.). 28S, 18S, and 5.8S ribosomal RNA is transcribed (by RNA polymerase I) from hundreds to thousands of tandemly arranged **rDNA genes** distributed (in humans) on 10 different chromosomes. The rDNA-containing regions of these 10 chromosomes cluster together in the nucleolus. Once formed, the rRNA molecules associate with the dozens of different ribosomal **proteins** used in the assembly of the large and small subunits of the ribosome. But all proteins are synthesized in the cytosol—and all the ribosomes are needed in the cytosol to do their work—so there must be a mechanism for the transport of these large structures in and out of the nucleus. This is one of the functions of the nuclear pore complexes.

4. Nucleosomes. Two copies of each of four kinds of histones—H2A, H2B, H3 and H4 form a core of protein, the **nucleosome core**. Around this is wrapped 147 base pairs of DNA to form the structures called **Nucleosomes on DNA** chain just like beads of a chain.

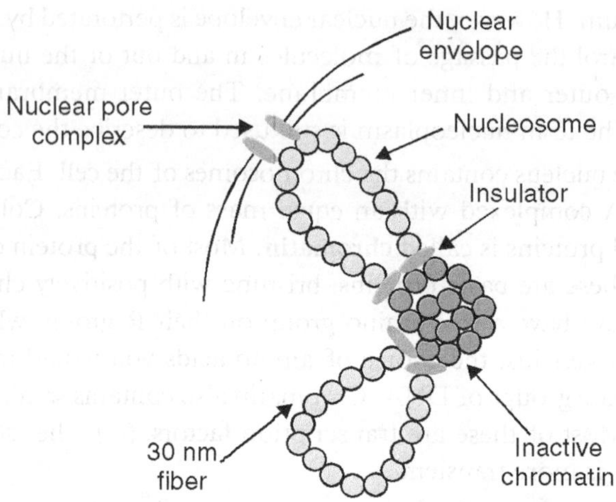

Figure 1.5 Nucleosomes in heterochromatin and euchromatin regions.

1.6 CYTOPLASM

The cytoplasm was defined earlier as the material between the plasma membrane (cell membrane) and the nuclear envelope. Fibrous proteins that occur in the cytoplasm, referred to as the cytoskeleton maintain the shape of the cell as well as anchoring organelles, moving the cell and controlling internal movement of structures. Microtubules function in cell division and serve as a "temporary scaffolding" for other organelles. Actin filaments are thin threads that function in cell division and cell motility. Intermediate filaments are between the size of the microtubules and the actin filaments.

1.7 ENDOPLASMIC RETICULUM

Endoplasmic reticulum is a network of tubules, vesicles and sacs that are interconnected. They may serve specialized functions in the cell including protein synthesis, sequestration of calcium, production of steroids, storage and production of glycogen, and insertion of membrane proteins. There are two types of endoplasmic reticulum (ER) rough endoplasmic reticulum, which gets its name from the presence of ribosomes on its surface during protein synthesis and smooth ER with out any ribosomes. Rough endoplasmic reticulum may either be vesicular or tubular. Or it may consist of stacks of flattened cisternae (like sheets) that may have bridging areas connecting the individual sheets. The Ribosomes sit on the outer surfaces of the sacs (or cisternae). They resemble small beads sitting in rosettes or in a linear pattern.

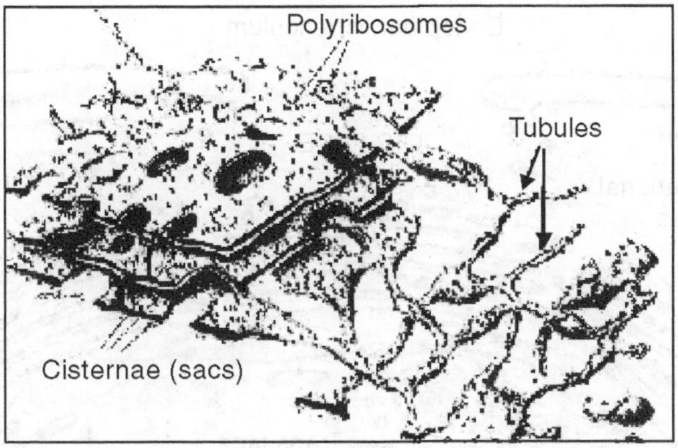

Figure 1.6 A part of rough endoplasmic reticulum showing interconnected network of tubules, cisternae with ribosomes attached to the surfaces giving a rough appearance.

1.8 VACUOLES AND VESICLES

Vacuoles are single-membrane organelles that are essentially part of the outside that is located within the cell. The single membrane is known in plant cells as a tonoplast. Many organisms will use vacuoles as storage areas. Vesicles are much smaller than vacuoles and function in transport within and to the outside of the cell.

1.9 GOLGI COMPLEX

It is another important membrane bound organelle present in eukaryotic cells, both in plants animals. It usually present in association with or near to ER. Golgi complex consists of pile of flattened membrane bound sacs called cisternae, which are formed by the fusion of vesicles. The vesicles are formed from ER. Golgi apparatus has an outer convex surface and an inner concave surface. The vesicles from ER fuse together and forms cisternae on the outer or convex surface and vesicles bud off constantly from the lower or concave side. Thus entire stack consists of a number of cisternae which is moving from the outer convex to the inner concave face of the Golgi complex.

The nascent protein transported from ER in the form of vesicles under goes post synthetic modifications inside the cisternae of Golgi complex. One of the important modifications on protein molecules is addition of carbohydrate moiety to the proteins, a process known as glycosilation. Glycoproteins function as the receptors on the cell surfaces along with phospholipids of the membranes. Another important function of Golgi complex is the production of lysosomes. Lysosomes are the vesicles containing hydrolytic enzymes.

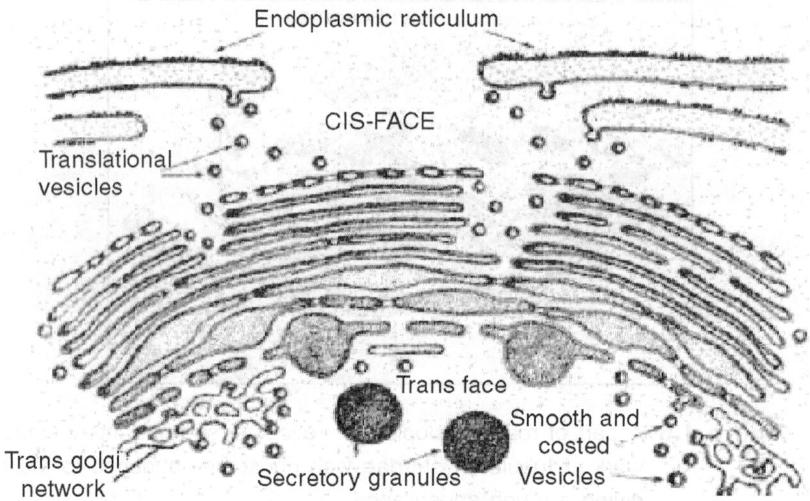

Figure 1.7 Diagrammatic sketch of Golgi complex showing the interphase with ER.

1.10 LYSOSOMES

These are single membrane bound vesicles present in eukaryotic cells containing hydrolytic enzymes such as nucleases, proteases, and lipases. These enzymes are synthesised in ER and are transported to Golgi complex through their cisternae and vesicles. From Golgi complex these are removed as vesicles known as primary lysosomes. The lysosomes are the defense units of cells and in unicellular eukaryotic organisms like protozoans they form the digestive enzymes for digesting the food materials engulfed by the organisms.

Lysosomes fuse with the phagocytic vesicle or the phagosomes (the material engulfed by the cells) to form the phagolysosomes. The enzymes of the lysosomes will digest the food material or the invading organisms present with in the phagocytic vesicle. Therefore lysosomes are abundant on those cells, which are specialised for phagocytosis.

1.11 MITOCHONDRIA

Mitochondria are another very important membrane bound organelle present in both animal and plant cells. They are short cylindrical structures. There are two membranes covering this organelle, an outer membrane and an inner membrane with a space in between. The inner membrane has got a number of infoldings into the matrix known as cristae. The matrix of mitochondria contains 70S ribosomes and a circular DNA in addition to a large number of enzymes. On the surfaces of the cristae there are a large number of granular structures known as oxysomes.

The main function of mitochondria is that they are the centres of power generation inside the cells. The final stage of respiration, the cellular respiration takes place in mitochondria. Cellular respiration or the oxidative break down of glucose occurs in three steps—glycolysis, Krebs cycle and electron transport system and oxidative phosphorylation (synthesis of ATP). The site of Glycolysis is outside the mitochondria in the cytoplasm. The other two steps take place inside the mitochondria. The site of Krebs's cycle is in the matrix and that of ETS is on the inner side of the cristae. The granular structures called oxysomes are the membrane bound enzyme assembly of ATPase on which the phosphorylation or ATP synthesis takes place. The membranes of cristae contain all the required components for operating ETS. Similarly the matrix contains all the enzymes and other components required for operating the Citric acid cycle.

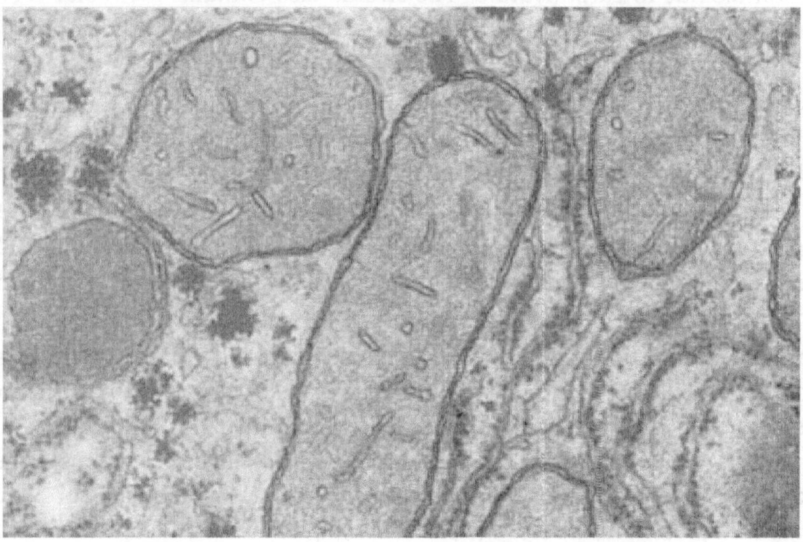

Figure 1.8 Electron micrograph of mitochondria.

1.12 PLASTIDS

These are the special organelles present only in plants. Plastids are of different shapes covered by two membranes similar to that of mitochondria. Depending on the functions plastids are classified into three groups. They are chloroplasts, chromoplasts and leucoplasts. Among these chloroplasts are the most important on the functional point of view.

Chloroplasts. These are the green plastids specialised for photosynthesis. They contain the photosynthetic pigments like chlorophylls and carotenoids. They generally have discoid shape with a double membrane covering enclosing the matrix known as the stroma. The stroma is traversed by system of double membranes known as lamellae. Some of these lamellae appear in the form of flattened and discoid sacs enclosing an inner space. These membrane structures are known as thylakoid membranes. Thylakoid membranes are present in piles of discs interconnected by the normal types of double membranes. These stacked thylakoids are called grana and the interconnecting membrane is called grana lamellae. There are membranes with out any grana traversing the stroma and they are known as stroma lamellae. The photosynthetic pigments are present on the thylakoid membranes as pigment protein complexes. They are organised in to light harvesting systems known as pigment systems I and II (PSI and PSII). Thylakoid membranes are the centres of light reaction, where the light energy is converted in to chemical energy. The components required for the light harvesting and its conversion into the reducing powers of NADPH2 and ATP are distributed on the thylakoid membrane in a very organised manner.

The second reaction of the photosynthesis –the dark reaction or the calving cycle occurs in the matrix or the stroma of the chloroplast. The stroma is a mixture containing all the required enzymes and other compounds including the intermediates of Calvin cycle essential for operating the absorption and reduction of carbon dioxide to carbohydrate. The stroma also contains a circular DNA molecule and 70S ribosomes needed for protein synthesis. The presence of these things indicates that chloroplasts like mitochondria are semi autonomous organelles.

The presence of a circular DNA molecule and 70S ribosomes and the absence of membrane bound organelles and the presence of membrane infoldings postulates the prokaryotic origin of chloroplasts and mitochondria. It is believed that chloroplasts are evolved from a symbiotic type of blue green algae (cyanobacteria) and mitochondria might have evolved from a symbiotic type of bacteria.

Chromoplasts. Chromoplasts are non photosynthetic the coloured plastids. They do not have non-photosynthetic pigments. They contain red, orange or yellow pigments and are present in the coloured parts of the plant like flowers, fruits etc.

Leucoplasts. These plastids do not contain any pigments and therefore are clourless. These plastids some times form the precursors of chloroplasts or change in to structures of for food storage. Depending on the type of stored food they are further classified into amyloplasts, lipidoplasts and proteoplasts. Amyloplasts contain starch and are also known as starch grains, which are present in the storage regions of the plants like tubers and grains. Lipidoplasts store lipids and proteoplasts or aleuron grains store proteins. These are also present in the storage parts of the plants like endosperm of seeds and pulses.

1.13 RIBOSOMES

Ribosomes are protein-synthesizing machines of the cell. They **translate** the information encoded in **messenger RNA (mRNA)** into a protein. They are not membrane-bound and thus occur in both prokaryotes and eukaryotes. Eukaryotic ribosomes are slightly larger than prokaryotic ones. The ribosomes of the prokaryotes are 80S and those found in bacteria and eukaryotic organelles like mitochondria and chloroplasts are 70S. The presence of bacterial (prokaryotic) ribosomes (70S) in chloroplasts and mitochondria shows the prokaryotic origin of these organelles.

Despite these differences, the basic structure and operations of bacterial and eukaryotic ribosomes are very similar. Structurally the ribosome consists of a small and larger subunit. These two units join together when the ribosome attaches to messenger RNA to produce a protein in the cytoplasm. Biochemically the ribosome are ribonucleoprotein complexes consisting of **ribosomal RNA** (rRNA) and some 50 structural proteins. In eukaryotes, Ribosomes that synthesize proteins for use within the cytosol (*e.g.*, enzymes of glycolysis) are suspended in the cytosol. Ribosomes that synthesize proteins destined for secretion (by exocytosis), the plasma membrane (*e.g.*, cell surface receptors), lysosomes etc are attached to the cytosolic face of the membranes of the **endoplasmic reticulum**. During protein synthesis, ribosomes form a complex with mRNA and moves along the molecule. During active periods of protein synthesis a number of ribosomes attach to the rRNA and such structures are known as polyribosomes.

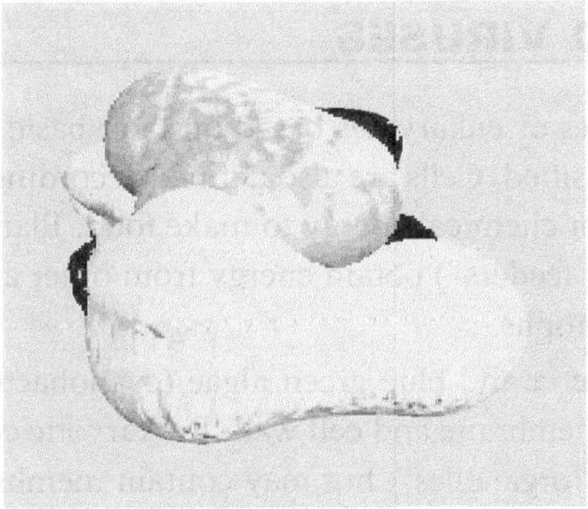

Figure 1.9 Model of ribosome, showing the small and large subunits together.

1.14 PEROXISOMES

These are small single membrane bound granular structures seen in eukaryotic cells. They are small lysosomes containing the enzyme catalase or Peroxidase. These structures are involved in a number of metabolic reactions, which involves oxidation. Peroxidase enzymes are needed for the oxidation of peroxides formed in the cells as the by products of other metabolic reactions.

1.15 CYTOSKELETON

There is a net work of fibers made up of fibrous proteins present in side the cells which gives mechanical support and shape to the cells. In addition to this there are some other specific functions for which it is modified. The fibers of cytoskeleton are of two types – microtubules and microfilaments. The other functions of these fibers of cytoskeleton include movements of cells, as the components of cilia and flagella and the movements of organelles and components with in the cells. For example movements of chromosomes by spindle fibers during cell division.

Microtubules. Microtubules are present in spindle fibers, cilia and flagella. These fibers are hollow cylindrical fibers consisting of proteins tubulin. In flagella, cilia and spindle the microtubules undergo sliding moments. They are responsible for the movement of chromosomes attached to the spindle fibers during cell division and for the rhythmic movements of cilia and flagella. They are also responsible for the movements of other organelles like lysosomes, Golgi complex etc. In addition to all these they give mechanical support for the cells, particularly in the case of animal cells, which does not have a rigid cell wall.

Microfilaments. These filaments are present in bundles or sheets just below the cell surface membranes. These are very thin filaments and are composed of the fibrous protein actin. They are supposed to be involved in the phagocytosis or endocytosis and exocytosis. In addition to this they are also involved in the movements of cells and cellular parts.

1.16 BACTERIA AND VIRUSES

Cells are classified as prokaryotes or eukaryotes based on their basic structure. and the way by which they obtain energy. Cells are classified. Cells are also defined according the need for energy. Autotrophs are "self feeders" that use light or chemical energy to make food. Plants are an example of autotrophs. In contrast, heterotrophs ("other feeders") obtain energy from other autotrophs or heterotrophs. Many bacteria and animals are heterotrophs.

Prokaryotes include bacteria and blue green algae (cyanobacteria). Simply stated, prokaryotes are molecules surrounded by a membrane and cell wall. Prokaryotic cells lack characteristic eukaryotic subcellular membrane enclosed "organelles", but may contain membrane systems inside a cell wall as an extension or infoldings of the cell membrane. The nucleus is not well organised and is with out any membrane.

Virus particle does not have any cells and therefore is termed as acellular. They are the macromolecular assembly formed primarily of proteins and nucleic acids. They are some times regarded as dormant nucleoprotein particles with out any life activities. They become active and show the signs of life activities only when they enter into specific types of cells like bacteria animal or plant cells.

• REVIEW QUESTIONS •

1. Plasma membrane has got the basic fluid mosaic structure, mainly consisting of phospholipids and proteins. Even then, plasma membrane at different regions of the cells like mitochondria chloroplast, cell surface carries out different types of functions. Explain.
2. What are the functions of Endoplasmic reticulum?
3. What are the similarities between chloroplasts and mitochondria?
4. Compare the structure of a plant and animal cell.
5. What is meant by rough endoplasmic reticulum? Explain its involvement in protein synthesis.
6. Explain the role cytoskeleton in a eukaryotic cell.
7. Compare the structure of a eukaryotic cell with that of a prokaryotic cell.
8. Explain the chemical composition and function of ribosomes.
9. Eukaryotic cells are compartmentalized in to a number of organelles. Explain its advantage in carrying out cellular metabolism with suitable examples.
10. What are the differences between prokaryotic cell and a virus?
11. Explain the structure and function of various types of plastids.
12. Explain how the ER, Golgi complex and nuclear membranes are interrelated.
13. What are the various types of tissues present in animals ?

• SUGGESTED READING •

1. Discover Biology., M.L. Cain, H. Damman, R.A. Lue and C.K. Yoon., W.W. Norton & Company, New York.
2. Biological Science, N.P.O. Green, G.W. Stout, and D.J. Taylor, Cambridge University Press. Cambridge.
3. The Diversity of Life, by E.O. Wilson, 1992, Harvard Belknap, Cambridge, Mass.
4. Molecular Biology of the Cell, third edition, Alberts et al, Garland Publishing, 1994.
5. Principle of Genetics E. J. Gardener , M.J. Simmons and D.P. Snusted; John Wiley & Sons. Inc., New York.

2 BIOMOLECULES

2.1 INTRODUCTION

The living cell of an organism is a chemical machine; its parts and its language are chemical. It is an assembly various types of biomolecules having diverse physico-chemical and biological properties. All these small and big biomolecules take part in a variety of biochemical reactions and the phenomenon of life is an out come these activities. Biopolymers or the biological macromolecules form the major class among the biomolecules both structurally and functionally. All cellular structures like cell wall, cell membranes, membranes of organelles, microtubules, flagella, and Celia etc. and there are a lot of biopolymers actively engaged in different metabolic reactions. For example biopolymers like cellulose

Biomolecules are the compounds synthesized by the living organisms. These biological compounds are of different types having different size, shape, chemical and physical properties and biological functions. These Biomolecules include different classes of compounds, which are broadly divided in to two, depending on size and nature. Those molecules, which are complex polymers and bigger in size are known as Macromolecules and other molecules are simple in complexity and small in size. It includes all the ordinary organic molecules synthesized in a living system.. There are four types of macromolecules in biological systems namely carbohydrates, proteins, lipids and nucleic acids. Out of these four types three are polymers composed of monomers or the building blocks and lipids are not a polymer similar to that of carbohydrates, proteins and nucleic acids.

This chapter deals with a brief account of the structure of the major macromolecules and their building blocks or the monomers. We first study about the small molecules including the building blocks followed by its polymers or the macromolecules. It includes monosaccharides or sugars, which forms the building blocks of carbohydrates like starch and cellulose; amino acids, which forms the building blocks of proteins; nucleotides, which forms the building blocks of nucleic acids like DNA and RNA.; and fatty acids form the components of various types of lipids. In addition to these there are the other small biomolecules like vitamins, coenzymes and cofactors carrying out very important metabolic activities and gene expression. In biological system all these molecules both macro and micro are in a state of flux or in a dynamic state. That is they are always subjected to chemical transformations in order to maintain the state of life. Actually the sum total of all these biochemical changes forms the metabolism of an organism. These metabolic reactions are interconnected and arranged in to specific metabolic pathways and they may be linear or cyclic interacting with each other, one regulating the other.

The macromolecules such as protein, nucleic acids, carbohydrates and lipids are very important in marinating the normal course of function and life activities of a cell. In proteins, there is a special focus on enzymes, which mediate all cellular biochemical reactions. Nucleic acids are mainly the genetic materials, carbohydrates form the primary energy sources and lipids are the energy reservoirs and they also form the barriers of cells and organelles. All these biomolecules organize in different manner and forms supramolecular assemblies, and that give rise to the organelles like cell membranes, ribosomes etc. These macromolecules are also known as biopolymers. They can be classified into four major groups—carbohydrates, Proteins, Lipids and Nucleic aids. All these biopolymers are composed of small repeating units or the monomers of the same type or different types. These structural units of biopolymers are called their building blocks. Even though lipids are not polymers they are macromolecules and put among the biopolymers. In this chapter we discuss about the structure and dynamics the building blocks of all major macromolecules of the biological system. The monomers of these biopolymers are characterized by the presence of certain specific functional groups. Therefore before going into the details of these monomers, a basic understanding about the functional groups found in the biomolecules is required. The following table 2.1 shows the major monomers and its corresponding polymer of the biological system. Basic principle of cell is that it acquire small molecules (**monomers**), then assemble these into much larger molecules (**polymers**).

TABLE 2.1 Major monomers and its polymers of biological system

Monomer	Polymer
amino acids	proteins
nucleotides	DNA, RNA
sugars	Polysaccharides
fatty acids	complex lipids

2.2 ELEMENTS TO FUNCTIONAL GROUPS

1. Elements of Life

An element is a pure substance that is made up of only one kind of atom. There are 92 naturally occurring elements. All elements that are present on earth are not represented in living system. Similarly the abundance of elements on earth's crust is not mirrored in living tissue. Some of earth's major elements are not found in cells (*e.g.* Aluminum) and some major elements in cells are not abundant on earth's crust (*e.g.* Carbon). There are about 25 elements observed in the living in cells.

The elements present in the cells can be identified as Macro and Micro elements.

Macro-elements—CHNOPS

The following 6 elements are the predominant elements present in all cells :

1. C = Carbon 2. H = Hydrogen
3. N = Nitrogen 4. O = Oxygen
5. P = Phosphorus6. S = Sulfur.

Some organisms have extra abundance of other elements (*e.g.*, vertebrates have lots of Calcium, mostly in bone)

Micro-elements and Trace Minerals

The remaining elements are required in comparatively fewer amounts, generally very small amounts relative to CHNOPS, and are used in a variety of ways. Many of these minerals are critical for the function of certain enzymes (*e.g.* Copper, Molybdenum, Zinc) and others required for certain special functions :

- Calcium for bone.
- Silicon for glassy shell in certain protists.
- Potassium and Sodium for generation and transmission of nerve impulse in animals.

2. Atoms & Isotopes

- Atom is the smallest unit of matter It contains electrons (– charge), protons (+ charge), and neutrons (no charge), with electron represented dynamically as "electron cloud".
- The number of protons and electrons in an atom is the same, unique for each **element.**
- Example: H = 1 proton; C = 6 protons; O = 8 protons. This is also called the **atomic number**.
- Electrons fit into specific geometrical patterns, called **orbitals** (not described further here — learn about this in chemistry). Each orbital holds only 2 electrons.
- As orbitals closest to nucleus become filled, additional electrons must occupy orbitals further from nucleus.
- Groups of orbitals can be approximated as **electron shells**.
- Some atoms lose or gain electrons and become charged **ions**.

  Examples : Na $\longrightarrow$ Na^+ ; Cl $\longrightarrow$ Cl^-
- Isotopes are the atoms having the same atomic number and different atomic weight. That is they differ in neutron number. Example: isotopes of Carbon.

  C^{12} = 'normal' (99 + %)

  C^{13} = less common (< 1%)

  C^{14} = 'unstable' isotope, weakly radioactive; useful in research, in biology and medicine as tracer element.

3. Molecules

- Elements adjacent to Noble Gases tend to ionize and can form ionic bond between them to form molecules.

- Elements further away from noble gases tend to form **covalent bonds** by electron sharing, and become stably connected into **molecules.**
- Each atom in a molecule tries to complete outer electron shell by sharing electrons.
- Number of electrons shared is called **valence.**

Valences of C, O, N, H, S, P :

H = 1, O = 2, N = 3, C = 4, P = 5 and S = 2

Some examples of molecules:

$$2H + O = \mathbf{H_2O\ (water)}$$

$$C + 4H = \mathbf{CH_4\ (methane)}$$

$$6C + 12H + 6O = \mathbf{C_6H_{12}O_6\ (glucose)}$$

4. The Nature of Chemical Bonds

1. Ionic Bonds. Ionic bonds are formed when there is a complete transfer of electrons from one atom to another, resulting in two ions, one positively charged and the other negatively charged. The electrostatic attractions between the positive and negative ions hold the compound together. For example, when a sodium atom (Na) donates the one electron in its outer valence shell to a chlorine (Cl) atom, which needs one electron to fill its outer valence shell, NaCl (table salt) results. The symbol for sodium chloride is Na^+Cl^-. The energy value or strength of electronic bonds is often 4-7 kcal/mol.

2. Covalent Bonds. Covalent Bonds are the strongest chemical bonds, and are formed by the sharing of a pair of electrons. Formed by elements such as C, H, N, O, P, S, which are usually not found immediately adjacent to noble gases in the periodic table. The energy of a typical single covalent bond is ~80 kilocalories per mole (kcal/mol). However, this bond energy can vary from ~50 kcal/mol to ~110 kcal/mol depending on the elements involved. Once formed, covalent bonds rarely break spontaneously. This is due to simple energetic considerations; the thermal energy of a molecule at room temperature (298 K) is only ~0.6 kcal/mol, which is much lower than the energy required to break a covalent bond.

There are single, double, and triple covalent bonds as shown in the table 2.2 below.

TABLE 2.2 Different types of covalent bonds and its energy values

<table>
<tr><th>Bond Number</th><th>Example</th><th>Energy (kcal/mol)</th></tr>
<tr><td>Single</td><td>H
|
H—C—H
|
H</td><td>~80</td></tr>
<tr><td>Double</td><td>H H
| |
H—C = C—H
| |
H H</td><td>~150</td></tr>
<tr><td>Triple</td><td>H
|
C = C
|
H</td><td>~200</td></tr>
</table>

Note that carbon-carbon bonds are unusually strong and stable covalent bonds. Usually covalent bonds are with out any charges. But in certain cases they can also develop partial charges when the atoms involved have different electro negativities. Water is perhaps the most obvious example of a molecule with partial charges. The symbols δ^+ and δ^- are used to indicate partial charges.

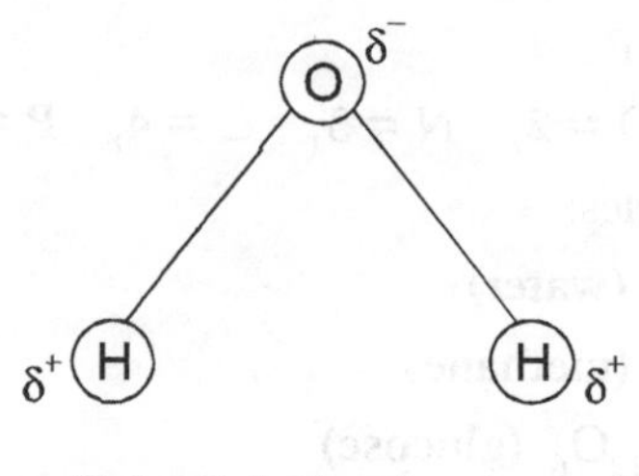

Figure 2.1 Polarity of water molecule.

Oxygen, because of its high electro negativity, attracts the electrons away from the hydrogen atoms, resulting in a partial negative charge on the oxygen and a partial positive charge on each of the hydrogen atoms. The possibility of hydrogen bonds (H-bonds) is a consequence of partial charges.

3. Hydrogen Bonds. In covalent linkages, if electrons are not evenly distributed, result is **polarity** and it results in hydrogen bonding. A hydrogen atom covalently attached to a very electronegative atom (N, O, or P) shares its partial positive charge with a second electronegative atom (N, O, or P). In H_2O; H carries fraction of + charge, O fraction of – charge. Therefore water molecules can form hydrogen bonding with each other very easily. These bonds are frequently found in proteins and nucleic acids, and by reinforcing each other serve to keep the protein (or nucleic acid) structure secure. **Hydrogen bond** is a weak interaction and is stable only for short periods of time. Its bond strength is ~ 3-5 kcal/mole.

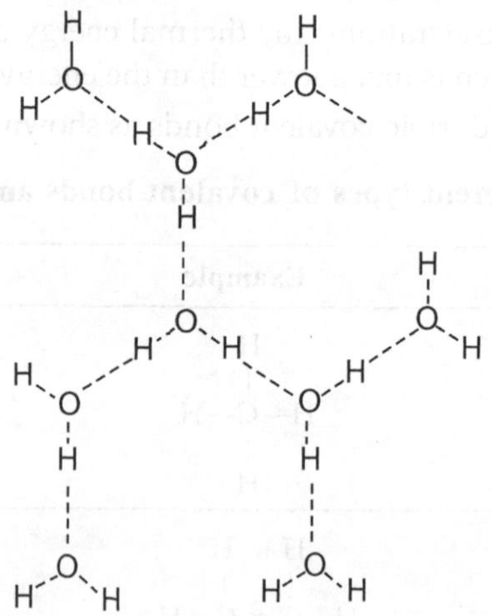

Figure 2.2 Hydrogen bonding between water molecules.

4. Van der Waals Bonds. Van der Walls interactions are very weak bonds (generally no greater than 1 kcal/mol) formed between nonpolar molecules or non-polar parts of a molecule. The weak bond is created because a C—H bond can have a transient dipole and induce a transient dipole in another C—H bond.

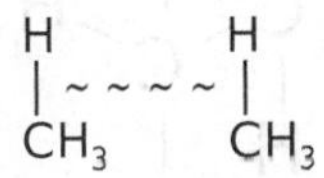

Figure 2.3 Representation of Van der Waals Bonds.

5. Hydrophobic interactions. Nonpolar molecules cannot form H-bonds with H_2O, and are therefore insoluble in H_2O. These molecules are known as hydrophobic (water hating), as opposed to water loving hydrophilic molecules, which can form H-bonds with H_2O. Hydrophobic molecules tend to aggregate together in avoidance of H_2O molecules; hydrophobic interactions are clearly demonstrated when you put an oil drop on water. This attraction/repulsion is known as the hydrophobic (fear of water) force. To understand the energetics driving this interaction, visualize the H_2O molecules surrounding a 'dissolved' molecule attempting to form the greatest number of hydrogen bonds with each other. The best energetic solution involves forcing all of the nonpolar molecules together, thus reducing the total surface area that breaks up the H_2O H-bond matrix.

5. Polar and Nonpolar Molecules

Some molecules have electrons asymmetrically distributed. That results in polar **molecules.** Because of polarity, molecule acquires a **dipole moment**, and will orient to one direction when placed in electric field.

Example : CH_3CH_2OH (ethanol); the oxygen atom attracts more of the electrons than other atoms.

All polar molecules interact by charge-charge interactions; interact well with water, form good hydrogen bonds. Sugars are good examples and are readily soluble molecules, due to many **—OH** groups. These —OH groups can easily form hydrogen bonds with water molecules.

By contrast, **non-polar molecules**, such as hydrocarbons, do not dissolve in water. Because to do so would have to break hydrogen bonds already present in water. Instead, hydrophobic or non-polar compounds form a layer that separate it from water (*e.g.* oil film on surface) or will form globules of fatty material inside water.

6. Functional Groups of Biomolecules

The functional groups are specific groups of atoms and molecules which, when substituted for one or more hydrogen atoms in a hydrocarbon, confer particular chemical properties to the new compound. The chemical property or the 'behavior' of a molecule can be attributed to the functional group present. For example all organic acids are characterized by the presence of —COOH group. The following are the important functional groups present in the biological molecules. Table 2.3 also gives a summary of the most common functional groups present in various biomolecules. Once you have learned the properties of these functional groups, the study of the major compounds of living organisms will be easy.

1. Hydrophobic Functional Groups

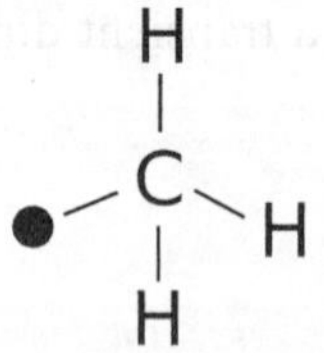

Methyl group

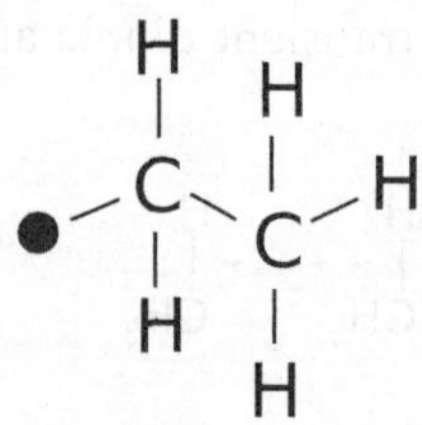

Ethyl group

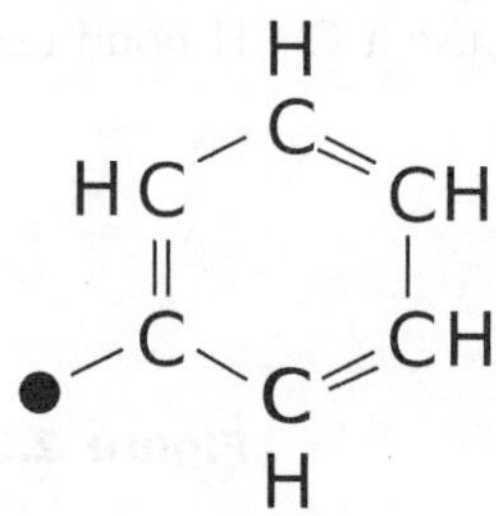

Phenyl group

2. Polar Functional Groups

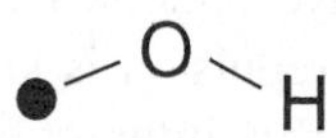

Hydroxyl group

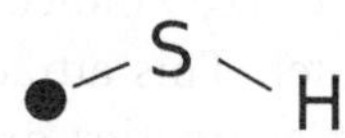

Sulfhydryl group

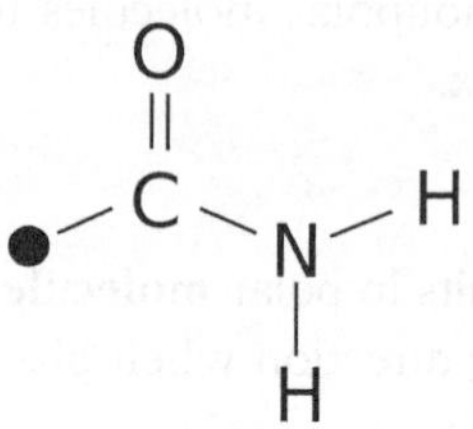

Amido group

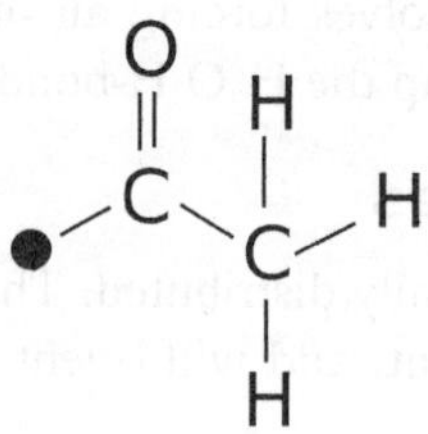

Acetyl group

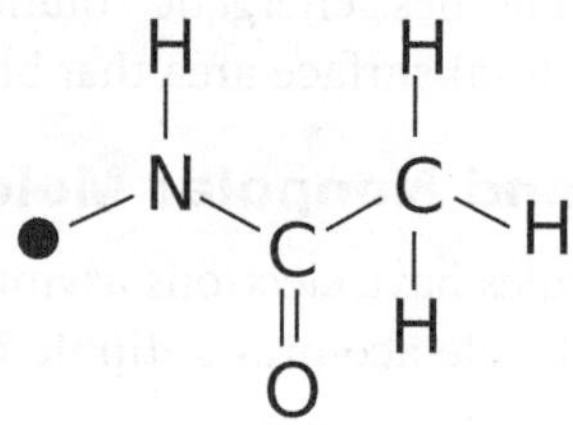

N-acetyl group

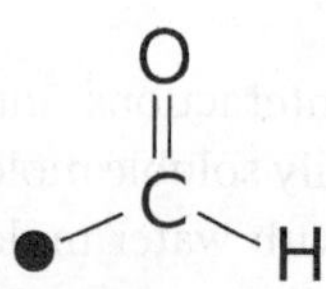

Carbonyl group: aldehyde

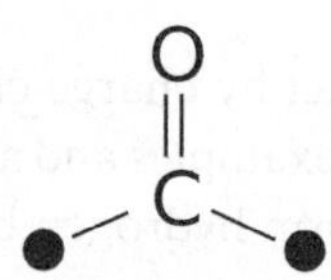

Carbonyl group: ketone

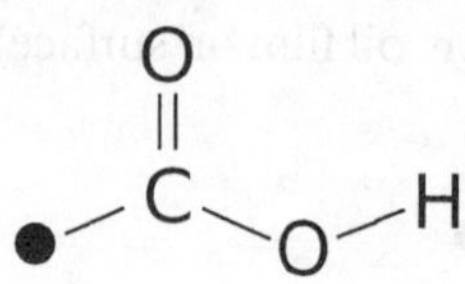

Carboxylic acid group

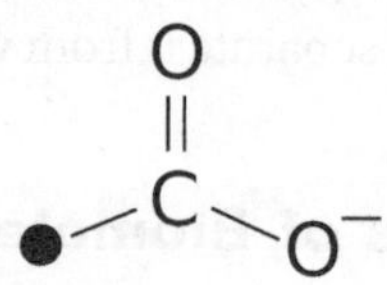

Carboxylate group

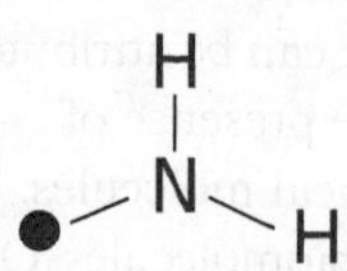

Amino group

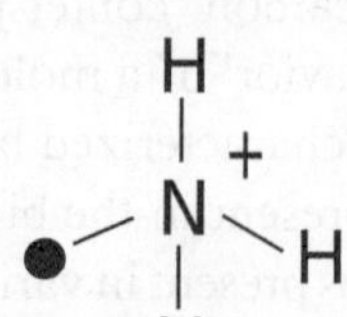

Ammonium group

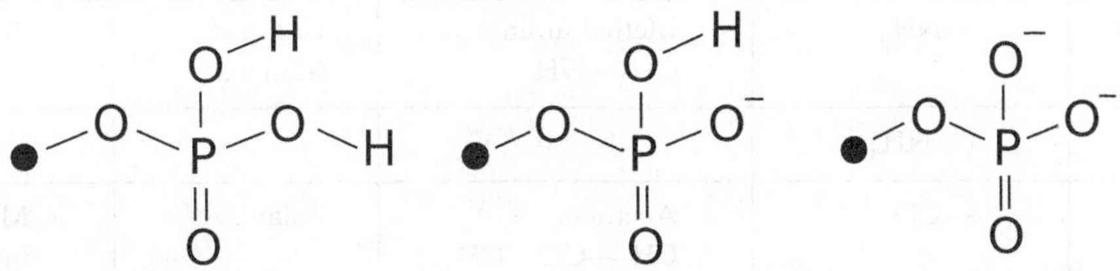

3. Linkages and Bridges

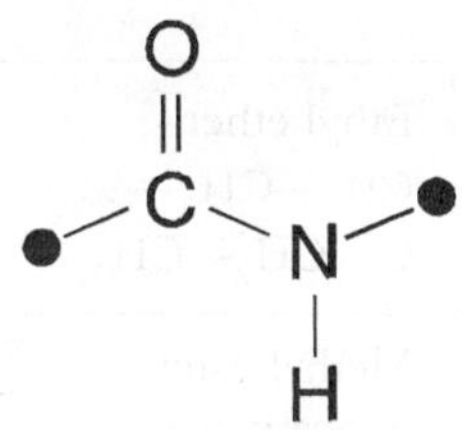

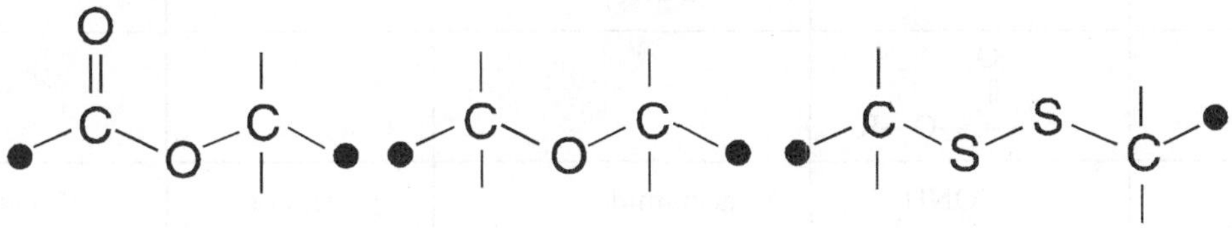

Figure 2.4 Structure of various Functional Groups present in Biomolecules.

TABLE 2.3 Functional groups of biomolecules

Name	Group	Chemical Example	Comment	Biological Example
Hydroxyl	—OH	Ethanol: CH_3 —CH_2 —OH	Polar	Sugars
Aldehyde	—CHO	Acetaldehyde: CH_3—CHO	Polar	Glucose
Detail	O ‖ —C—H			
Carboxylic acid	—COOH	Acetic acid: CH_3—COOH	Charged (ionized)	Fatty acid
Detail	O ‖ —C—H			

Amine	$—NH_2$	Methyl amine: $CH_3—NH_2$	Charged (ionized)	Amino acid
Detail	$(—NH_3^+)$			
Ketone	—CO—	Acetone: $CH_3—CO—CH_3$	Polar	Metabolic intermediate (pyruvic acid)
Detail	O ‖ —C—			
Ether	—O—	Ethyl ether: $CH_3—CH_2—$ $O—CH_2—CH_3$	Not so polar (symmetric)	Some lipids
Ester	—COOR	Methyl ester of acetic acid: $CH_3—CO$ $—OCH_3$	Polar. Not charged	Fats
Detail	O ‖ —C—O—R			
Amide	$—CONH_2$	acetamide: $CH_3—CONH_2$	polar, not charged	Proteins (asparagine)
Detail	O ‖ $—C—NH_2$			
Sulfhydryl	$—SH_2$	2-mercaptoethanol: $HO—CH_2$ $—CH_2—SH$	Reducing agent	Protein (cysteine)
Disulfide	—S—S—	R—S—S—R	Crosslinks in proteins	Protein (cystine)
Phenyl	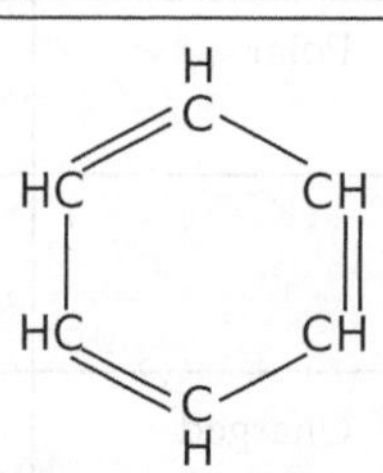	Phenol: $C_6H_5—OH$, Ø—OH	Hydrophobic	Proteins (phenylalanine)

2.3 THE STRUCTURE OF BIOMOLECULES

1. CARBOHYDRATES

Building Blocks of Carbohydrates—Monosaccharides and Disaccharides

Carbohydrates are carbon compounds that contain large quantities of hydroxyl groups. The simplest carbohydrates or commonly known as sugars, also contain either an aldehyde group or a ketone group. The aldehyde containing sugars are termed **polyhydroxyaldehydes** and ketone-containing simple sugars are termed as **polyhydroxyketones**. All carbohydrates can be classified as **monosaccharides, oligosaccharides** or **polysaccharides**. Monosaccharides are the monomers or the building blocks of carbohydrates. Anywhere from two to ten monosaccharide units, linked by glycosidic bonds, make up an oligosaccharide. Polysaccharides are much larger, containing hundreds of monosaccharide units. The presence of the hydroxyl groups allows carbohydrates to interact with the aqueous environment and to participate in hydrogen bonding, both within and between chains. Derivatives of the carbohydrates can contain nitrogens, phosphates and sulfur compounds. Carbohydrates also can combine with lipid to form glycolipids or with protein to form glycoproteins.

Monosaccharides

The monosaccharides are commonly known as simple sugars such as glucose, fructose, galactose etc. The various types of sugars present in nature are classified according to the number of carbons they contain in their backbone structures. Their basic molecular formula is $C_nH_2O_n$ and therefore is considered as the hydrates of carbon. The monosaccharides are classified in to two based on the type of functional group it posses. Those sugars having aldehyde groups are called aldoses and those having keto groups are termed as ketoses. The major monosaccharides both ketoses and aldoses, contain four to seven carbon atoms. Accordingly monosaccharides classified in to seven groups depending on the number of carbon atoms. They are triose with 3 carbon atoms, tetrose with 4 carbon atoms, pentose with 5 carbon atoms, hexose with 6 carbon atoms and heptulose with 7 carbon atoms. Each of these groups includes both aldoses and ketoses.

The trioses having aldehyde groups are termed as aldotriose and aldoketose. Similarly Tetrose may be aldotetrose or ketotetrose, Pentose sugar includes aldopentose and ketopentose; and the hexose sugar includes aldohexose and ketohexose and so on. Heptulose or the seven-carbon sugar is not common and is not present in free form. Sedoheptulose, a keto form of heptose is present in plants as an intermediate in the biosynthetic pathway of glucose in photosynthesis. Table 2.4 gives a summary of classification of monosaccharides.

TABLE 2.4 Classifications of monosaccharides with examples

Number of Carbons	Category Name	Relevant examples
3	*Triose* Aldotriose Ketoriose	Glyceraldhyde Dihydroxyacetone
4	*Tetrose* Aldotetrose Ketotetrose	Erythose Erythrulose

5	*Pentose* Aldopentose Aldoketose	Ribose, Xylose Ribulose, Xylulose
6	*Hexose* Aldohexose Ketohexose	Glucose, Galactose, Mannose Fructose
7	*Heptose* Aldoheptose Ketoheptose	No naturally occurring aldo heptose Sedoheptulose

Nomenclature of Monosaccharides—D and L Forms

The predominant monosaccharides encountered in the body are structurally related to the aldotriose **glyceraldehyde** and to the ketotriose **Dihydroxyacetone**. All carbohydrates contain at least one asymmetrical (chiral) carbon and are, therefore, optically active. In addition, carbohydrates can exist in either of two conformations, as determined by the orientation of the hydroxyl group about the asymmetric carbon farthest from the carbonyl. With a few exceptions, those carbohydrates that are of physiological significance exist in the **D-conformation**. The mirror-image conformations, called **enantiomers**, are in the **L-conformation**.

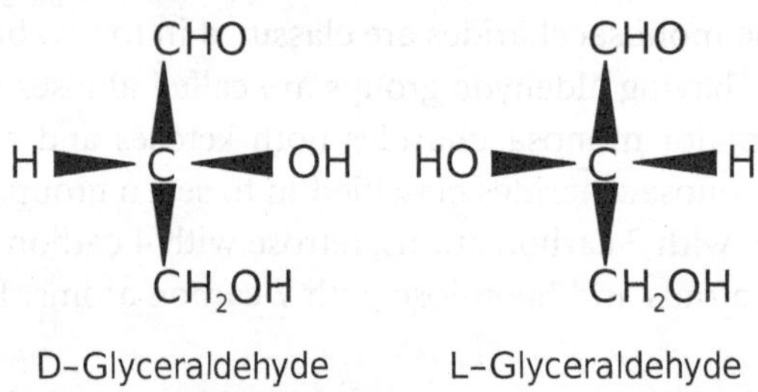

Figure 2.5 Structures of glyceraldehyde enantiomers.

The simple sugars or the monosaccharides have two types of functional groups :

- —OH (Hydroxyl)
- =O (Carbonyl), includes both aldehyde and ketone.

The aldehyde and ketone moieties (carbonyl group) of the monosaccharides with five and six carbons will spontaneously react with alcohol groups present in neighboring carbons to produce intramolecular **hemiacetals** or **hemiketals**, respectively. This results in the formation of five- or six-membered rings. Because the five-membered ring structure resembles the organic molecule **furan**, derivatives with this structure are termed **furanoses**. Those with six-membered rings resemble the organic molecule **pyran** and are termed **pyranoses**. Such structures can be depicted by either **Fischer** or **Haworth** style diagrams. The numbering of the carbons in monosaccharides proceeds from the carbonyl carbon, for aldoses, or the carbon nearest the carbonyl, for ketoses.

The rings can open and re-close, allowing rotation to occur about the carbon bearing the reactive carbonyl yielding two distinct configurations (α and β) of the hemiacetals and hemiketals. The carbon about which this rotation occurs is the **anomeric carbon** (the Carbon No.1 in glucose and Carbon No. 2 in fructose) and the two forms are termed anomers. Carbohydrates can change spontaneously between α and β configurations—a process known as **mutarotation**. When drawn in the Fischer projection, the α configuration places the hydroxyl attached to the anomeric carbon to the right, towards the ring. When drawn in the Haworth projection, the α configuration places the hydroxyl downward as shown in Fig. 2.6.

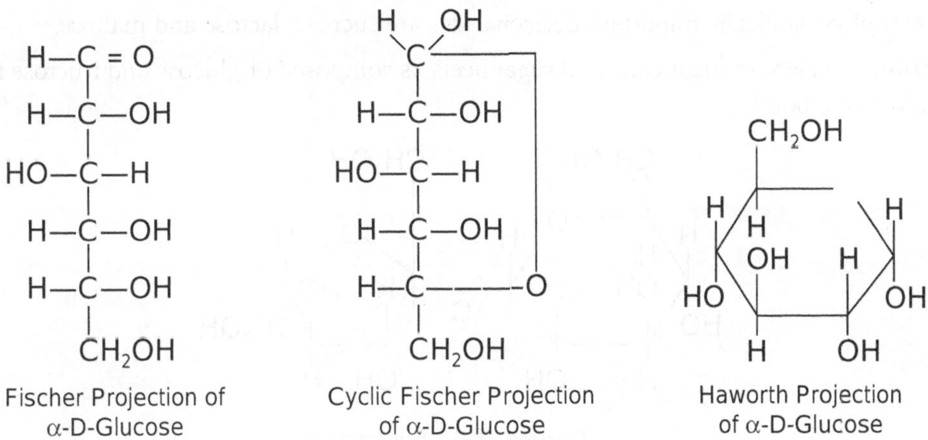

Figure 2.6 Structure of glucose, different structural representations.

The spatial relationships of the atoms of the furanose and pyranose ring structures are more correctly described by the two conformations identified as the **chair form** (Fig. 2.7) and the **boat form**. The chair form is the more stable of the two. Constituents of the ring that project above or below the plane of the ring are **axial** and those that project parallel to the plane are **equatorial**. In the chair conformation, the orientation of the hydroxyl group about the anomeric carbon of α-D-glucose is axial and equatorial in β-D-glucose.

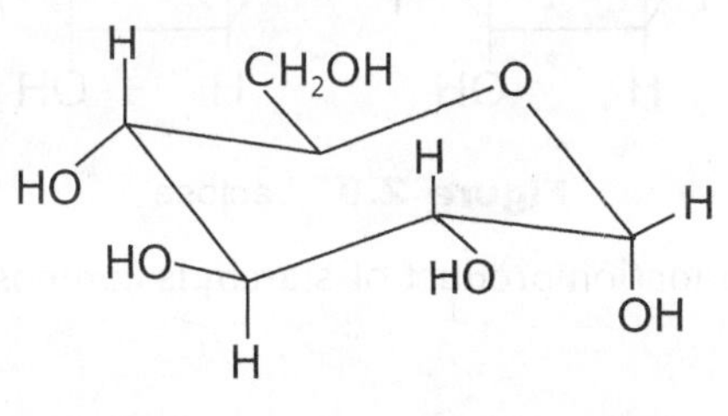

Figure 2.7 Chair form of α-D-Glucose.

Disaccharides

The hydroxyl group of anomeric carbon of a cyclic monosaccharide can react with the hydroxyl group of another monosaccharide to form disaccharide. Covalent bonds between the anomeric hydroxyl of a cyclic sugar and the hydroxyl of a second sugar (or another alcohol containing compound) are termed **glycosidic bonds**, and the resultant molecules are **glycosides**. The linkage of two monosaccharides to form disaccharides involves a glycosidic bond. Various numbers of monosaccharides can be joined in this manner into a chain to form carbohydrate polymers. (Disaccharides, trisaccharides, oligosaccharides (a few), or polysaccharides (many).

Several physiologically important disaccharides are sucrose, lactose and maltose.

Sucrose prevalent in sugar cane and sugar beets, is composed of glucose and fructose through an α-(1, 2) β-glycosidic bond.

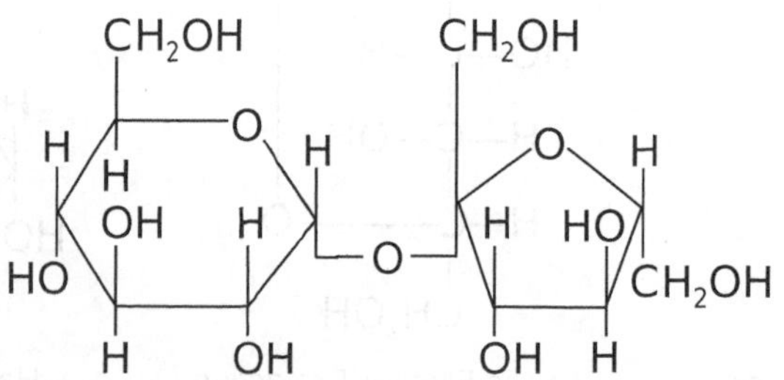

Figure 2.8 Sucrose

Lactose is found exclusively in the milk of mammals and consists of galactose and glucose in a β-(1, 4) glycosidic bond.

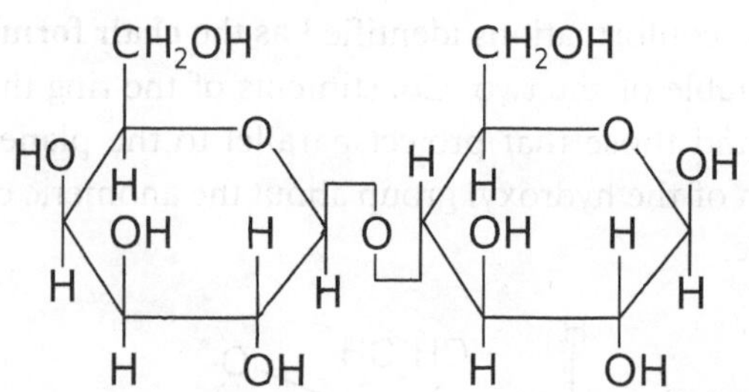

Figure 2.9 Lactose

Maltose is the major degradation product of starch, is composed of 2 glucose monomers in an α-(1, 4) glycosidic bond.

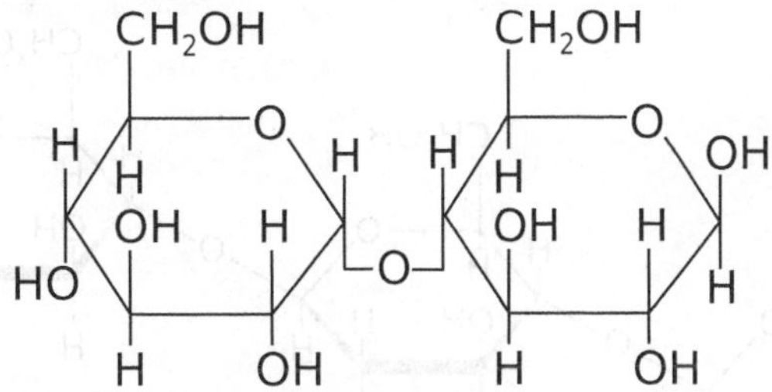

Figure 2.10 Maltose

Chemical Properties of Sugars

Monosaccharides are called reducing sugars. The presence of hemiacetal group or the aldehyde group in monosaccharides make them a reducing agent. It can reduce alkaline solutions of copper salts giving rise to red precipitate of cuprous oxide. This is the basic reaction of the qualitative tests like Fehling's and Benedict's test used to identify the presence of glucose in various samples like urine blood etc. Both Fehling's and Benedict's reagents contain copper sulphate in an alkaline state. These test are routinely conducted in pathological laboratories to find out the presence of sugar in clinical samples like blood and urine. All monosaccharides and disaccharides except sucrose are positive to these tests and are therefore reducing sugars. In sucrose, the glycosidic linkage is between the anomeric carbon (C-1) of glucose and anomeric carbon (C-2) of fructose. Thus both the anomeric carbons are involved in the glycosidic linkage and therefore there is no other active group to participate in reaction. Hence sucrose is nor a reducing sugar. Aldoses and Ketoses are distinguished by Seliwanoff's test. Ketoses and aldoses undergo dehydration with concentrated sulfuric acid and form furfural derivatives. In the case of ketones, furfural derivatives condense with resorcinol and give a red complex.

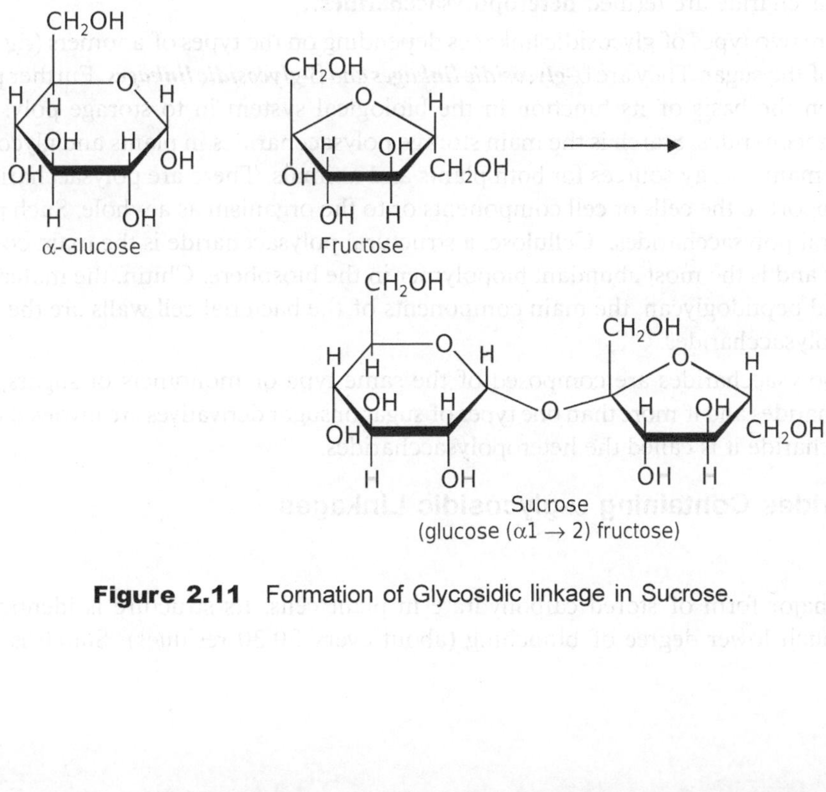

Figure 2.11 Formation of Glycosidic linkage in Sucrose.

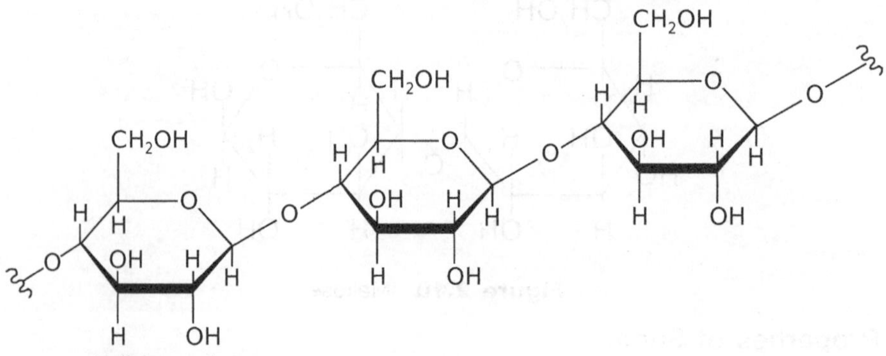

Figure 2.12 Starch and cellulose–the alpha and beta linkages.

Note. To distinguish between alpha and beta linkages, examine the position of the hydrogen on the first carbon molecule (Hemiacetal group). In an alpha linkage, the hydrogen is pointing up, and in a beta linkage it is pointing down.

Polysaccharides

Most of the carbohydrates found in nature occur in the form of high molecular weight polymers called **polysaccharides**. The polysaccharides are defined, as the polymers of sugars and its various derivatives, like glucosamine and galactosamine linked together by **glycosidic bonds**. The monomeric building blocks used to generate polysaccharides can be varied. In all cases, however, the predominant monosaccharide found in polysaccharides is D-glucose. When polysaccharides are composed of a single monosaccharide building block, they are termed **homopolysaccharides**. Polysaccharides composed of more than one type of monosaccharide are termed **heteropolysaccharides**.

There are two types of glycosidic linkages depending on the types of anomers (*e.g.* α-D glucose or β-D-Glucose) of the sugar. They are *α-glycosidic linkages and β-glycosidic linkages*. Further polysaccharides are classified on the basis of its function in the biological system in to storage polysaccharides and structural polysaccharides. Starch is the main storage polysaccharides in plants and glycogen in animals. They form the main energy sources for both plants and animals. There are polysaccharides, which give mechanical support to the cells or cell components or to the organism as a whole. Such polysaccharides are the structural polysaccharides. Cellulose, a structural polysaccharide is the main component of the plant cell walls and is the most abundant biopolymer in the biosphere. Chitin, the material of the insect exoskeleton and peptidoglycan, the main components of the bacterial cell walls are the other examples of structural polysaccharides.

If the polysaccharides are composed of the same type of monomers or sugars, it is termed as homopolysaccharides and if more than one types of sugar or sugar derivatives are involved in the formation of the polysaccharide it is called the heteropolysaccharides.

Polysaccharides Containing α-glycosidic Linkages

Starch

Starch is the major form of stored carbohydrate in plant cells. Its structure is identical to glycogen, except for a much lower degree of branching (about every 20-30 residues). Starch is the polymer of glucose units linked together by α-1 > 4 linkages and/or α-1 > 6 linkages. Starch is the main storage form of polysaccharides in plants. In most of the cases starch is a mixture of two types of glucose chains– **amylose and amylopectin.** Amylose molecules are straight chains of glucose units linked by α-1 > 4 linkages, unbranched and water-soluble. They are, like most polysaccharides, polydispers, meaning that the length of the molecule is not exactly defined, the number of glucosyl residues is indefinite and it can range from some hundreds to few thousands.

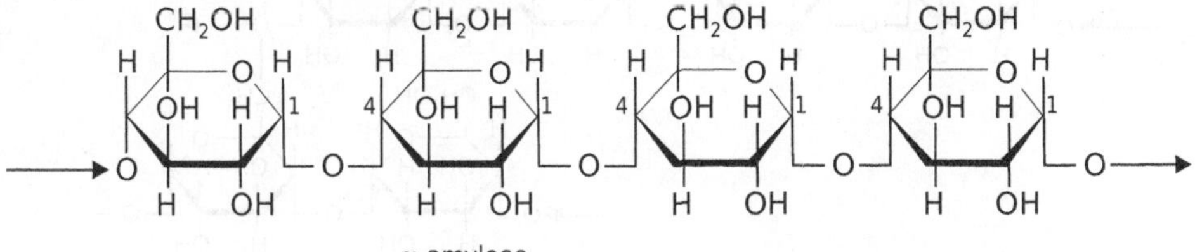

(A)

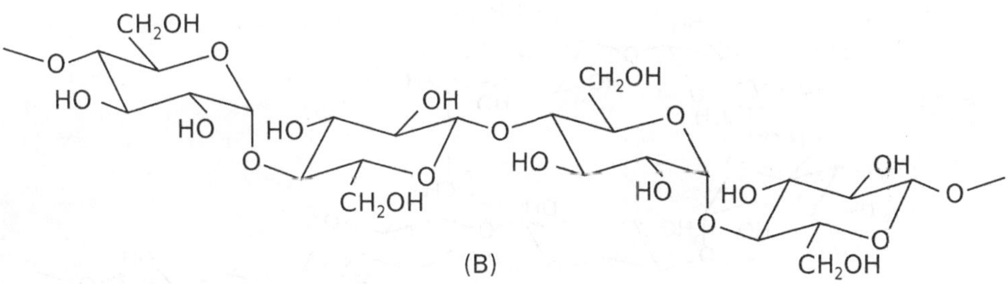

(B)

Figure 2.13 (A) α-1 > 4 Glycosidic bonds in α- amylose and (B) Glucose units in Chair configuration.

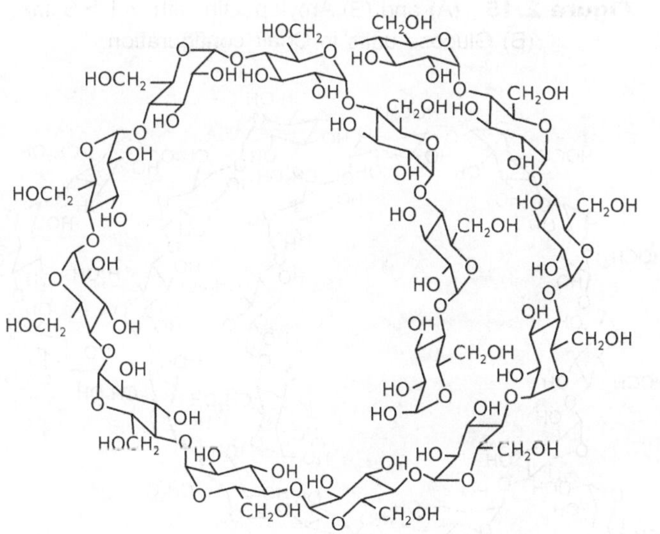

Figure 2.14 Unbranched chain of glucose (amylose) can orient in the form of helix known as Starch helix .

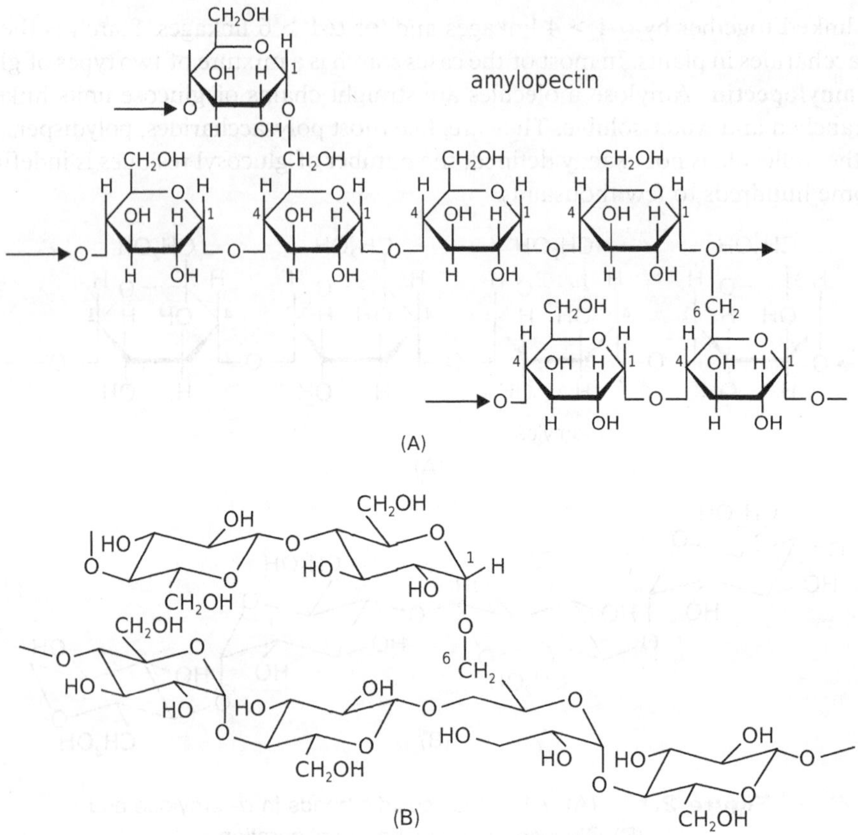

Figure 2.15 (A) and (B) Amylopectin with α-1 > 6 links. (B) Glucose units in Chair configuration.

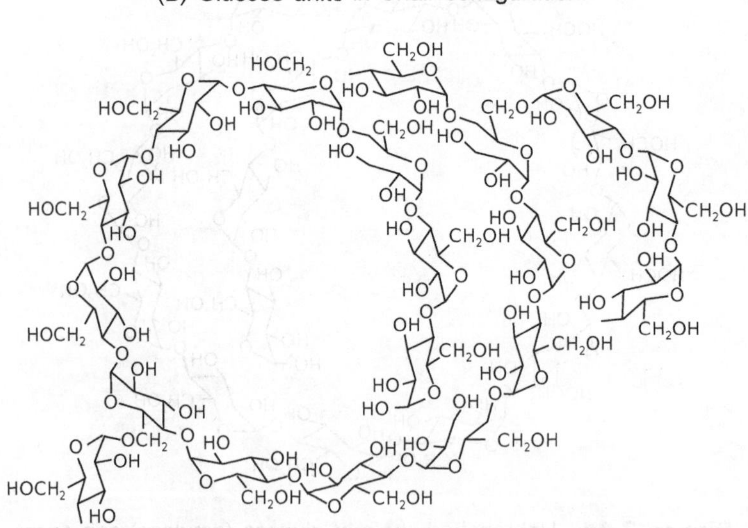

Figure 2.16 Helix formation in the case of amylopectin.

The pyranose ring occurs usually in the boat conformation. The α-1 > 4 linkages cause the helical structure of the polysaccharide chain. The inner diameter of the helix is big enough for elementary iodine to become deposited forming a blue complex (evidence for starch). **Amylopectin** is characterized by branchings. About 2000 to 200,000 glycosyl residues form a molecule. Branching occurs by **α-1 > 6 links** on average at every 20th residue. Starch occurs in layered starch grains within cells. The structure of starch grains is specific for the respective species.

Glycogen also has the same structure as that of Amylopectin except that the rate of branching is more in glycogen than that of Amylopectin. Branching occurs at every 8th to 10th glucose residue by **α-1 > 6 links.** Because of this glycogen molecules are more compact and can accommodate more glucose units per molecule than starch. Glycogen is the storage form of carbohydrates in animals and is stored in liver and muscles.

Glycogen

Glycogen is the major form of stored carbohydrate in animals. This crucial molecule is a homopolymer of glucose in α-(1, 4) linkage; it is also highly branched, with α-(1, 6) branch linkages occurring every 8-10 residues. Glycogen is a very compact structure that results from the coiling of the polymer chains. This compactness allows large amounts of carbon energy to be stored in a small volume, with little effect on cellular osmolarity.

Polysaccharides Containing β-glycosidic Linkages

Cellulose

No branchings occur in cellulose, a linear molecule. Neighbouring cellulose chains may form hydrogen bonds leading to the formation of microfibrils with partially crystalline parts (micelles). Cellulose is the most important structural component of nearly all green plants' cell walls. The glucose units are linked by β-1 > 4- glycosidic linkages.

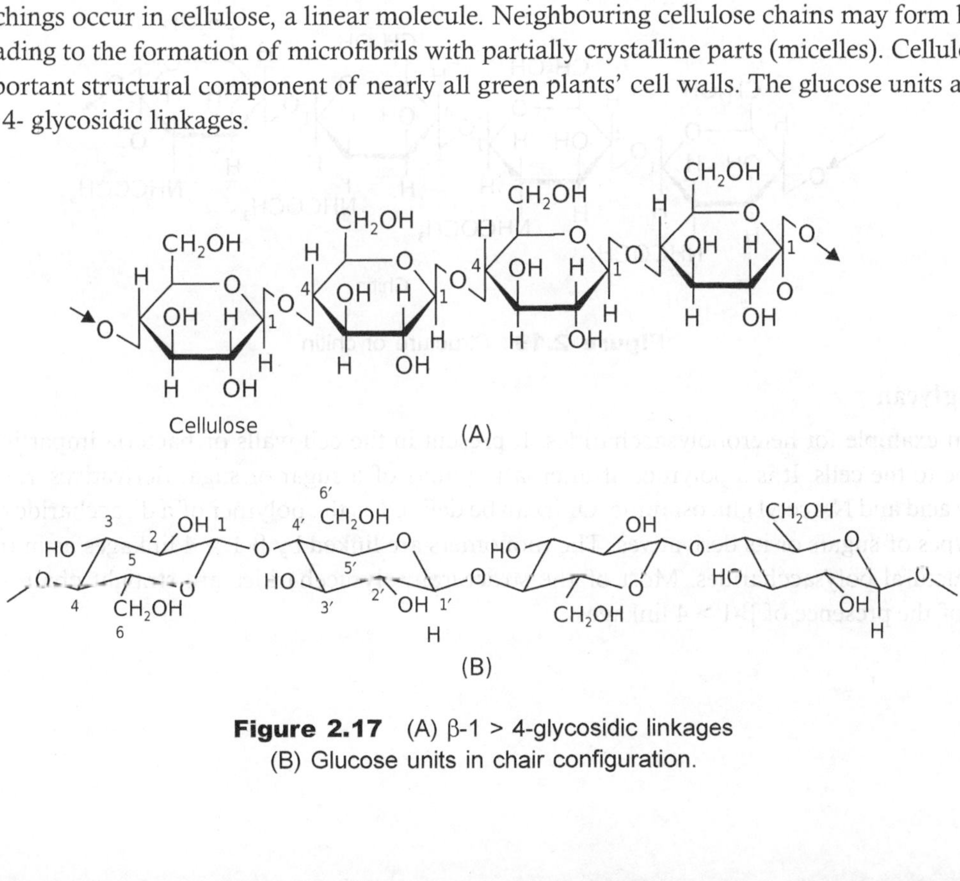

Figure 2.17 (A) β-1 > 4-glycosidic linkages
(B) Glucose units in chair configuration.

Callose

This is branched cellulose present in cell walls in very small quantity. In callose, the glucose units are linked via 1, 3 glycosidic linkages. This type of bond causes the polymer to arrange in a helix. The helix binds aniline blue ; the resulting complex leads to yellow fluorescence. Callose contains also 1 > 4 and 1 > 6 linkages.

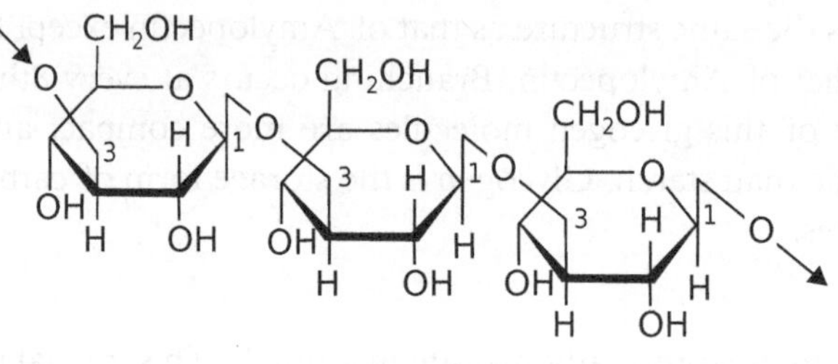

Figure 2.18 Structure of callose.

Chitin

Chitin is a polymer that consists of glucose derivatives ; N-acetyl glucosamine units connected by β-1 > 4 linkages. It is only exceptionally found in plants, like in some algae (see theme 44) but it is the main structural component of the cell walls of molst fungi.

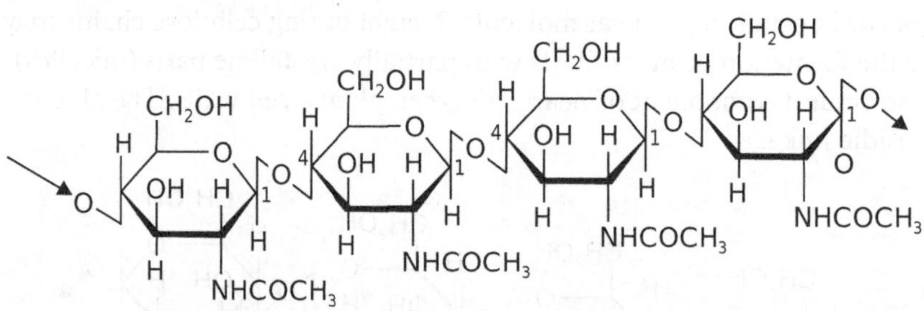

Figure 2.19 Structure of chitin.

Peptidoglycan

This is an example for heteropolysaccharides. It present in the cell walls of bacteria imparting rigidity and shape to the cells. It is a polymer of alternating units of a sugar or sugar derivatives, *e.g.* N-acetyl muramic acid and N-acetyl glucosamine. Or it can be defined as the polymer of a disaccharide consisting of two types of sugars or its derivatives. The monomers are linked by β-1 > 4 linkages as in the case of other structural polysaccharides. Most of the structural polysaccharides are straight chain molecules because of the presence of β-1 > 4 linkages.

Synthesis of Carbohydrates (Polysaccharides)

The characteristic biosynthetic pathway of which leads to the synthesis of polymeric carbohydrates such as starch, cellulose, etc in plants and glycogen in animals occurs as a part of photosynthesis in plants and carbohydrate metabolism in animals.

Glucose and other phosphorylated sugars and its derivatives are the starting compound for the synthesis of polymeric carbohydrates. For example, the biosynthetic pathways of many polysaccharides that are the structural elements of the plant cells, the cellulose and storage compounds such as starch are synthesised using the phosphorylated glucose as the precursors. If the Glucose is not already activated, like all other starting compounds of biosynthetic pathways at first it has to be activated. This occurs either by phosphorylation or by binding of a sugar residue to a nucleotide (like ATP). Thus the formation of glucose-6-phosphate.

Fructose-6-phosphate is an important intermediate of both photosynthesis and glycolysis. It is at equilibrium with glucose-6-phosphate and this again is at equilibrium with glucose-1-phosphate. Glucose-1-phosphate and UTP Uridine Triphosphate) react to UDP-glucose that again polymerizes with fructose-6-phosphate to form saccharose phosphate. Upon cleavage of the phosphate, saccharose (sucrose) is produced. The reaction takes part in chloroplasts. Similarly other types of disaccharides are produced. UDP-glucose (but not ADP-glucose) can also be incorporated into glycolipids and glycoproteins.

In the case of starch, glucose-1-phosphate is coupled to ATP to form ADP-glucose. ADP-glucose that again polymerizes with Glucose-6-phosphate to form a disaccharide, which undergoes further polymerization to from the polysaccharide phosphate. Phosphate is cleaved off from the polysaccharide to generate the starch. This reaction also takes place in part in chloroplasts.

Glucose-1-phosphate + ATP + H_2O • ADP-glucose + 2 Pi

ADP-glucose + Glucose-1-phosphate • ADP-glucose – glucose + Pi

Two enzymes are necessary for starch production: one for the start and the chain elongation (α-1 > 4 glycosidic linkage) with monosaccharide units the other for the introduction of α-1 > 6 glycosidic linkages in amylopectins, as well as for linking together different chains (via alpha 1 > 4 glycosidic bonds).

2. PROTEINS

Building Blocks of Proteins—Amino Acids

Proteins are very large molecules composed of combinations of 20 different amino acids. The precise physical shape of a protein is very important for its function. There are many, many different proteins essential for the functioning of each cell in a living organism.

All proteins (peptides and polypeptides) are polymers of alpha-amino acids. There are 20 α-amino acids that are relevant to the make-up of all proteins. Several other amino acids are found in the body free or in combined states (*i.e.* not associated with peptides or proteins). These non protein associated amino acids perform specialized functions. Several of the amino acids found in proteins also serve functions distinct from the formation of peptides and proteins, *e.g.*, tyrosine in the formation of thyroid hormones or glutamate acting as a neurotransmitter.

The α-amino acids in peptides and proteins consist of a carboxylic acid (**—COOH**) and an amino (**—NH_2**) functional group (except in proline, in which there is **—NH** group instead of **—NH_2** group) attached to the same tetrahedral carbon atom. This carbon is the α-carbon. Distinct R-groups, that distinguish one amino acid from another, also are attached to the alpha-carbon (except in the case of glycine where the R-group is hydrogen). The fourth substitution on the tetrahedral α-carbon of amino acids is hydrogen. All twenty amino acids have the same general formula as given below.

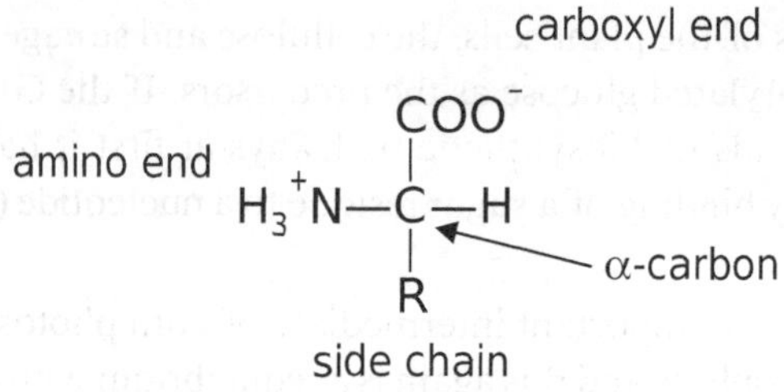

Figure 2.20 General Structure of Amino acid.

The R-group in glycine is a H atom and in proline the α carbon atom has an imino group (—NH) instead of an amino (—NH_2) group. All naturally occurring amino acids have L-configuration with respect α-carbon position. There is no naturally occurring D-amino acid except in bacterial cell walls.

Properties of Amino acids

Hydrophobicity and Hydrophilicity of Amino Acids

Each of the 20 α-amino acids found in proteins can be distinguished by the R-group substitution on the α-carbon atom. There are two broad classes of amino acids based upon whether the R-group is hydrophobic or hydrophilic. The hydrophobic amino acids tend to repel the aqueous environment and, therefore, reside predominantly in the interior of proteins. This class of amino acids does not ionize nor participate in the formation of H-bonds. The hydrophilic amino acids tend to interact with the aqueous environment, are often involved in the formation of H-bonds and are predominantly found on the exterior surfaces proteins or in the reactive centres of enzymes.

Acid-Base Properties of the Amino Acids

The α—COOH and α—NH_2 groups in amino acids are capable of ionizing (as are the acidic and basic R-groups of the amino acids). As a result of their ionizability the following ionic equilibrium reactions may be written :

$$R—COOH \leftrightarrow R—COO^- + H^+$$

$$R—NH_3^+ \leftrightarrow R—NH_2 + H^+$$

The equilibrium reactions, as written, demonstrate that amino acids contain at least two weakly acidic groups. However, the carboxyl group is a far stronger acid than the amino group. At physiological pH (around 7.4) the carboxyl group will be unprotonated and the amino group will be protonated. An amino acid with no ionizable R-group would be electrically neutral at this pH. This species is termed a **zwitterion**.

Like typical organic acids, the acidic strength of the carboxyl, amino and ionizable R-groups in amino acids can be defined by the association constant, K_a or more commonly the negative logarithm of K_a, the pK_a. The **net charge** (the algebraic sum of all the charged groups present) of any amino acid, peptide or protein, will depend upon the pH of the surrounding aqueous environment. As the pH of a solution of an amino acid or protein changes so too does the net charge. This phenomenon can be observed during the titration of any amino acid or protein. When the net charge of an amino acid or protein is zero the pH will be equivalent to the **isoelectric point, pI**.

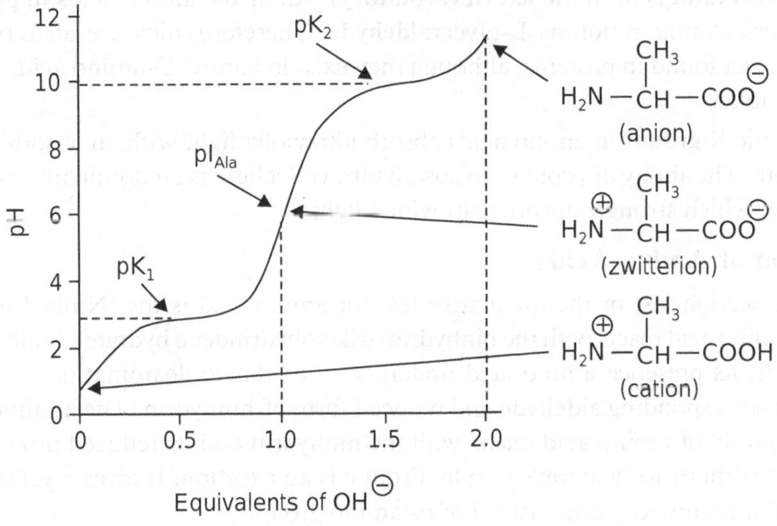

Figure 2.21 Titration curve for Alanine.

Functional Significance of Amino Acid R-Groups

In solution it is the nature of the amino acid R-groups that dictate structure-function relationships of peptides and proteins. The hydrophobic amino acids will generally be encountered in the interior of proteins shielded from direct contact with water. Conversely, the hydrophilic amino acids are generally found on the exterior of proteins as well as in the active centers of enzymatically active proteins. Indeed, it is the very nature of certain amino acid R-groups that allow enzyme reactions to occur.

The imidazole ring of histidine allows it to act as either a proton donor or acceptor at physiological pH. Hence, it is frequently found in the reactive center of enzymes. Equally important is the ability of histidines in hemoglobin to buffer the H^+ ions from carbonic acid ionization in red blood cells. It is this property of hemoglobin that allows it to exchange O_2 and CO_2 at the tissues or lungs, respectively.

The primary alcohol of serine and threonine as well as the thiol (—SH) of cysteine allow these amino acids to act as nucleophiles during enzymatic catalysis. Additionally, the thiol of cysteine is able to form a disulfide bond with other cysteines :

Cysteine—SH + HS—Cysteine ↔ Cysteine—S—S—Cysteine

This simple disulfide is identified as cystine. The formation of disulfide bonds between cysteines present within proteins is important to the formation of active structural domains in a large number of proteins. Disulfide bonding between cysteines in different polypeptide chains of oligomeric proteins plays a crucial role in ordering the structure of complex proteins, *e.g.* the insulin receptor.

Optical Properties of the Amino Acids

A tetrahedral carbon atom with 4 distinct constituents is said to be **chiral**. The one amino acid not exhibiting chirality is glycine since its 'R-group' is a hydrogen atom. Chirality describes the handedness of a molecule that is observable by the ability of a molecule to rotate the plane of polarized light either to the right (**dextrorotatory**) or to the left (**levorotatory**). All of the amino acids in proteins exhibit the same absolute steric configuration as **L-glyceraldehyde**. Therefore, they are all L-α-amino acids. D-amino acids are never found in proteins, although they exist in nature. D-amino acids are often found in polypeptide antibiotics.

The aromatic R-groups in amino acids absorb ultraviolet light with an absorbance maximum in the range of 280nm. The ability of proteins to absorb ultraviolet light is predominantly due to the presence of the tryptophan, which strongly absorbs ultraviolet light.

Colour Reaction of Amino Acids

One of the identification test or the qualitative test for amino acid is the 'Ninhydrin' test. The alpha amino group of amino acid reacts with the ninhydrin (triketohydrindene hydrate). Ninhydrin is a powerful oxidizing agent. In its presence amino acid undergoes oxidative deamination liberating ammonia, carbon dioxide, a corresponding aldehyde and reduced form of ninhydrin. The ammonia liberated from the alpha amino group of amino acid reacts with the ninhydrin and its reduced product to form a blue substance, diketohydrin or Ruhemann's purple. Proline is an excetion. It gives a yellow colour because of the presence of an α-imino group instead of α- amino group.

The Peptide Bond

Peptide bond formation is a condensation reaction leading to the polymerization of amino acids into peptides and proteins. Peptides are small consisting of few amino acids. A number of hormones and neurotransmitters are peptides. Additionally, several antibiotics and antitumor agents are peptides. Proteins are polypeptides of greatly divergent length. The simplest peptide, a **dipeptide**, contains a single peptide bond formed by the condensation of the carboxyl group of one amino acid with the amino group of the second with the concomitant elimination of water. The presence of the carbonyl group in a peptide bond allows electron resonance stabilization to occur such that the peptide bond exhibits rigidity not unlike the typical —C=C— double bond. The peptide bond is, therefore, said to have **partial double-bond character**.

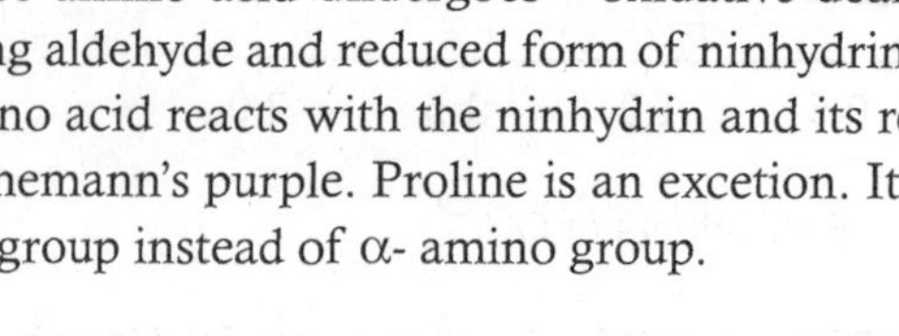

Figure 2.22 Resonance stabilization forms of the peptide bond.

TABLE 2.5 The twenty protein amino acids classified according to the nature of their R groups. The structures and names, both full names and three-letter abbreviations, and dissociation constants (pK_a Value) of each group (—COOH, —NH_2, and the R-group) in the amino acids are shown in separate columns.

Amino Acid	Symbol	Structure	pK_1	pK_2	pKR Group
		Hydrophobic Amino Acids (with Aliphatic R-Groups)			
Glycine	Gly-G	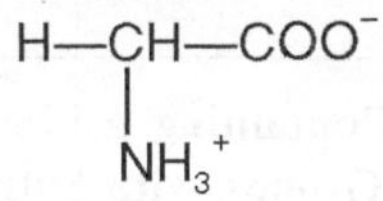	2.4	9.8	
Alanine	Ala-A	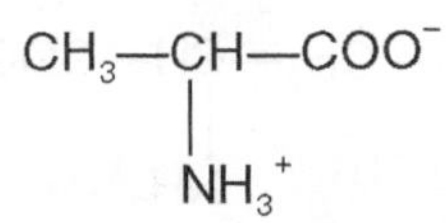	2.4	9.9	
Valine	Val-V	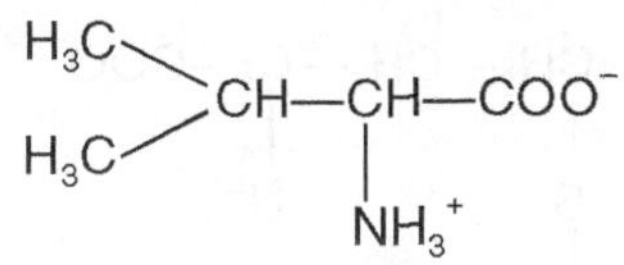	2.2	9.7	
Leucine	Leu-L	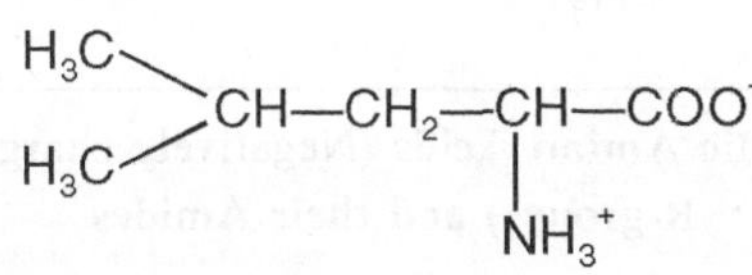	2.3	9.7	
Isoleucine	Ile-I	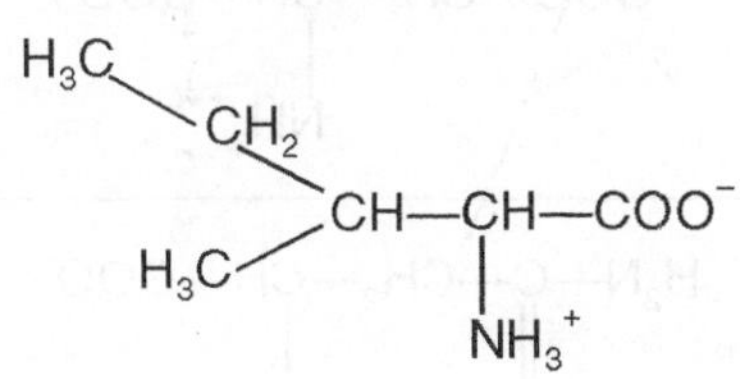	2.3	9.8	
Proline	Pro-P	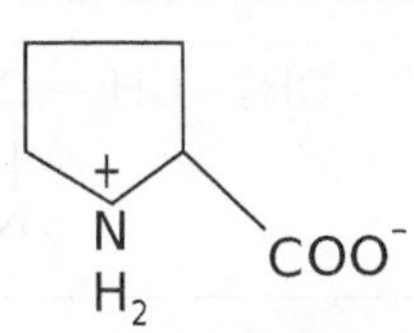	2.0	10.6	

		Hydroxilated aliphatic Amino Acids (with Hydroxilated R-Groups)			
Serine	Ser-S	$HO—CH_2—CH(NH_3^+)—COO^-$	2.2	9.2	~ 13
Threonine	Thr-T	$(H_3C)(HO)CH—CH(NH_3^+)—COO^-$	2.1	9.1	~ 13
		Sulfur-Containing Amino Acids (R-Groups with Sulfur)			
Cysteine	Cys-C	$HS—CH_2—CH(NH_3^+)—COO^-$	1.9	10.8	8.3
Methionine	Met-M	$CH_3—S—CH_2—CH_2—CH(NH_3^+)—COO^-$	2.1	9.3	
		Acidic Amino Acids (Negatively charged R-groups) and their Amides			
Aspartic Acid	Asp-D	$^-OOC—CH_2—CH(NH_3^+)—COO$	2.0	9.9	3.9
Asparagine	Asn-N	$H_2N—C(=O)—CH_2—CH(NH_3^+)—COO$	2.1	8.8	
Glutamic Acid	Glu-E	$^-OOC—CH_2—CH_2—CH(NH_3^+)—$	2.1	9.5	4.1

Glutamine	Gln-Q	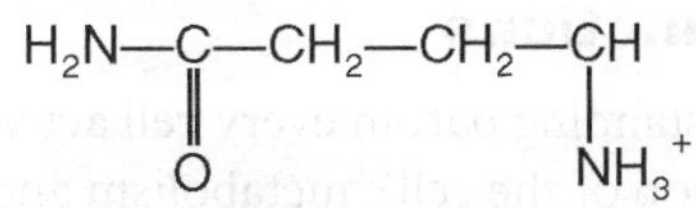	2.2	9.1	
		Basic Amino Acids (with positively charged R-groups)			
Arginine	Arg-R	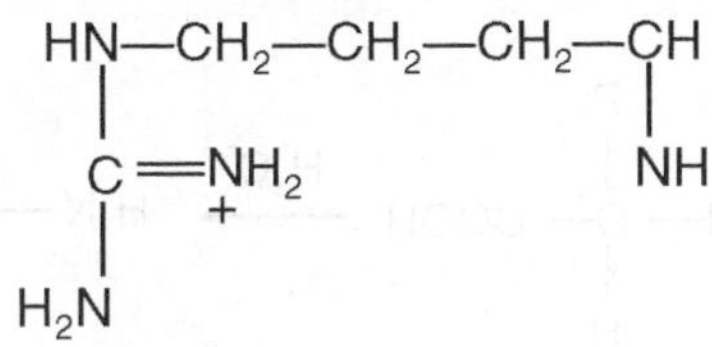	1.8	9.0	12.5
Lysine	Lys-K	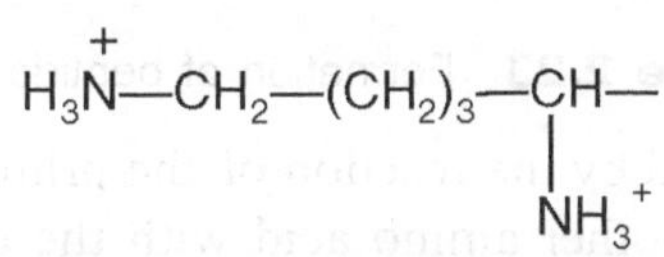	2.2	9.2	10.8
Histidine	His-H	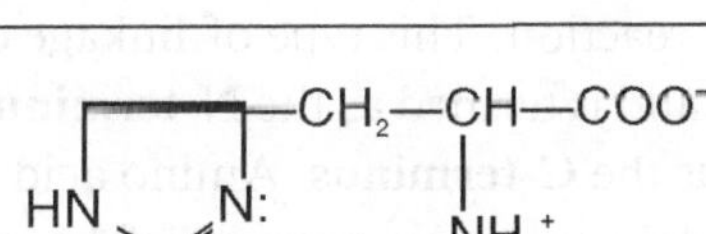	1.8	9.2	6.0
		Aromatic Amino Acids (with aromatic R-groups)			
Phenylal anine	Phe-F	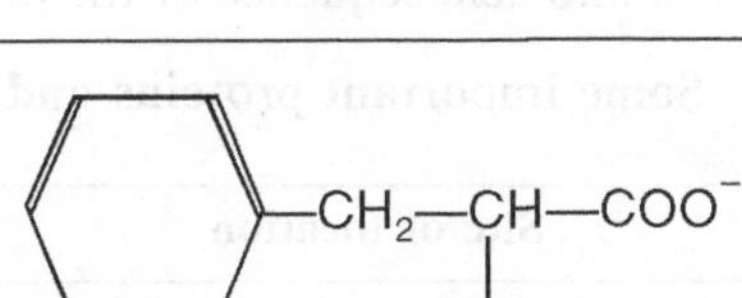	2.2	9.2	
Tyrosine	Tyr-Y	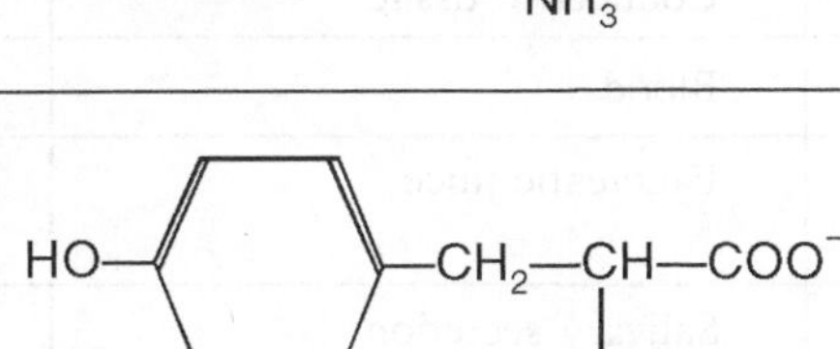	2.2	9.1	10.1
Tryptophan	Trp W	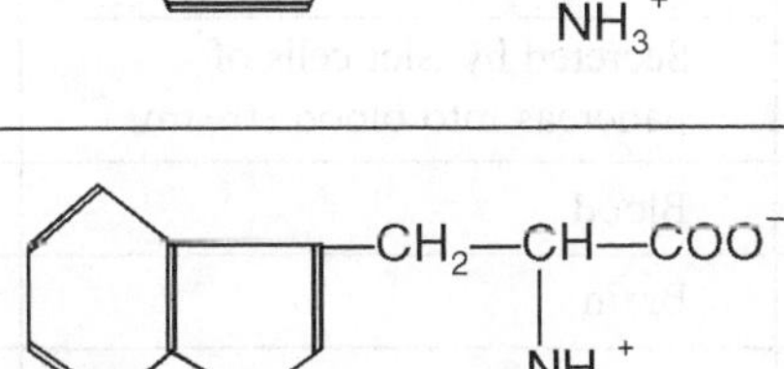	2.4	9.4	

Proteins-Structure and Functions

Proteins or polypeptides play an outstanding part in every cell activities. They act as biological catalysts (= enzymes), take part in the regulation of the cell's metabolism and in the interaction between cells and are required for the generation of specific structures. They are linear chains consisting of a sequence of 20 amino acids in different combinations linked exclusively by peptide bonds. The different combination and sequence of amino acids are responsible for the diverse nature and functions of proteins (Table 2.6).

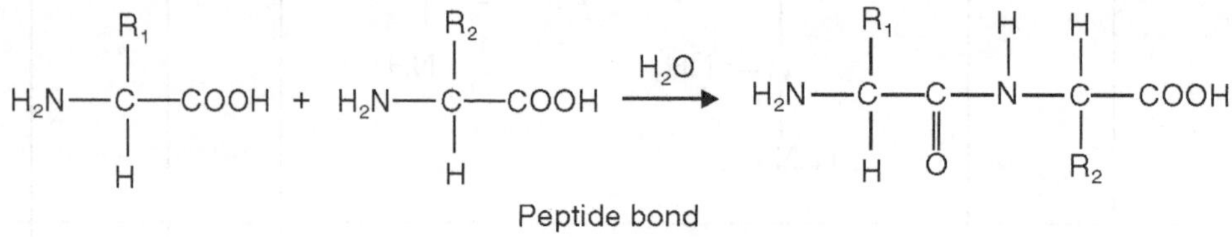

Figure 2.23 Formation of peptide bond.

The peptide bonds are formed by the reaction of the primary amino group of one amino acid with primary carboxyl group of another amino acid with the elimination of a molecule of water (Fig. 2.23). Thus it is a condensation reaction. This type of linkage causes a polarity for the polypeptide chain. One end has the amino group and is termed as the **N-terminus**, while the other end terminates by a free carboxyl group and is termed as the **C-terminus**. Amino acid sequences are written from N- to C-terminus, the direction in which protein synthesis proceeds. The exact sequence of amino acids (also called the protein's primary structure) is determined by the nucleotide sequence of the gene, the part of the DNA strand, which codes for the protein. The 3-dimensional structure and the function of the protein very closely depended on the amino acid sequence of the polypeptide.

TABLE 2.6 Some important proteins and its function

Protein	Site of location	Function
Collagen	Connective tissue	Gives tensile strength
Thrombin	Blood	Blood coating
Trypsin	Pancreatic juice	Cleaves the polypeptide during protein digestion.
Amylase	Salivary secretion	Starch hydrolysis
Insulim	Secreted by islet cells of pancreas into blood stream	Controls the blood sugar level. Immunity
Immunoglobulins	Blood	Regulates the brain activities
Endorphins	Brain	Vision
Rhodopsin	Retinal cells of eye	Cellular respiration
Cytochromes	Mitochondrial membranes	

There are three types of proteins based on their complexity. Some proteins are made up of a single polypeptide chain and are known as simple proteins. Those proteins having two or more polypeptide chains are termed as complex proteins. In some cases the protein molecule is associated with a non-protein component known as the prosthetic group. Such proteins are known as conjugated proteins. The non protein component may be metallic ions like Zn^+ in the case of Carbonic anhydrase, haem part of haemoglobin enclosing Fe^+ ion in it or organic molecules like vitamin derivatives like NAD and NADP, nucleotides like ATP and GTP or may be sugars, oligosaccharides, or various types of lipids.

Amino Acid Composition and Protein Sequencing

Amino Acid Composition

To determine the amino acid composition the peptide is first hydrolyzed into its constituent amino acids by heating in 6 N HCl at 110°C for 24 hours. The amino acids in the hydrolysate can be separated by ion-exchange chromatography and quanititated by reacting then with ninhydrin. Alpha amino acids treated this way give an intense bluc color, whereas imino acids, such as proline, give a yellow color. The concentration of amino acid in a solution is proportional to the optical absorbance of the solution after heating it with ninhydrin. This technique can detect a microgram (10 nmol or nanograms) of an amino acid. After getting the information about the amino acid composition and relative quantity of each amino acid, one can proceed to do the sequencing of amino acids for a particular protein or polypeptide.

Amino Acid Sequencing

The amino acid sequence of a protein is very important because it is essential to know about the structure and function of that protein; and also it can help in identifying and isolating the gene that code for the protein. So obtaining at least a partial amino acid sequence is a critical first step in studying many proteins.

The first protein to be sequenced was the peptide hormone insulin, which controls the glucose level in blood, and its deficiency can leads to the metabolic error, diabetes. Nobel Laureate, Frederic Sanger showed for the first time that proteins have a definite amino acid sequence and a specific 3-D structure that is determined by the amino acid sequence. He found out that certain reagents like fluoro-dinitro-benzene (FDNB) known as Sanger's reagent can react specifically with the free NH_2 group of the amino acid at the N-terminal of a polypeptide. Fluoro-DNB reacts with the N-terminal amino acid and on acid hydrolysis it yields a yellow DNP derivative of the amino acid, which can be separated and identified by ion exchange chromatography. This procedure cannot be used on the same sample of polypeptide because the peptide is totally hydrolyzed on the acid hydrolysis step. But Sanger managed to use this technique with new samples of peptide in each cycle of experiments to sequence insulin. He has used more than a gram of insulin for completing this task. Now this method is used only for identifying the N-terminal amino acid of a polypeptide and not for sequencing because of the above-mentioned demerit.

Edman Degradation Reaction

The most popular direct protein-sequencing technique in use today is the Edman degradation procedure. The Edman reaction is a series of chemical reactions, which remove one amino acid at a time from amino terminus of a protein, releasing an amino acid derivative, phenylthiohydantoin (PTH) that may be chromatographically identified (reversed phase). The Edman reagent, Phenylthiocyanate reacts with the NH_2 group of the terminal amino acid and forms an intermediate, phenylthiocarbomyle derivative. A simplified schematic diagram of the process is shown in figure 2.24. Release of amino acids from the amino terminus in a sequential manner is possible because the Edman procedure consists of three chemical reactions, which proceed under different pH conditions (Fig. 2.24).

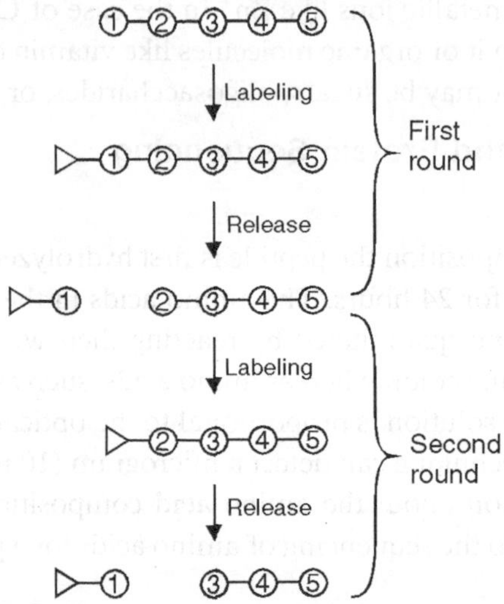

Figure 2.24 Sequential removal of amino acids in Edman's degradation reaction.

The first step is the coupling step (Fig. 2.25), which occurs a high pH values and results in formation of phenylthiocarbamoylated (PTC) amino groups on the protein. The second reaction is the cleavage step, which occurs at low pH, resulting in release of an anilionothiazolinone (ATZ) form of the amino acid and regeneration of a free amino terminus on the protein. The ATZ-amino acid is converted to the phenylthiohydantoin (PTH) derivative in a separate reaction generally exposure to strong acid. The PTH-amino acids are more stable and a bit more amenable to chromatographic resolution. The PTH-amino acid can be separated and identified by HPLC (High Performance Liquid Chromatography). This leaves the intact peptide shorten by one amino acid. The Edman procedure can be repeated on the shortened peptide obtained in the previous cycle for identifying the second amino acid of the polypeptide from the N-terminal.

The whole method including the reaction steps and the identification of PTH-amino acid derivative is automated and the instrument that can carryout these reactions is known as 'Sequenator'. Using this automated instrumentation, more than 100 amino acids of a polypeptide can be sequenced efficiently. Polypeptides and proteins of high molecular weight have to be fragmented in to smaller polypeptides of 50 to 100 amino acids before carrying out the sequencing by a Sequenator. Amino acid sequencing by Edman degradation reaction can be carried out with very minute quantities of protein to the level of micro grams. There are a number of physical techniques like Mass spectrometry, NMR (Nuclear Magnetic Resonance spectrometry) and X-ray diffraction techniques used in the protein analysis and structural elucidation. There are various types of Mass spectrometric techniques for the analysis of proteins. For example, Matrix assisted Laser Desorption Ionization (MALDI) Mass Spectrometry and Tandem Mass spectrometry (2D-MS or MS-MS) are efficient techniques developed to identify and sequence proteins rapidly.

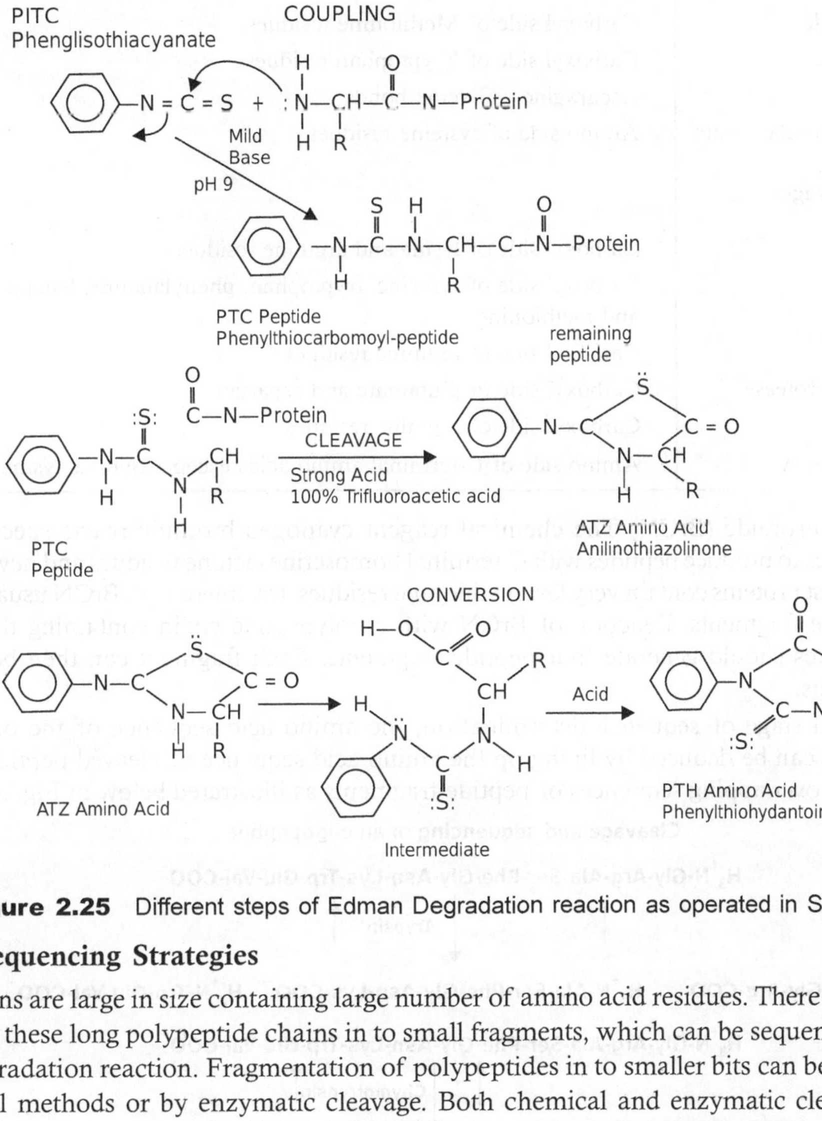

Figure 2.25 Different steps of Edman Degradation reaction as operated in Sequenator.

Protein Sequencing Strategies

Most proteins are large in size containing large number of amino acid residues. There fore it is necessary to fragment these long polypeptide chains in to small fragments, which can be sequenced completely by Edman degradation reaction. Fragmentation of polypeptides in to smaller bits can be carried out either by chemical methods or by enzymatic cleavage. Both chemical and enzymatic cleavage of are very specific and the cleavage profile of a polypeptide by a specific chemical or enzyme can be used for the identification of an unknown protein. Proteases or certain chemical reagents are used to selectively cleave some of the peptide bonds of a protein. The smaller peptides fragments formed are then isolated and subjected to sequencing by the Edman degradation procedure. A list of chemicals and protease used to cleave polypeptides at selected points are given in table 2.7.

TABLE 2.7 Specific cleavage of polypeptides

Reagents	Cleavage site
Cyanogenbromide	Carboxyl side of Methionine residues
O-Idosobenzoate	Carboxyl side of Tryptophan residues
Hydroxylamine	Asparagine—Glycine bonds
2-Nitro-5-thiocyanobenzoate	Amino side of cysteine residues
Enzymatic cleavage	
Trypsin	Carboxyl side of lysine and arginine residues
Chymotrypsin	Carboxyl side of tyrosine, tryptophan, phenylalanine, leucine and methionine
Clostripain	Carboxyl side of arginine residues
Staphylococcal protease	Carboxyl side of glutamate and aspartate
Thrombin	Carboxyl side of arginine residues
Carboxypeptidase A	Amino side of C-terminal amino acids except arginine, lysine and proline

Cyanogen bromide (BrCN) The chemical reagent cyanogen bromide reacts specifically with methionine residues to produce peptides with C-terminal homoserine lactone residues and new N-terminal residues. Since most proteins contain very few methionine residues, treatment with BrCN usually produces only a few peptide fragments. Reaction of BrCN with a polypeptide chain containing three internal methionine residues should generate four peptide fragments. Each fragment can then be sequenced from its N-terminus.

In the final stage of sequence determination, the amino acid sequence of the original large polypeptide chain can be deduced by lining up the amino acid sequence of cleaved peptide fragments and matching the overlapping sequences of peptide fragments as illustrated below in Fig. 2.26.

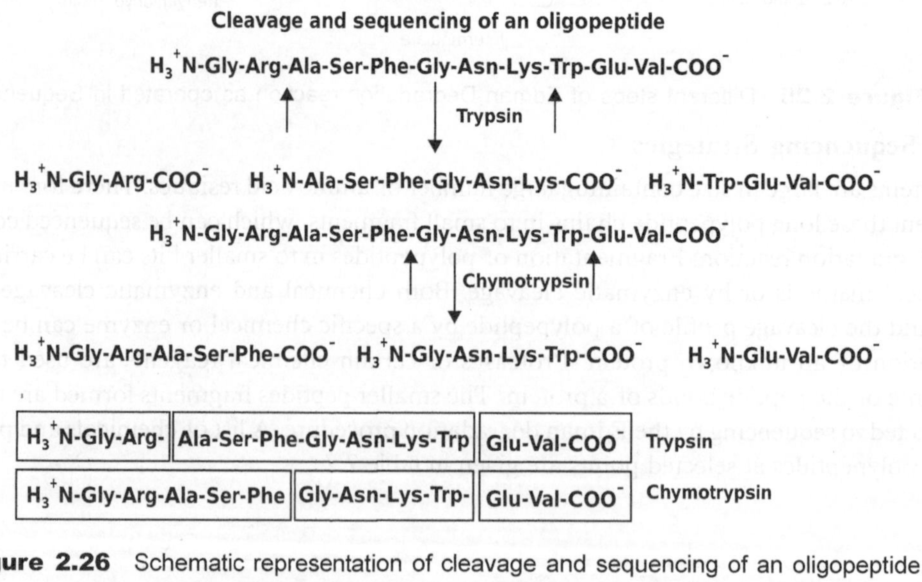

Figure 2.26 Schematic representation of cleavage and sequencing of an oligopeptide.

Three Dimensional Structure of Proteins

All proteins exhibit a specific molecular shape, which is determined by the amino acid sequence or the primary structure of the protein. This was first demonstrated by Christian Anfinsen on enzyme pancreatic Ribonuclease A, which hydrolyses RNA (ribonucleic acids). He has shown that a pure sample of Ribonuclease A lost its 3-D structure and most of its properties including catalytic activity, when treated with strong chemicals like urea or by heating. The enzyme could regain its most of its properties and 3-D structure, when the denaturing agent (high temperature or the chemicals like urea) is removed. This demonstrated mainly two things : (1) The function of a protein is closely linked to the 3-D structure of the protein. (2) All the information necessary for folding the polypeptide in to 3-D structure is present in the sequence of amino acid or the primary structure of the protein.

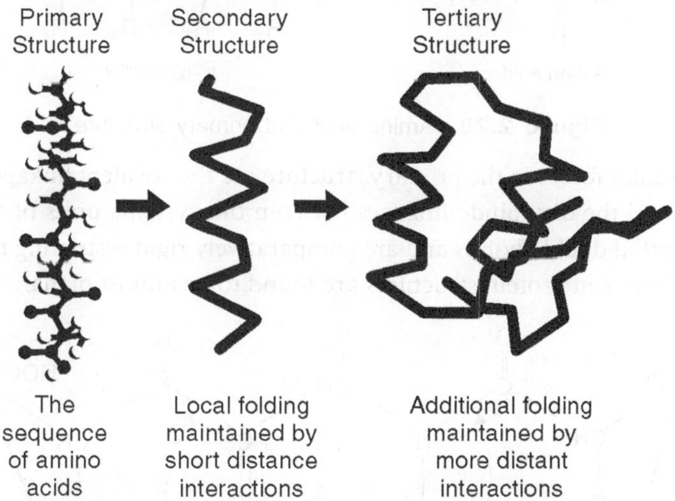

Figure 2.27 Three stages of protein structure.

Based on the molecular shape at its 3-D structure proteins are classified into two—Globular proteins and Fibrous proteins. Globular proteins have globular 3-D shape and are therefore soluble proteins. Fibrous proteins have a long thin fibre like structure and are insoluble proteins.

The structure of protein includes three stages in its transition from amino acid sequence to the functional 3-D structure. In multimeric proteins where there are more than one polypeptide chains, there are four stages in the process of folding and conversion to the functional 3-D stage. They are primary structure, secondary structure tertiary structure and quaternary structure (applicable in multimeric proteins).

The sequence of R-groups along the chain is called the primary structure (Fig. 2.27 and 2.28). Secondary structure refers to the local folding of the polypeptide chain. Tertiary structure is the arrangement of secondary structure elements in 3 dimensions and quaternary structure describes the arrangement of a protein's subunits.

Primary Structure of Proteins

The primary structure of a protein or polypeptide is the sequence of amino acids joined by the peptide bonds formed between the COOH group of one amino acid with the NH_2 group of the other. The primary structure of protein can be determined by Edman Degradation Reaction or it can be obtained from cloning and sequencing the gene responsible for the production of the protein. Size of the protein molecule can be determined by SDS-PAGE (Sodium dodecyle sulphate-Polyacrylamide gel electrophoresis) or Electrospray-Mass Spectrometry (ES-MS).

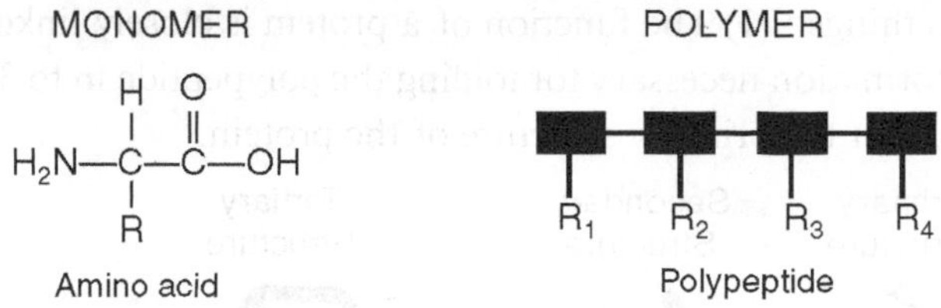

Figure 2.28 Amino acid and primary structure.

The main molecular forces in the primary structure are the covalent linkages – the peptide bonds between amino acids and the disulphide linkages between the cysteine units of the polypeptides. The peptide linkages are partial double bonds and are comparatively rigid restricting the free rotation about the bond. All peptide bonds in protein structures are found to be almost planar.

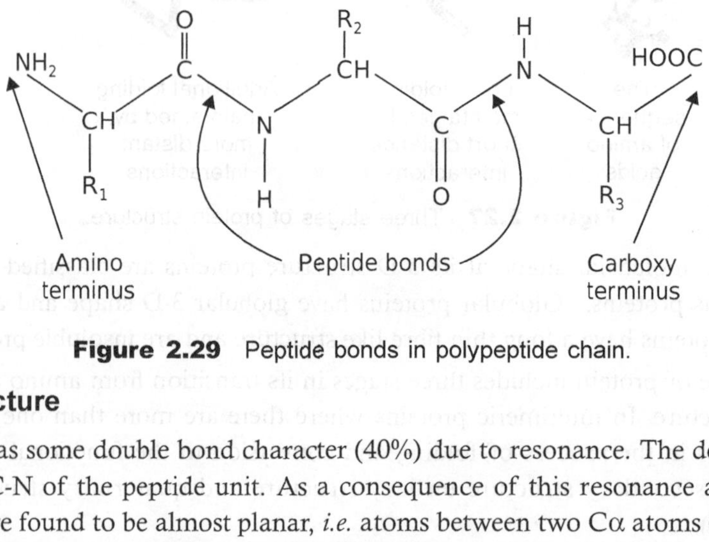

Figure 2.29 Peptide bonds in polypeptide chain.

Secondary Structure

The peptide bond has some double bond character (40%) due to resonance. The double bond of C=O alternate with the C-N of the peptide unit. As a consequence of this resonance all peptide bonds in protein structures are found to be almost planar, *i.e.* atoms between two Cα atoms (C, O, N and H) are approximately co-planar. This rigidity of the peptide bond reduces the degrees of freedom of the polypeptide during folding. One can visualize the polypeptide chain consisting of rigid planes of peptide units interconnected by Cα atoms of each amino acid residue.

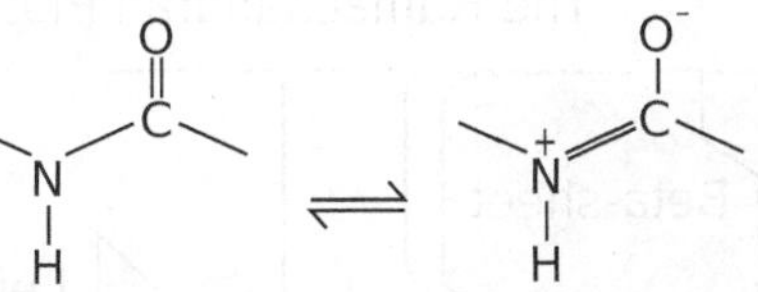

Figure 2.30 Resonance of double bond in peptide linkage.

The peptide bond nearly always has the *trans* configuration since it is more favorable than *cis*, which is sometimes found to occur with proline residues. Movement in the polypeptide chain is possible only by rotation in the bond angles of Cα—NH— and Cα—CO—. The angles are called the torsion angles. There are three main chain torsion angles of a polypeptide. These are phi (ϕ), psi (ψ) and omega (ω) (Fig. 2.31). To make the peptide bond planar omega is nearly always 180 degrees. Rarely it is 0 degrees. The ϕ and ψ angles are present on either side of the Cα atoms. The angle N— Cα bond is the ϕ angle and that of Cα—C linkage is the y angle. Polypeptide chain is flexible only at these linkages and therefore movements are possible by rotation about ϕ and ψ angles. These movements are also limited by the nature of the side chain (R group) of the amino acids. The R-groups attached to the Cα— atoms interfere with the values of the torsion angles and only certain values are permitted for the ϕ and ψ angles, to avoid steric hindrance (collision of atoms during bond rotation).

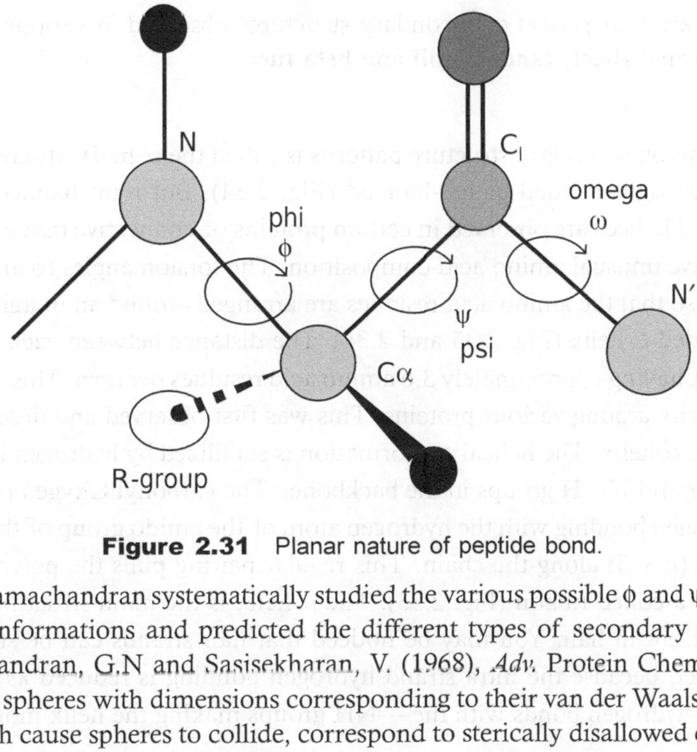

Figure 2.31 Planar nature of peptide bond.

Prof. G.N. Ramachandran systematically studied the various possible ϕ and ψ angles via computer to finding stable conformations and predicted the different types of secondary structures in protein molecules (Ramachandran, G.N and Sasisekharan, V. (1968), *Adv.* Protein Chem., 23, 283.). Atoms were treated as hard spheres with dimensions corresponding to their van der Waals radii. Therefore, phi and psi angles, which cause spheres to collide, correspond to sterically disallowed conformations of the polypeptide backbone. A graphical representation in which the ϕ values plotted against ψ values is known as the Ramachandran plot (Fig. 2.32).

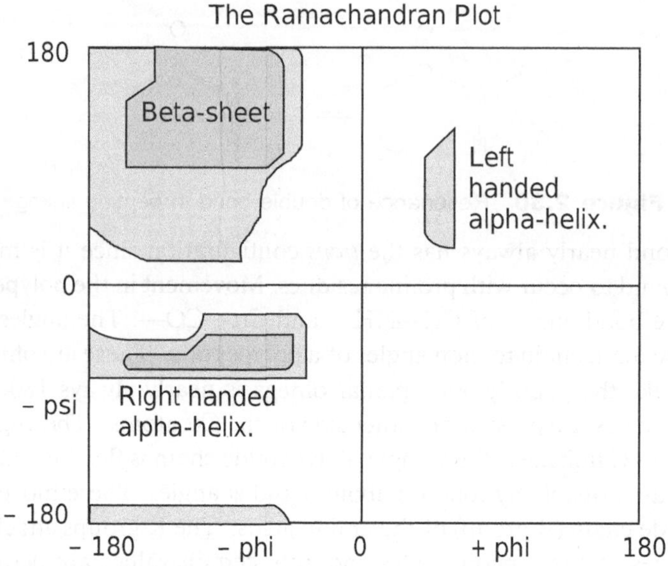

Figure 2.32 Ramachandran plot.

Mainly there are four patters of secondary structures observed in various protein molecules—**alpha helix, beta pleated sheet, random coil** and **beta turn.**

Alpha Helix

One of the most common secondary structure patterns is called the α-**helix**, discovered by Pauling and Corey. α-helix can be right handed or left-handed (Fig. 2.34). But right handed α-helices are more common. Left-handed helices are observed in certain proteins of connective tissue such as collagen and are also known to have unusual amino acid composition. The torsion angles (ϕ and ψ angles) have the characteristic values so that the amino acid residues are arranged around an imaginary axis and form a helix—the right handed α-helix (Fig. 2.33 and 2.34). The distance between each turn in an α-helix is calculated to be 5.4 Å having approximately 3.6 amino acid residues per turn. This is the most frequently observed widely α-helix among various proteins. This was first observed and described in α-keratin of hair, and so the name α-helix. The helical conformation is stabilized by hydrogen bonds formed readily between C=O groups and N—H groups in the backbone. The carbonyl Oxygen of one amino acid (*n*) residue make a hydrogen bonding with the hydrogen atom of the amido group of the residue four amino acid residues further ($n + 3$) along the chain. This regular pairing pulls the polypeptide into a helical shape that resembles a coiled ribbon (Fig. 2.33). The α-helix is the main structural component of the protein α-keratin present in hair. You may be noticed that hair strands can be stretched and is more flexible when it is wet, because the intra strand hydrogen bonding is reduced as water molecules are available for making hydrogen bonds with the —NH groups making the helix more flexible.

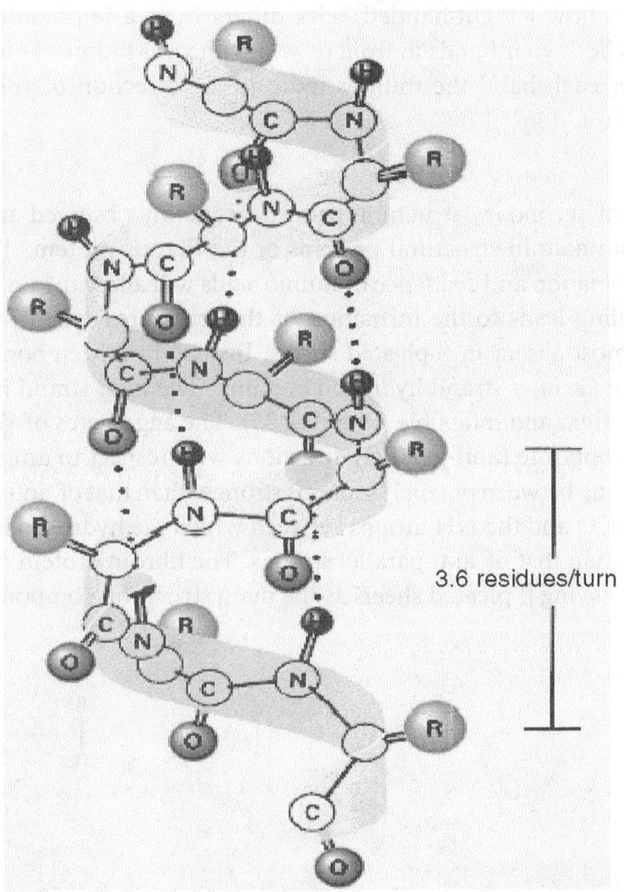

Figure 2.33 Diagrammatic representation of a right handed α-helix of a polypeptide.

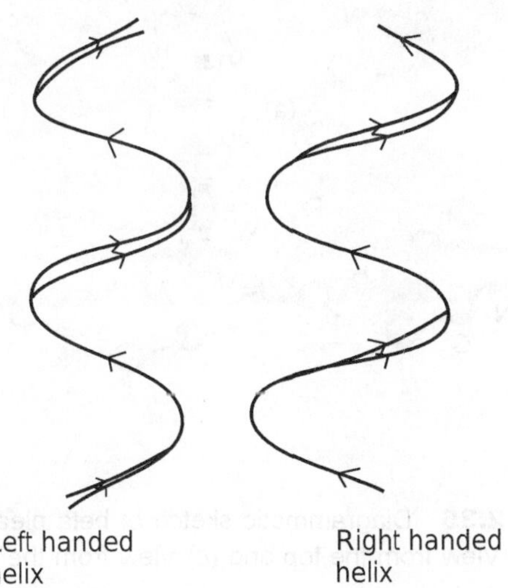

Figure 2.34 Right handed and left handed helix.

The figure shows how a right-handed helix differs from a left-handed one. An easy way to remember this is to hold both your hands in front of you with your thumbs pointing up and your fingers curled towards you. For each hand the thumbs indicate the direction of translation and the fingers indicate the direction of rotation.

Beta Pleated Sheet

This is the second type of secondary structural pattern frequently observed among different classes of proteins. But it is predominant in structural proteins or the fibrous protein. The characteristic torsion angles for a certain combination and sequence of amino acids will allow the polypeptide to form straight chains with out any coiling leads to the formation of the structures called β-sheets. The intra strand hydrogen bonding is almost absent in β-pleated sheets. Instead hydrogen bonding is possible between β-strands in an aggregate *i.e.* inter strand hydrogen bonding. The inter strand hydrogen bonding makes the protein more strong, rigid and inflexible (Figure 2.35). The aggregates of β-strands may be running in the same (parallel) or opposite (anti-parallel) directions with respect to amino terminal and carboxy terminal. Hydrogen bonding between parallel strands is stronger than that of anti-parallel strands. Because, in parallel strands the —CO and the NH groups between which the hydrogen bond is formed are closer and are aligned straight than that of anti-parallel strands. The fibroin protein of natural silk and spider web are fibrous proteins having β-pleated sheets as the main structural components and has high tensile strength.

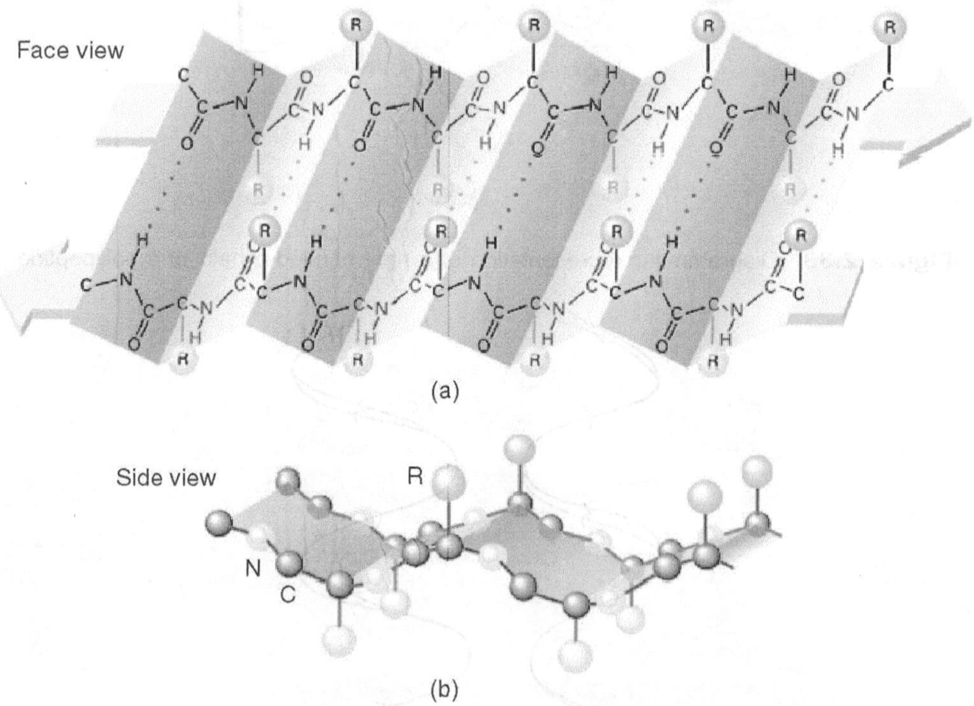

Figure 2.35 Diagrammatic sketch of beta pleated sheet. (*a*) View from the top and (*b*) View from the side.

Random Coil

These secondary structure patterns consist of unspecified coils, loops and sheets, which do not have any constant pattern. They are usually resulted by a mixture of amino acid sequence, which cannot form either a α-helix or a β-sheet. Random coils are present along with α-helices and β-sheets in most of the globular proteins.

Beta Turns

To combine helices and sheets in their various combinations, protein structures must contain turns that allow the peptide backbone to fold back. Two turn structures will be discussed here using their Ramachandran plot coordinates. These turns can be found almost always on the surface of proteins and often contain Pro and/or Gly. Proline gives the backbone a special rigidity (fixed Phi torsion angle at —60° , Cα —N) and glycine has a high flexibility because of its hydrogen substituent. Turn structures are also stabilized through H-bond formation.

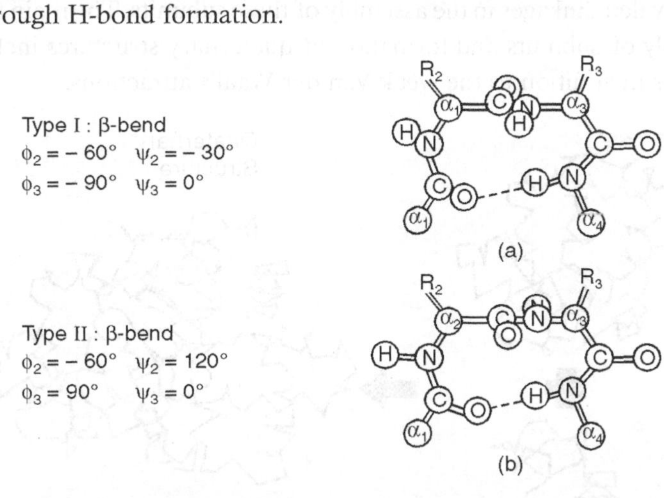

Figure 2.36 Two types of β-turns.

Tertiary Structure of Proteins

The regular polypeptide structures or the secondary structures are assembled or grouped together by specific folding of the polypeptide units like helices and sheets are grouped together to from a specific shape. It is only with the complete, compact folding into tertiary structure that the proteins attain their "native conformation" and become active proteins (as a result of the creation of active sites). Forces that contribute to tertiary folding include: hydrogen bonds hydrophobic interactions and ionic bonds. Sulfhydryl bonds (—S—S— bonds) were earlier considered, as a force contributing to the tertiary structure, but now it is considered as molecular interaction contributing to the primary structure of polypeptide along with peptide bonds. These are especially important, because thcy are covalent bonds and quite strong compared to H-bonds. It was observed that certain combination of secondary structures like helix and sheets are conserved in certain classes of proteins and such groups of secondary structures are known as **motifs**. The motifs form the structural elements of proteins since they are conserved. Domains certain other conserved assembly of secondary structures of protein molecules they are functional units. The presence of certain conserved domains and motifs in the tertiary structure of protein can be used for the predicting the function of a protein or its identification.

Quaternary Structure of Protein

Some proteins are made of more than one polypeptide chains and such proteins are referred to as multimeric proteins. The polypeptides at its tertiary structures are assembled to form the native conformation and become functional proteins. The individual polypeptides are known as the subunits, which must be assembled together after each individual polypeptide has reached its tertiary structure. Examples: **Hemoglobin** (blood protein involved in oxygen transport) has **four subunits** (Fig. 2.37). **Pyruvate dehydrogenase** (mitochondrial protein involved in energy metabolism) has **72 subunits.** There is no involvement of covalent linkages in the assembly of these subunits. The main molecular interactions involved in the assembly of subunits and formation of quaternary structures include hydrophobic and electro static attractions in addition to the weak Van der Waal's attractions.

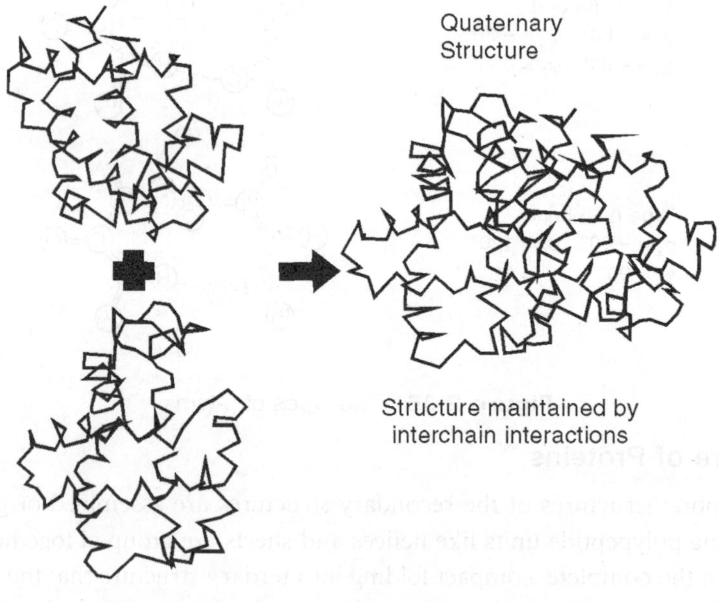

Figure 2.37 Tertiary Structure of protein.

When two subunits or polypeptides are linked through a covalent bond like disulphide linkages, then they cannot be considered as subunits. For example, in the case hemoglobin there are four polypeptides and each one is a subunit. Whereas in the case of the peptide hormone insulin there are two polypeptides, which are connected together by disulphide linkages and therefore they are not subunits.

RIBBON MODEL OF HEMOGLOBIN

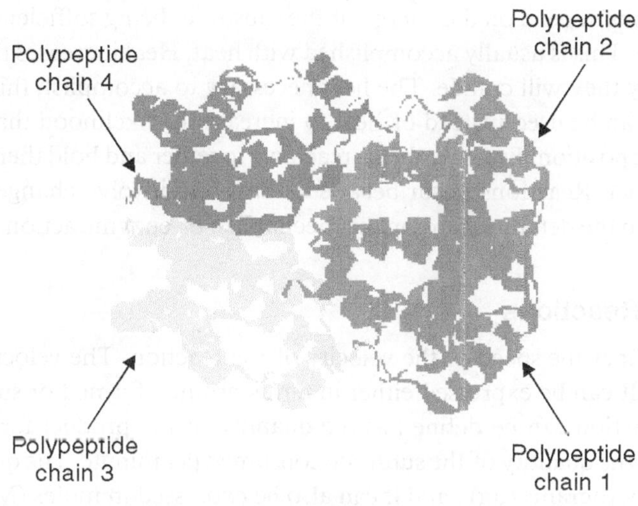

Figure 2.38 Tertiary structure of hemoglobin.

3. ENZYMES

Introduction

Enzymes are the biocatalysts that control almost all cellular reactions. A catalyst is a compound, which accelerates a chemical reaction with out undergoing any change in quality or quantity. Enzymes are globular proteins, each with a specific structure (native conformation), function, distribution of electrical charges, and surface geometry whose specificity depends on their tertiary structure. The tertiary structure determines the three-dimensional shape. They are responsible for control of a single reaction and are thus responsible for control of metabolism. There are about 700 enzymes in a typical prokaryotic cell and there are thousands in a eukaryotic cell.

Enzymes function as **catalysts**, which are substances that facilitate (speed up) reactions without actually entering into the reaction. They are used over and over and a single enzyme molecule may mediate thousands of reactions in a single second. Even simple reactions like dissolution of carbon dioxide in water will not take place to an appreciable extent by itself. But we can make it to dissolve in water in higher concentrations under high pressure. The carbonated drinks have CO_2 under high pressure. On releasing the pressure by removing the cap, lots CO_2 bubbles will release out. But in biological systems the dissolution of CO_2 takes place under normal conditions at a rate more than 10^6 times than that of uncatalysed reactions. This is possible because of an enzyme known as carbonic anhydrase is mediating the reaction. Similarly all reactions in the biological systems are mediated by one or more enzyme and so that reactions takes place at higher speed.

Enzymes operate on reactants, which are known as substrates, and convert them into products. The reaction may require energy or it may release energy. The enzyme is unaffected by the reaction.

$$\text{A + B + ENZYME} \longrightarrow \text{AB + ENZYME}$$

Enzyme + substrates ⟶ products + enzyme

Chemical reactions depend on the energy of the substrates being sufficient to cause the molecules to move and thus collide. This is usually accomplished with heat. Heat increases the motion of molecules and makes it more likely they will collide. The heat necessary to accomplish this is often inappropriate inside a cell. Enzymes can be used instead of heat to increase the likelihood that reactants will collide, and collide in the correct position. Enzymes bring reactants together and hold them in the correct position with respect to each other. Reactions occur between atoms and involve changes in the distribution of electrons. Before going to the details regarding the mechanism of enzyme action, let us discuss about the rate of a reaction.

Rate of Enzymatic Reactions

The rate of reaction defines the speed or the velocity of the reaction. The velocity of a reaction can be expressed in two ways. It can be expressed either in terms product formed or substrate consumed. The rate or velocity of a reaction can be defined as the quantity of the product formed per unit time (per second or per minute) or the quantity of the substrate consumed per minute. The quantity can be expressed in milligrams (mg) or micrograms (μg) ; and it can also be expressed in moles (M), millimoles (mM), or micromoles (μM).

Rate of a reaction = Quantity of product formed per minute

In the case of enzyme mediated reactions the rate of reaction is usually represented as the **enzyme activity,** which is related to the quantity of enzyme. The quantity of enzyme is expressed as units. It is the international unit (I.U.) for expressing the quantity of enzymes. One unit of enzyme is defined as the amount of enzyme required to convert 1 μM of substrate to product or to produce 1 μM product in unit time under physiological conditions (room temperature, normal pressure and physiological pH).

Enzymes are Highly Specific

Enzyme molecules, compared to substrate molecules are very big. The substrate molecules bind to the enzyme molecule at certain specific sites on the surface of the enzyme molecule known as substrate binding sites or actives sites to form a complex, the enzyme substrate complex (ES complex). The ES complex finally ends up in products with the release of free and intact enzyme. The active site of the enzyme is responsible for the reactions leading to the formation of the products. The active site in the enzyme molecule is only a very small portion and is complementary to the molecular shape of the substrate. This the main reason for the substrate specificity of an enzyme. The chemical reactions normally occur because of the molecular collisions between the reactants or the substrates. In enzyme mediated reactions the substrate molecules are brought to the proximity by taking them in to the active sites of the enzyme. Thus the enzyme molecule facilitates the reaction between the substrates with out any collision between them. The ability of an enzyme molecule to accelerate a reaction is several orders of magnitude greater than normal inorganic catalysts.

Mechanism of Enzyme Action

Energy of Activation

Reactions involve making and breaking bonds. Some bonds must be broken at the start of a reaction and this requires energy. No matter how exergonic (reactions, which proceeds with the release of energy) the overall reaction may be, some energy must be added initially to break the necessary bonds and get the reaction started. (For example, match to paper, spark to cylinder). This is the **energy of activation** (E_a). The activation energy can be defined as the energy needed by a molecule to take part in a reaction. This energy often supplied as heat but this may not be practical in a cell.

All organic molecules contain energy that could be released by the breakdown of the molecule. Such a reaction would be strongly exergonic and it does not happen spontaneously because the activation energy must be provided to get the reaction started and this energy is usually not available. To start the reaction it is necessary to provide the activation energy, or else reduce the amount of activation energy needed. Enzymes accelerate reactions more likely by reducing the energy of activation required. Enzymes bring down the activation energy by holding the reactants (substrates) in exactly the right orientation to each other so that contact will result in a reaction each time.

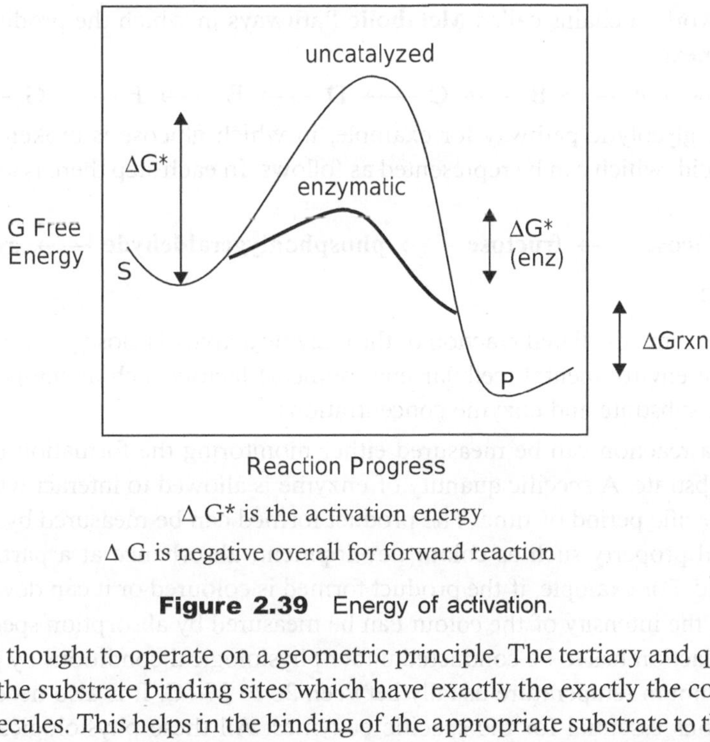

Δ G* is the activation energy

Δ G is negative overall for forward reaction

Figure 2.39 Energy of activation.

Enzymes are thought to operate on a geometric principle. The tertiary and quaternary structures of an enzyme have the substrate binding sites which have exactly the exactly the complementary shape of the substrate molecules. This helps in the binding of the appropriate substrate to the active centres just like a key fits in the keyhole. This hypothesis that explains the mechanism of enzyme substrate interaction is known as **Lock and key hypothesis**. Some enzymes change shape of the active centre slightly to accommodate the substrate molecules, a process known as **induced fit**. This theory is known as induced fit theory. **The Enzyme substrate complex thus formed lowers the energy of activation by either stressing an existing bond or correctly orienting two molecules to favor a reaction.** The enzyme holds the substrate molecules in exactly the right position relative to each other to facilitate the reaction due to geometric and electrical configuration. The characteristic functional groups of the amino acid residues present in the active sites of the enzyme molecule can interact with the various bonds of the substrate molecule. The reaction occurs and the new product molecule leaves the enzyme due to diffusion gradients, or to new repulsive electrical forces or the shape changes in either the enzyme or the product. New substrate molecules move into position.

Anabolic reactions are endergonic. Endergonic reactions do not occur spontaneously. In them the system moves from low to high potential energy

$$A + B \longrightarrow C$$

Catabolic reactions are exergonic

$$C \longrightarrow A + B$$

Exergonic reactions usually do not happen spontaneously because the activation energy is not available in the cell. If an enzyme is present to lower the activation energy there may then be enough energy present for the reaction to occur.

Metabolic Pathways

Enzymes typically work in chains called **Metabolic Pathways** in which the product of one reaction is the substrate of the next.

substrate ⟶ A ⟶ B ⟶ C ⟶ D ⟶ E ⟶ F ⟶ G ⟶ product

Consider the glycolytic pathway for example, in which glucose is broken down in a stepwise fashion to pyruvic acid, which can be represented as follows. In each step there is an enzyme facilitating the reaction.

Glucose ⟶ fructose ⟶ phosphoglygeraldehyde ⟶ pyruvate

Enzyme Kinetics

The velocity of an enzyme mediated reaction or the enzyme activity is closely controlled by a number of factors including the environmental (cellular environment) factors such as temperature and pH. The other factors are the substrate and enzyme concentrations.

The rate of a reaction can be measured either monitoring the formation of product or by the disappearance of substrate. A specific quantity of enzyme is allowed to interact with an excess amount of substrate for a specific period of time. The product formed can be measured by exploiting any of its physical or chemical property such as colour development, absorbance at a particular wavelength of light radioactivity etc. For example, if the product formed is coloured or it can develop any colour with some other reagent, the intensity of the colour can be measured by absorption spectrophotometer. The enzyme-mediated reaction has to be conducted under physiological conditions by providing a buffer of appropriate pH, optimum temperature usually between 20 to 37 °C. It is also necessary to provide co-factors and metal ions essential for the specific enzymes. All these physiological factors have to be maintained *in vitro* to keep the protein in its specific 3-D conformation and active state. Change in any of the parameters particularly the factors like temperature, pH and ionic strength can destroy the enzyme conformation or the 3-D structure. This change of conformation in the 3-D structure of an enzyme followed by the loss of activity is called denaturation.

Effect of Substrate Concentration

Substrate concentration is one of the factors that affect the rate of an enzyme-catalysed reaction. Velocity of an enzyme-catalysed reaction at different substrate concentrations can be measured and the values can be plotted on a graph with substrate concentration on the X- axis and its corresponding velocity on the Y-axis. It can be observed from the graph that as the concentration of substrate increases there is a corresponding increase in the velocity of reaction. However, beyond a particular concentration of substrate concentration, the velocity remains constant with out any further increase. This maximum velocity of an enzyme-catalysed reaction under substrate saturation is called the V_{max}. A reaction attains its V_{max} when all the active sites of the enzyme molecules are saturated with its substrate molecules and so that the rate of product formation will be at its maximum. There is a substrate concentration at which the reaction attains its V_{max}. Similarly there is a substrate concentration at which the velocity of the reaction is half of the V_{max}. This substrate concentration is a constant for a particular enzyme and that is known as K_M or Michaelis -Menton constant. K_M value of an enzyme is inversely proportional to affinity of the enzyme towards its substrate. If the Km value of an enzyme is low it indicates its high affinity towards the substrate. The relationship between the V, K_M, Vmax and substrate concentration [S] can be represented by an equation, the Michaelis -Menton equation :

$$V = \frac{V_{max}\,[S]}{K_M\ [S]}$$

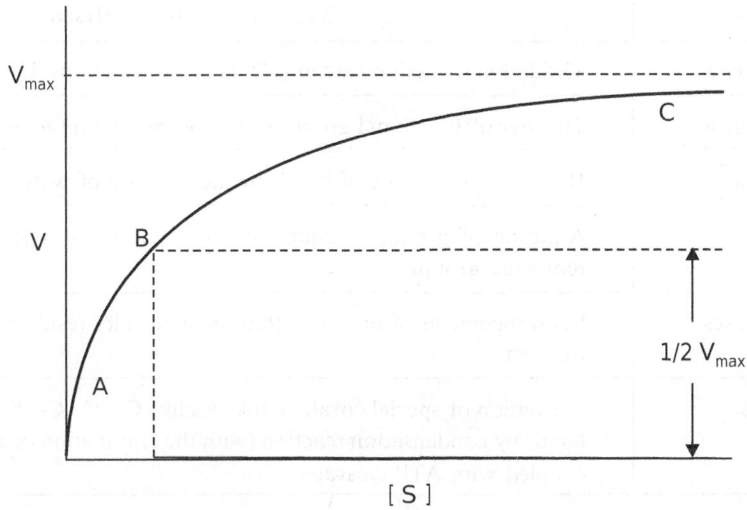

Figure 2.40 Effect of substrate concentration [S] on the velocity of a reaction.

Turnover Number

It is the number of substrate molecules converted to product per second by one molecule of enzyme. It can be calculated from the V_{max}. It indirectly indicates the catalytic power of an enzyme. Higher values of turnover number point toward better catalytic power of the enzyme.

Effect of Enzyme Concentration

The effect of enzyme concentration on the rate of reaction is more or less similar to that of substrate concentration. Under excess of substrate concentration, if you increase enzyme concentration there is a corresponding increase in the rate of reaction, but only up to a particular point. After this point there is no further increase in the rate of reaction corresponding to the increase of enzyme concentration.

Effect of Temperature and pH

Enzymes always exhibit typical temperature and pH optima that are characteristics of that enzyme and appropriate for its function. Most of the enzyme works in the physiological pH, which is 7.5, even though their optimal pH is different. The enzyme activity decreases when temperature and pH changes from the optimum value. At extreme temperature and pH the enzyme activity may drop to zero because of the complete denaturation of the enzyme.

Mammalian enzymes for example always show temperature optima around 37 °C while those of the ice fish are around 0°C. Most enzymes are denatured at 45°C or so, but not those of thermophilic organisms.

Classification of Enzymes

Enzymes are classified into six major groups or classes based on the reactions they catalyze. They are listed below in the table along with the reaction they catalyze.

TABLE 2.8 Different classes of enzymes and the reactions they catalyze

No.	Class	Type of reaction catalysed
1	Oxidoreductase	Oxidation reduction reactions (Transfer of electrons, Hydrogen or oxygen)
2	Transferase	Transfer of functional groups from one compound to another
3	Hydrolases	Hydrolysis (Breaking of bonds by the addition of water)
4	Lyases	Addition of groups to double bonds, or formation of double bonds by removing groups
5	Isomerases	Rearrangements of groups within the molecules resulting in the formation isomers
6	Ligases	Formation of special covalent linkages like C—C, C—S, C—O and C—N bonds by condensation reaction (with the elimination of a water molecule) coupled with ATP cleavage.

Enzyme Inhibition

Enzymes may be regulated in several ways. It may be by temperature and pH as mentioned earlier or by control of its synthesis or locally *i.e.* after its synthesis. We are interested in the latter here and are by certain compounds called enzyme inhibitors, which inactivates the enzyme permanently or temporarily. There are many types of inhibitors. Some are deleterious (poisons) some are important regulatory mechanisms. Some inhibitions are deliberate control mechanisms; some are detrimental to the organism. Enzymes are often inactivated by some molecule (an inhibitor) that changes their shape or blocks the active site.

A. Competitive Inhibitors

These are molecules that bind reversibly or irreversibly to the active site. They compete with the substrate for space in the active site. Most naturally occurring competitive inhibitors are irreversible.

Reversible Competitive. These competitors are not the substrate molecules but molecules similar to the substrate molecules. They thus compete with the substrate to occupy the active centre of the enzyme molecule. If they are in high concentration they may essentially inactivate the enzyme. These are of limited importance in natural systems. The competition is reversible and the competitor can be overwhelmed by high concentration of substrate. For example Fumaric acid is the competitive inhibitor of the enzyme, succinic dehydrogenase. The structure of fumaric acid is similar to that of succinic acid, the actual substrate of the enzyme.

Irreversible Competitive. The competitor and substrate both compete for the active site but the competitor occupies the active site permanently thus deactivating the enzyme. Nonclassical competitive inhibition is the binding of S at the active site and prevents the binding of I at a different site. The reverse is also true. Carbon monoxide is an irreversible competitive inhibitor of hemoglobin. Oxygen is the substrate.

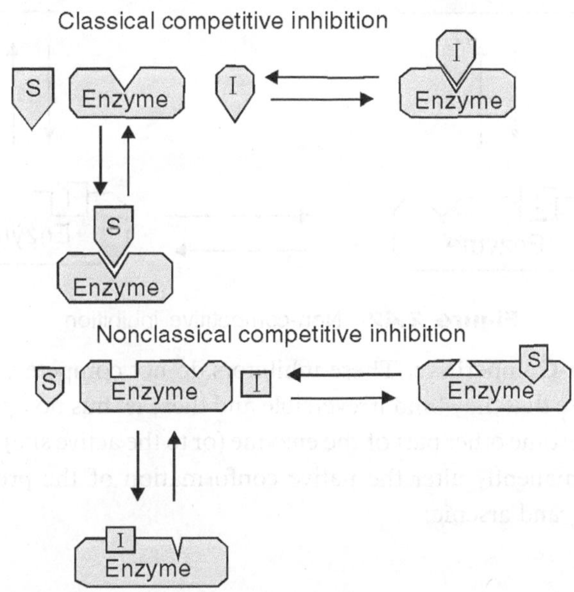

Figure 2.41 Diagrammatic illustration of competitive inhibition.

B. Non-Competitive Inhibition

In non-competitive inhibition the inhibitor and substrate do not compete for space in the active site. The substrate enters the active site but the inhibitor reacts with some other part of the enzyme molecule. It may be reversible or non-reversible. Reversible non-competitive inhibition is a major metabolic control mechanism.

1. Reversible Non-Competitive. This type of inhibition involves two binding sites on the enzyme molecule. The usual active site and a second **regulatory site** where the inhibitor binds. The inhibitor and the substrate do not compete for the same site. If the regulatory site is occupied by the inhibitor then the shape of the active site is changed so that the substrate molecules cannot fit and react. This type of inhibition, often called **allosteric control**, is very important in regulating metabolic pathways in the cell.

In allosteric control the final product molecule often acts as an inhibitor of one of the enzymes in the pathway, typically the first.

When high concentrations of final product accumulate, some of them react with the first enzymes at its regulatory site and render it inactive. This eventually results in a decrease in the concentration of the product. When this happens the inhibitor will drop out of the regulatory site and the enzyme will become active again. This is an example of **negative feedback**. This is a type of reaction that we will see over and over again in biology...at all levels from cell to ecosystem. Negative feedback prevents runaway reactions. The thermostat on ACs, furnace and incubators is a negative feed back mechanism. When temperature is raised over the control point the heating system of the incubator is turned off.

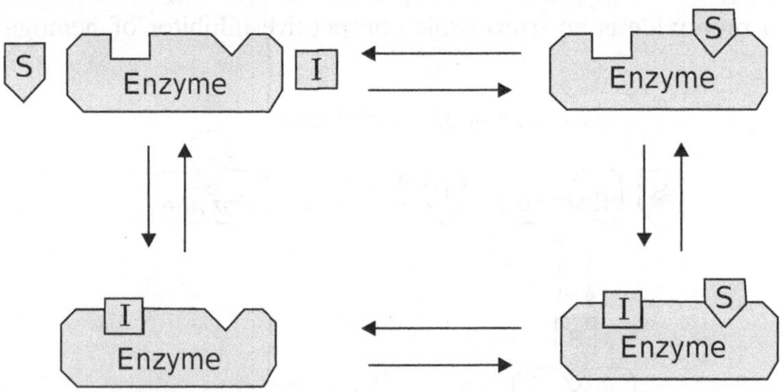

Figure 2.42 Non-competitive inhibition.

2. Irreversible Non-Competitive. These inhibitors do not compete with the substrate for space in the active site in the sense that they bind irreversible and there is thus no opportunity for competition. They bind irreversible with some other part of the enzyme (or to the active site) and permanently denature or inactivate it. They permanently alter the native conformation of the protein. They include heavy metals, cyanide, nerve gas, and arsenic.

Cofactors

Proteins are often associated with other chemical groups known as cofactors or prosthetic groups. Enzymes are no exception. Often molecules, atoms or ions other than proteins are necessary for the functioning of enzyme molecules. These may be a permanent part of the molecule, as are most ions (Zn, Mg, Fe) or may be only temporarily associated with the protein. Small organic prosthetic groups are often called **coenzymes**. Most of the coenzyme is the derivatives of **vitamins**. In addition to enzymes some non-enzyme proteins also have prosthetic groups or the non-protein parts. They are :

Nucleoproteins : have nucleic acid as the prosthetic group.

Glycoproteins : have carbohydrate as prosthetic group.

Lipoproteins : have lipid prosthetic group.

4. NUCLEIC ACIDS

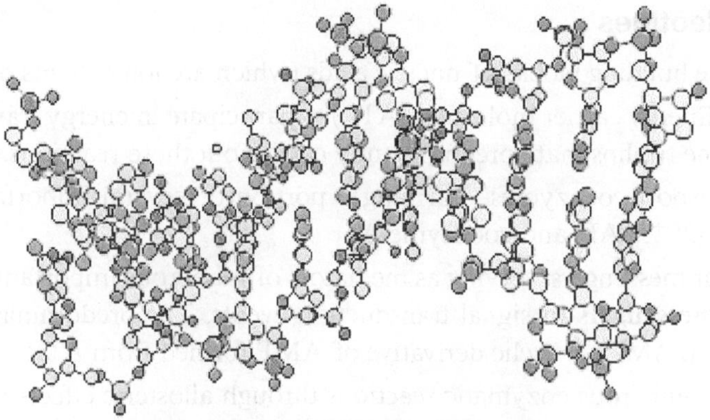

Friedrich Miescher, in 1869 discovered the nucleic acids, the molecular substrate of the genetic material. He also demonstrated that the regulation of breathing depends on the CO_2-concentration in the blood.

Building Blocks of Nucleic acids—The Nucleotides

Introduction

Nucleic acids are our information carrying molecules – our genetic material. As with many of our other compounds, the nucleic acids are composed of the repeating units or the monomers termed as nucleotides. Nucleotides are the building blocks of Nucleic acids. As a class, the nucleotides may be considered one of the most important metabolites of the cell. Nucleotides are found primarily as the monomeric units comprising the major nucleic acids of the cell, RNA and DNA. However, they also are required for numerous other important and independent functions within the cell.

Functions of Nucleotides

1. They are the building blocks of nucleic acids (which are long chains of nucleotides).
2. Serving as Energy carrier molecules (ATP)— participate in energy transaction reactions. ATP or Adenosine triphosphate predominantly carries out these reactions.
3. Energy transport coenzymes—forming a portion of several important coenzymes such as NAD^+, $NADP^+$, FAD and coenzyme A.
4. Intra cellular messengers- serving as mediators of numerous important cellular processes such as **second messengers** in signal transduction events. The predominant second messenger is cyclic-AMP (cAMP), a cyclic derivative of AMP formed from ATP.
5. Controlling numerous enzymatic reactions through allosteric effects on enzyme activity.
6. Serving as activated intermediates in numerous biosynthetic reactions. These activated intermediates include S-adenosylmethionine (S-AdoMet) involved in methyl transfer reactions as well as the many sugar-coupled nucleotides involved in glycogen and glycoproteins synthesis.

Nucleoside and Nucleotide Structure and Nomenclature

A nucleotide consists of three components :

1. 5- Carbon sugar or Pentose sugar (ribose)
2. Phosphate group
3. Nitrogen Base

The 5 Carbon sugar in Nucleic acid is Ribose sugar in Ribonucleic acids (RNA) and Deoxyribose sugar in Deoxyribose nucleic acids (DNA). The phosphoric acid in the form of phosphate residue is attached to the 5′ Carbon of the pentose sugar by an ester linkage. The nitrogen base is attached to the 1′ carbon of the pentose sugar by N-glycosidic linkage. Both the reactions, the formation of ester linkage between pentose sugar and phosphoric acid and formation N-glycosidic linkage between C-1 of the sugar and nitrogen bases are condensation reactions and results in the elimination of a molecule of water. The nucleotides found in cells are derivatives of the heterocyclic highly basic, compounds, **purine** and **pyrimidine**.

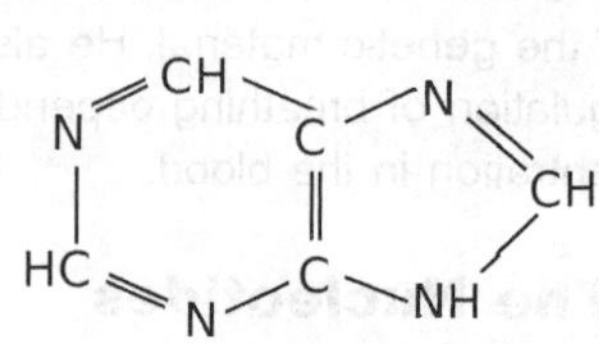

Purine

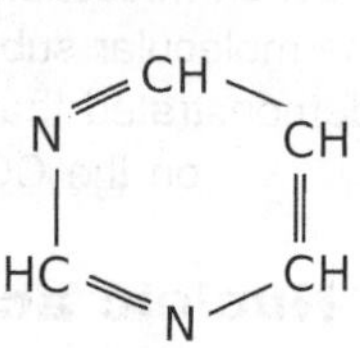

Pyrimidine

Figure 2.43 Nitrogen bases—purine and pyrimidine.

It is the chemical basicity of the nucleotides that has given them the common term "bases" as they are associated with nucleotides present in DNA and RNA. There are five major bases found in cells. The derivatives of purine are called **adenine** and **guanine**, and the derivatives of pyrimidine are called **thymine**, **cytosine** and **uracil**. The common abbreviations used for these five bases are, **A, G, T, C** and **U**. The purine and pyrimidine bases in cells are linked to D-ribose or 2′-deoxy-D-ribose through a β-N-glycosidic bond between the anomeric carbon (C—1) of the ribose and the N^9 of a purine or N^1 of a pyrimidine resulting in a nucleoside. The nucleoside condenses with phosphoric acid by forming an ester linkage with 5′ terminal OH group resulting in a nucleotide. Since there are five types of nitrogen bases there are five types of nucleoside and nucleotides as shown in the figure below.

In nucleosides and nucleotides the base can exist in 2 distinct orientations about the N-glycosidic bond. These conformations are identified as, ***syn*** and ***anti*** (Fig. 2.45). It is the anti conformation that predominates in naturally occurring nucleotides.

Nucleosides are found in the cell primarily in their phosphorylated form. These are termed **nucleotides**. The most common site of phosphorylation of nucleotides found in cells is the hydroxyl group attached to the 5′-carbon of the ribose The carbon atoms of the ribose present in nucleotides are designated with a prime (′) mark to distinguish them from the backbone numbering in the bases. Nucleotides can exist in the mono-, di-, or tri-phosphorylated forms.

Nucleotides are given distinct abbreviations to allow easy identification of their structure and state of phosphorylation. The monophosphorylated form of adenosine (adenosine-5′-monophosphate) is written as, **AMP**. The di- and tri-phosphorylated forms are written as, **ADP** and **ATP**, respectively. The use of these abbreviations assumes that the nucleotide is in the 5′-phosphorylated form. The di- and tri-phosphates of nucleotides are linked by acid anhydride bonds. Acid anhydride bonds have a high free energy ($\Delta G^{0'}$) for hydrolysis imparting upon them a high potential to transfer the phosphates to other molecules. It is this property of the nucleotides that results in their involvement in group transfer reactions for energy transactions in the cell.

The nucleotides found in DNA are unique from those of RNA in that the ribose exists in the 2′-deoxy form and the abbreviations of the nucleotides contain a **'d'** designation. The monophosphorylated form of adenosine found in DNA (deoxyadenosine-5′-monophosphate) is written as **dAMP**.

The nucleotide uridine is never found in DNA and thymine is almost exclusively found in DNA. Thymine is found in tRNAs but not rRNAs nor mRNAs. There are several less common bases found in DNA and RNA. The primary modified base in DNA is 5-methylcytosine. A variety of modified bases appear in the tRNAs. Many modified nucleotides are encountered outside of the context of DNA and RNA that serve important biological functions.

Base Formula	Base	Nucleoside (with ribose or deoxyribose)	Deoxyribose or ribose Nucleotide phosphate
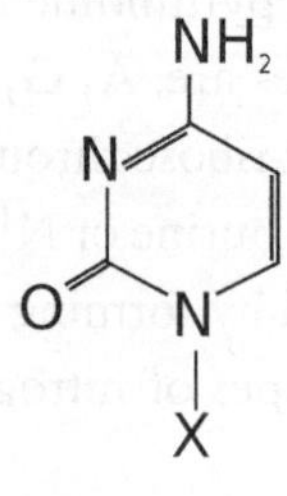	Cytosine, C	Cytidine	Cytidine monophosphate CMP
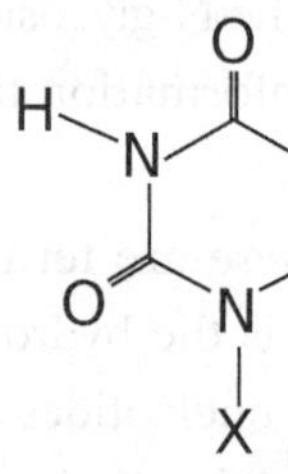	Uracil, U	Uridine	Uridine monosphate UMP
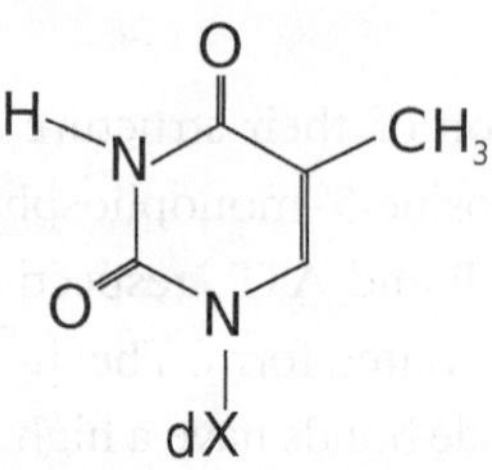	Thymine, T	Thymidine	Thymidine monophosphate TMP
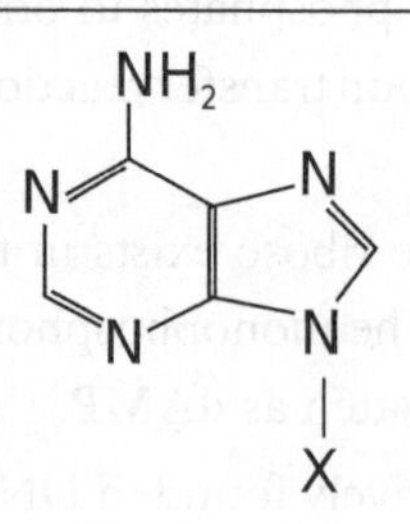	Adenine, A	Adenosine	Adenosine monophosphate AMP
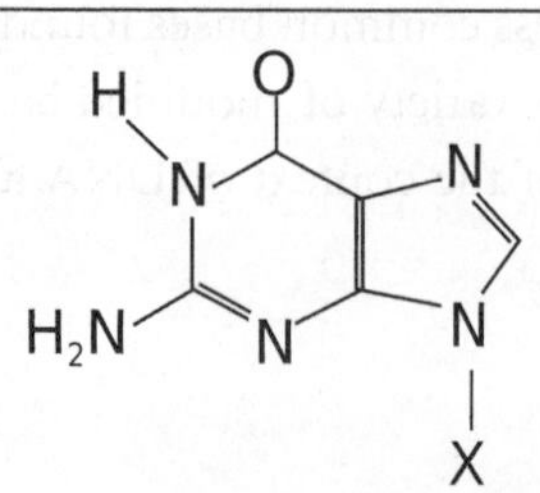	Guanine, G	Guanosine	Guanosine monophosphate GMP

Figure 2.44 Nitrogen bases and their corresponding nucleoside and nucleotides.

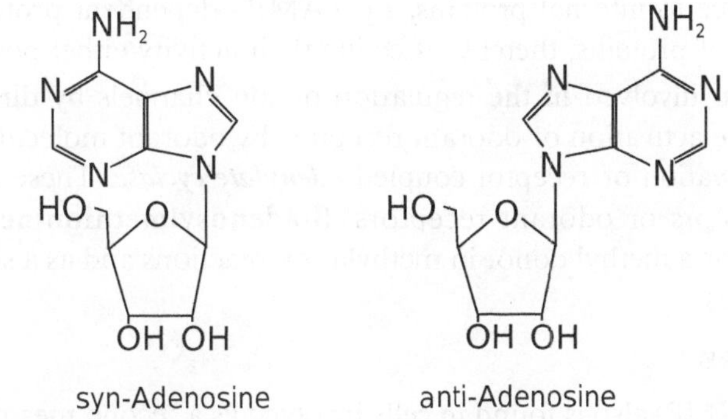

Figure 2.45 *syn* and *anti* conformations of Nucleoside, taking adenosine nucleoside as an example. *anti* conformation predominates in nature.

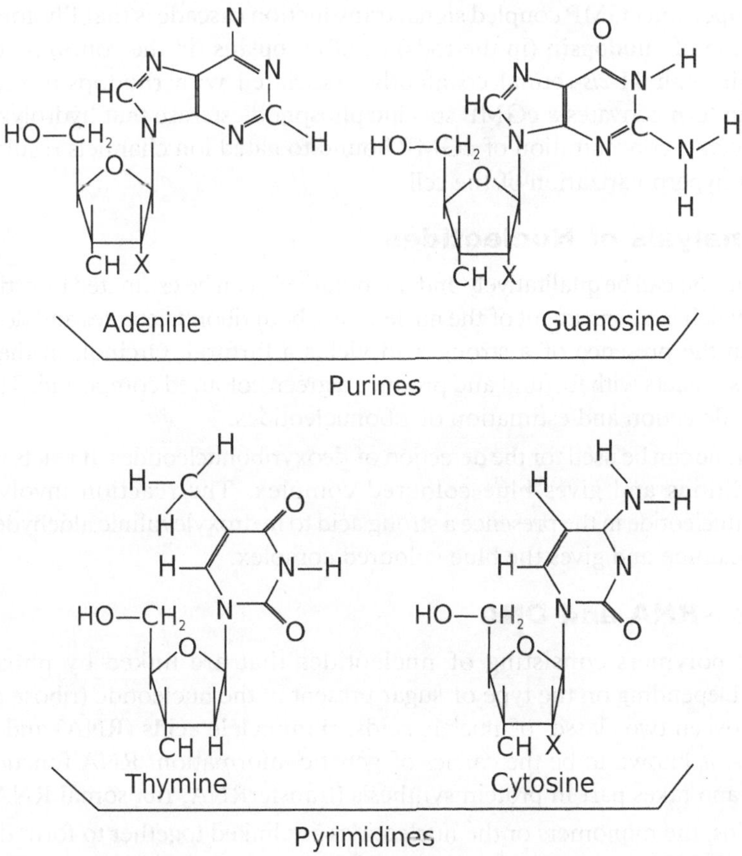

Figure 2.46 The types of nucleotides in DNA.

Adenosine Derivatives

The most common adenosine derivative is the cyclic form, **3′-5′-cyclic adenosine monophosphate, cAMP**. This compound is a very powerful **second messenger** involved in passing signal transduction events from the cell surface to internal proteins, *e.g.* cAMP– dependent protein kinase (**PKA**). PKA phosphorylates a number of proteins, thereby, affecting their activity either positively or negatively.

Cyclic-AMP is also involved in the regulation of ion channels by direct interaction with the channel proteins, *e.g.* in the activation of odorant receptors by odorant molecules. Formation of cAMP occurs in response to activation of receptor-coupled ***adenylate cyclase***. These receptors can be of any type, *e.g.* hormone receptors or odorant receptors. **S-adenosylmethionine** is a form of ***activated*** methionine, which serves as a methyl donor in methylation reactions and as a source of propylamine in the synthesis of polyamines.

Guanosine Derivatives

A cyclic form of GMP (**cGMP**) also is found in cells involved as a second messenger molecule. In many cases its role is to antagonize the effects of cAMP. Formation of cGMP occurs in response to receptor-mediated signals similar to those for activation of adenylate cyclase. However, in this case it is ***guanylate cyclase*** that is coupled to the receptor.

The most important cGMP coupled signal transduction cascade is that Photoreception. However, in this case activation of **rhodopsin** (in the rods) or other **opsins** (in the cones) by the absorption of a photon of light (through 11-*cis*-retinal covalently associated with rhodopsin and opsins) activates transducin which in turn activates a cGMP specific phosphodiesterase that hydrolyzes cGMP to GMP. This lowers the effective concentration of cGMP bound to gated ion channels resulting in their closure and a concomitant hyperpolarization of the cell.

Qualitative Analysis of Nucleotides

Nucleotides in a solution can be qualitatively and quantitatively can be estimated by estimating or detecting the pentose sugar, which is a constituent of the nucleotides, both ribonucleotides and deoxyribonucleotides. The ribose sugar in the presence of a strong acid yields a furfural. Orcinol, in the presence of ferric chloride as a catalyst, reacts with furfural and produce a green coloured compound. This Orcinol reaction can be used for the detection and estimation of ribonucleotides.

Diphenylamine can be used for the detection of deoxyribonucleotides. It reacts with diphenylamine under acidic conditions and gives blue coloured complex. The reaction involves dehydration of deoxyribose of the nucleotide in the presence a strong acid to hydroxylevulinic aldehydes. These aldehydes react with diphenylamine and gives the blue coloured complex.

Nucleic Acids—RNA and DNA

Nucleic acids are polymers consisting of nucleotides that are linked by phosphodiester bonds (polynucleotides). Depending on the type of sugar present in the nucleotide (ribose or desoxyribose), it is distinguished between two classes of nucleic acids: **ribonucleic acids (RNA)** and **desoxyribonucleic acids (DNA)**. DNA is known to be the carrier of genetic information, RNA functions as a messenger (messenger RNA) and takes part in protein synthesis (transferRNA, ribosomal RNA).

As in proteins, the monomers or the nucleotides are linked together to form the polynucleotides, a long chain of nucleotides connected together by the **phosphodiester linkages**. To distinguish the C-atoms of the nitrogen bases and those of the sugars from each other, the latter are marked by a′. It is thus spoken of as 1′, 2′, 3′, 4′ and 5′ positions. Polynucleotides are formed by the condensation of two or more nucleotides. The condensation most commonly occurs between the alcohol of a 5′-phosphate of one nucleotide and the 3'-hydroxyl of a second, with the elimination of H_2O, forming a **phosphodiester**

bond. This polymerization reaction involves the breaking of a pyrophosphate residue in each addition of nucleotide. The polynucleotide chain grows (during synthesis) from the 5′ to the 3′ terminus. Since the phosphodiester bond is formed between a 5′ phosphate group of one nucleotide and 3′ hydroxyl group of another nucleotide, it is known as 5′ ⟶ 3′ phosphodiester bond. The primary structure of DNA and RNA (the linear arrangement of the nucleotides) proceeds in the 5′ ⟶ 3′ direction. The common representation of the primary structure of DNA or RNA molecules is to write the nucleotide sequences from left to right synonymous with the 5′ ⟶ 3′ direction as shown :

5′-pGpApTpC-3′ ('p' is the phosphodiester linkage)

Ribonucleic Acids–RNA

RNA is the non-genetical nucleic acids except in some viruses like HIV, where it is the genetic material. They are comparatively small and present chiefly as linear single stranded molecules. There are four major forms of RNA in cells. They are messenger RNA (mRNA), transfer RNA (tRNA), ribosomal RNA (rRNA) and small nuclear RNA (snRNA). All these RNA molecules are synthesised based on the nucleotide sequence of the genetic material, DNA for the purpose of protein synthesis. In contrast to DNA, RNA is normally single-stranded. But it displays also some double-stranded, helical sections that are caused by folding of the single strand. These sequences are mirror images of each other and develop palindromes.

Messenger RNAs are the copies of the coding regions of the genes hand have the information to make the protein. It is very unstable and comprises about 5% of the total cellular RNA. It is highly variable in length and has a range of 75-3000 nucleotides.

Transfer RNAs are the different types of small molecules having a length in the range of 90 to 120 nucleotides. They also can fold and form secondary structures and have a clover leaf like model and can again fold to form a L-shaped structure as given in figure 2.47. They form about 15% of the total cellular RNA. During protein synthesis they carry the respective amino acids to the ribosomes where the translation of mRNA or the protein synthesis takes place. They contain a number of **rare and modified bases** that stabilizes its structure.

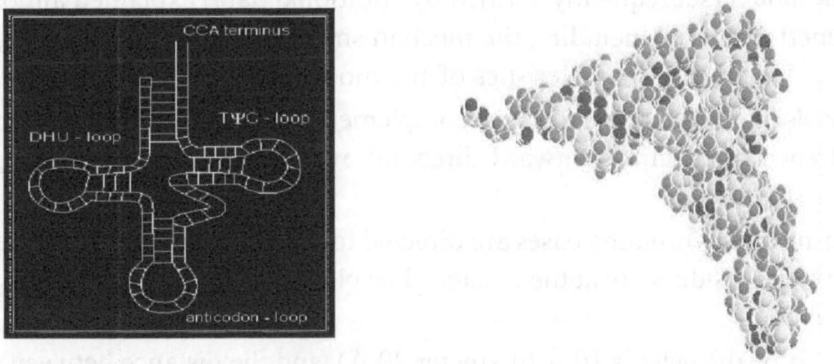

(A) (B)

Figure 2.47 The secondary structure and tertiary structure of tRNA molecule :
(A) The clover leaf structure of tRNA (secondary structure)
(B) L-shaped 3-D model (Tertiary structure).

Ribosomal RNAs are a group of RNA molecules having different sizes. Complex forms of secondary structures can be found in ribosomal RNA (rRNA). They form the structural components of the subcellular particles called ribosomes. Ribosomes are one of the important components needed for protein synthesis. The size of the molecules varies from 120 –3000 or more nucleotides. Approximately 80 % of the total cellular RNA is rRNA.

There is another class of RNA molecules mostly present in the nucleus of eukaryotic cells known as small nuclear RNA or **snRNA**. As the name indicates they are small molecules, but shows complicated secondary structures. They form 2% of the total cellular RNA molecules and are involved in the processing or modification called splicing of mRNA. They are organized in to small nucleoprotein particles called splicisomes. Most of them have catalytic activity and therefore are called **Ribozymes.** Ribozymes are the RNA molecules having catalytic (enzyme like) activity.

Deoxyribonucleic Acid–DNA

Deoxyribonucleic acid or the DNA forms the genetic material of almost all organisms except some viruses, which have RNA as the genetic material and therefore are named as RNA viruses. DNA is a very long polymer and always exists as double stranded helical molecule.

DNA contains the nitrogen bases adenine (A), Thymine (T), Cytosine (C) and Guanine (G). Erwin Chargaff isolated DNA samples from different organisms and analyzed nitrogen base composition and observed that A and T as well as C and G are present in equal amounts. The ratio A+T/C+G is specific for every species. This is known as Chargaff rule. All double stranded DNA samples irrespective of its source obey the Chargaff rule of base equivalence. There are some DNA viruses, where the genetic material is single stranded DNA. Single stranded DNA does not obey the Chargaff rule of base equivalence. Based on these data and X-ray diffraction data developed by Rosalind Franklin, obtained from crystals of DNA, James Watson and Francis Crick proposed a model, the **Watson** and **Crick model** for the structure of DNA. They published their findings in 1953 in the journal, Nature.

3-D Structure of DNA

The **Watson-Crick model** (subsequently verified by additional data) explained almost all the properties of DNA as the genetic material including the mechanism of its replication by using both of the DNA strands as matrices. The further characteristics of the model are :

- DNA molecule exist as a helix of two complementary antiparallel polynucleotide chains, wound around each other in a rightward direction and stabilized by H-bonding between bases in adjacent strands.
- The purine and pyrimidine bases are directed towards the interior of the helix, the phosphates and the sugar residues are at the outside. The planes of the bases are at a right angle to the axis of the helix.
- The radius of the helix is 10 Å (diameter 20 Å), and the distance between neighbouring bases is 3.4 Å.
- Both chains are linked by hydrogen bonds that develop between a pyrimidine and a purine base. An adenine is thus always linked with a thymine, a guanine with a cytosine. Between adenine and thymine exist two hydrogen bonds, between cytosine and guanine three (Fig. 2.48).

- The specific base pairing causes complementarity. If the sequence of one strand is known, then that of the other can be predicted.
- The antiparallel nature of the helix stems from the orientation of the individual strands. From any fixed position in the helix, one strand is oriented in the 5′ ⟶ 3′ direction and the other in the 3′ ⟶ 5′ direction.
- On its exterior surface, the double helix of DNA contains two deep grooves between the ribose-phosphate chains. These two grooves are of unequal size and termed the **major** and **minor grooves**. The difference in their size is due to the asymmetry of the deoxyribose rings and the structurally distinct nature of the upper surface of a base pair relative to the bottom surface.

The mechanism of replication was elucidated in 1958 **by M. Meselson** and **F.W. Stahl**, who could experimentally prove a prediction made by **Watson** and **Crick** and thus showed that DNA duplicated in a semiconservative way.

The sequence of nucleotides within a nucleic acid may seem arbitrary at first, but we know that it is this sequence that encodes the genetic information. Though it doesn't mean that every nucleotide sequence contains information. Long, repetitive DNA sequences exist and their function is not known. Since around 1975, methods for the sequencing of nucleotides have been developed.

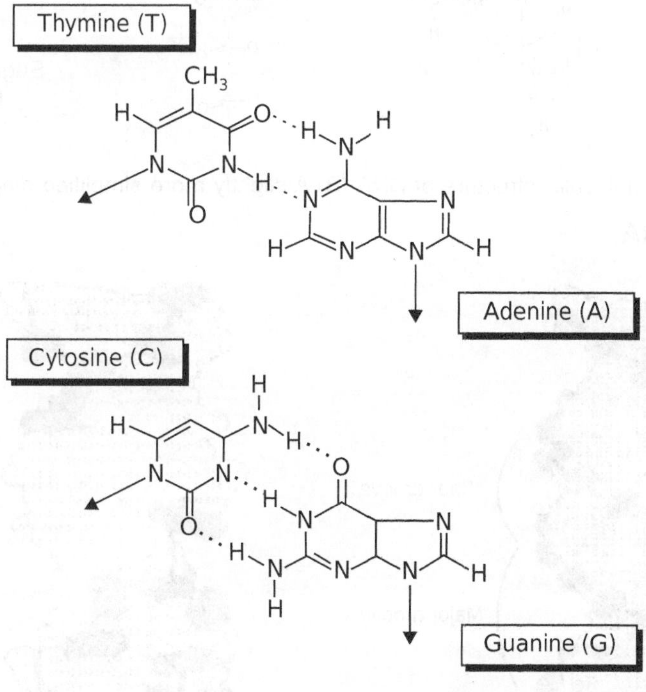

Figure 2.48 Watson-Crick base pairing in DNA strands.

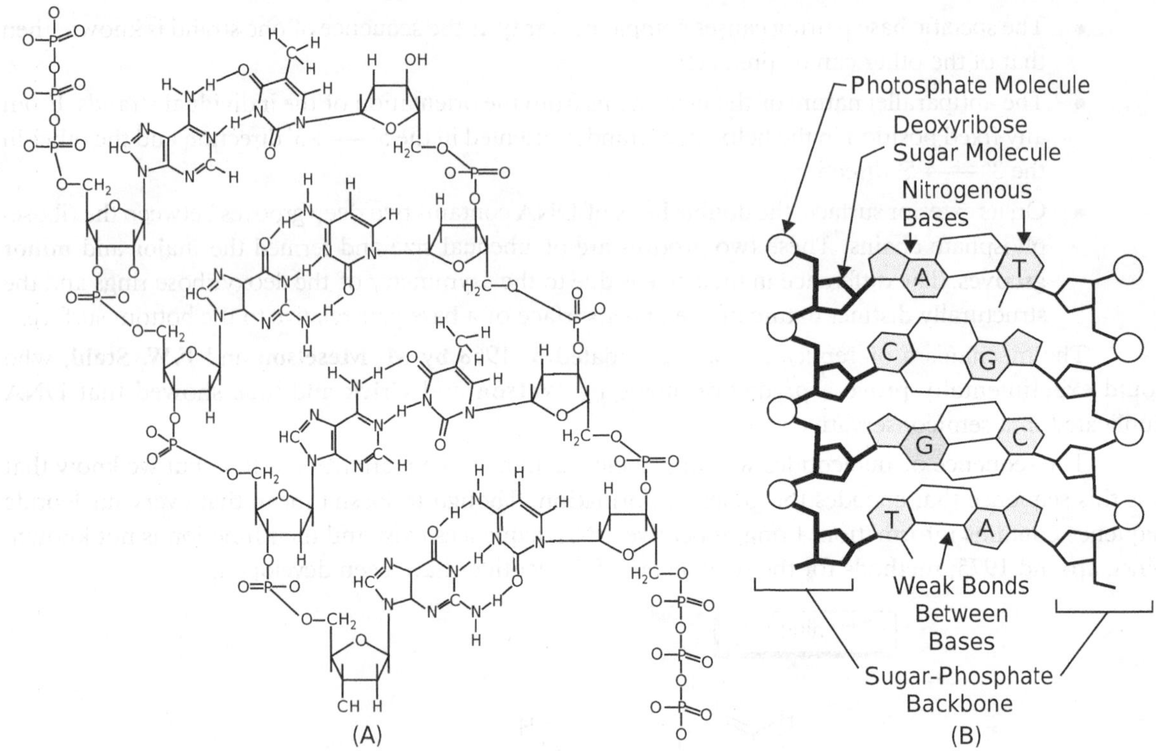

Figure 2.49 (A) Molecular structure of DNA (B) A slightly more simplified diagram is shown in B.

Other Forms of DNA

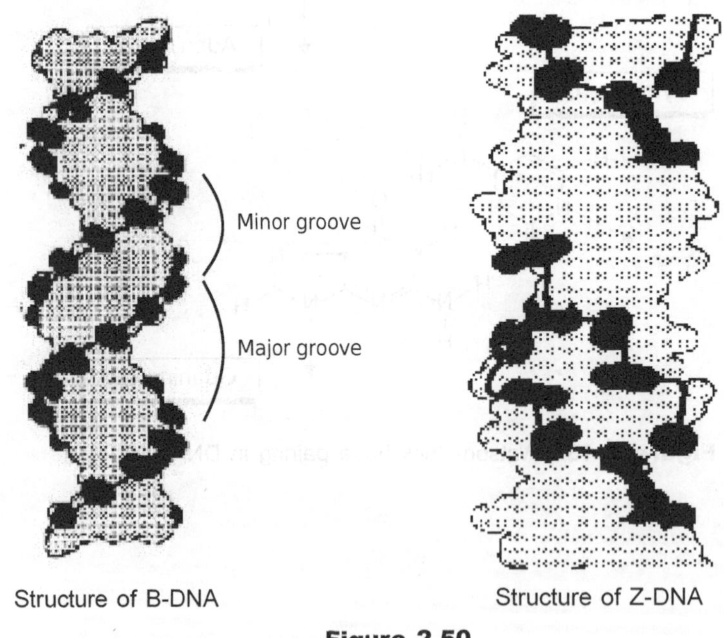

Structure of B-DNA

Structure of Z-DNA

Figure 2.50

The double helix of DNA has been shown to exist in several different forms, depending upon sequence content and ionic conditions of crystal preparation. The **B-form** of DNA prevails under physiological conditions of low ionic strength and a high degree of hydration. Regions of the helix that are rich in pCpG dinucleotides can exist in a novel left-handed helical conformation termed **Z-DNA**. This conformation results from a 180 degree change in the orientation of the bases relative to that of the more common A- and B-DNA.

TABLE 2.9 Difference in the properties between the major helical forms of DNA

Parameters	A-Form	B-Form	Z-Form
Direction of helical rotation	Right	Right	Left
Residues per turn of helix	11	10	12 base pairs
Rotation of helix per residue (in degrees)	33	36	– 30
Base tilt relative to helix axis (in degrees)	20	6	7
Major groove	narrow and deep	wide and deep	Flat
Minor groove	wide and shallow	narrow and deep	narrow and deep
Orientation of N-glycosidic Bond	Anti	Anti	Anti for Py, Syn for Pu
Comments		most prevalent within cells	occurs in stretches of alternating purine-pyrimidine base pairs

Chromosome structure, circular and supercoiled DNA

The DNA, the genetic material is a long molecule. The length of DNA of the bacteria *Escherechia coli* is approximately 1.44 mm. In higher forms of life DNA molecules are very long measuring to the range of centimetres. But the diameter is only 20 Å. These long and thin molecules are packaged in to small distinct subcellular structures called chromosomes, particularly in eukaryotes. The chromosomes can be visualized by special staining techniques under microscopes. The chromosomes contain DNA packaged with special type of proteins called histone proteins. Each species of organism has a characteristic number of chromosomes and is constant for that species. In Human cells there are 23 pairs of chromosomes and in bacteria like *E. coli* there is only a single chromosome. In most cases the bacterial DNA is circular with out any free end where as in higher forms of organisms the DNA molecule of the chromosomes are linear with free ends. These chromosomes are confined to the nucleus in the case of eukaryotic cells. Since the DNA molecules are very long, they are pachaged in to chromosomes having microns of length in a very specific manner. This packaging involves the process known as supercoiling. Supercoiling in DNA is promoted and maintained by enzymes known as topoisomerases, which keep the DNA in a highly supercoiled state. The supercoiling and packaging with histone proteins produces a beaded state for the DNA molecule, which is an intermediate stage in the formation of chromosomes known as nucleosomes.

It is very interesting to see that how the cells have managed to package the enormous length of DNA in to chromosomes having of microns of length. A typical human cell has 20 mm in diameter and the nucleus has about 5mm. The total length of DNA from all the 23 pairs of chromosomes will be corresponding to approximately 2 m (total base pairs = 6×10^9. length of 1 bp = 3.3 Å) If 2 m have to be packaged in to the confines of a 5 mm diameter nucleus the DNA molecules have to be condensed by a factor of 4×10^5. The fragile and long DNA molecules are packaged in chromosomes in such a way that the information stored in the DNA molecules can be retrieved with in a very short period. The microchips are very large when it is compared to this level of miniaturization.

In addition to the circular chromosomal DNA most of the bacterial cells contain certain type of one or more types of small extra chromosomal circular DNA molecules which can replicate independently. These circular DNA molecules are known as plasmids. The plasmids are responsible for certain unusual characters of organisms like antibiotic resistance. Plasmids have great importance in gene cloning and recombinant DNA experiments. The vectors commonly used in these experiments the modified forms of plasmids. Plasmids used for gene cloning and recombinant experiments are known as vectors.

5. LIPIDS

Building Blocks of Lipids—Fatty Acids, Glycerol

Introduction

Biological molecules that are insoluble in aqueous solutions and soluble in organic solvents are classified as lipids. The lipids of physiological importance for humans have four major functions :

1. They serve as structural components of biological membranes.
2. They provide energy reserves, predominantly in the form of triacylglycerols.
3. Both lipids and lipid derivatives serve as vitamins and hormones.
4. Lipophilic bile acids aid in lipid solubilization.

The building blocks or the chemical constituent of lipids are the long chain hydrocarbon acids known as fatty acids, the C_{18} amino alcohol or the sphingosine, glycerol and cholesterol.

Major Types of Lipids

- Triglycerides commonly known as the fats and oils
- Waxes (very similar to triglycerides)
- Phospholipids – the main constituent of membranes.
- Sterols (or steroids)
- Terpenes

Structure of Lipids

In chemical composition, Lipids are similar to carbohydrates in having carbon, hydrogen and oxygen as the main elemental components. However the proportion of oxygen is low, so lipids are mostly

hydrocarbons. The chemical structure of fats and oils, the most common lipids, is based on **fatty acid** building blocks and an alcohol, **glycerol.** The terms fats and oils are terms of convention. Fats are 'hard' or solid at room temperature and oils are liquids at room temperature.

Fatty Acids

Fatty acids fill two major roles in the body of plants and animals.

- As the components of more complex membrane lipids.
- As the major components of stored fat in the form of triacylglycerols.

Fatty acids are long-chain hydrocarbon molecules containing a carboxylic acid moiety at one end. The numbering of carbons in fatty acids begins with the carbon of the carboxylate group. At physiological pH, the carboxyl group is readily ionized, rendering a negative charge onto fatty acids.

Fatty acids are of two types—**saturated fatty acids** and **unsaturated fatty acids**.

Fatty acids that contain no carbon-carbon double bonds are termed saturated fatty acids; those that contain double bonds are unsaturated fatty acids. The numeric designations used for fatty acids come from the number of carbon atoms, followed by the number of sites of unsaturation (eg, palmitic acid is a 16-carbon fatty acid with no unsaturation and is designated by 16 : 0). The site of unsaturation in a fatty acid is indicated by the symbol Δ and the number of the first carbon of the double bond (*e.g.* palmitoleic acid is a 16-carbon fatty acid with one site of unsaturation between carbons 9 and 10, and is designated by 16 : $1^{\Delta 9}$).

Saturated fatty acids of less than eight carbon atoms are liquid at physiological temperature, whereas those containing more than ten are solid. The presence of double bonds in fatty acids significantly lowers the melting point relative to a saturated fatty acid. The majority of body fatty acids are acquired in the diet. However, the lipid biosynthetic capacity of the body (fatty acid synthase and other fatty acid modifying enzymes) can supply the body with all the various fatty acid structures needed. Two key exceptions to this are the highly unsaturated fatty acids know as **linoleic acid** and **linolenic acid**, containing unsaturation sites beyond carbons 9 and 10. These two fatty acids cannot be synthesized from precursors in the body, and are thus considered the **essential fatty acids**; essential in the sense that they must be provided in the diet. Since plants are capable of synthesizing linoleic and linolenic acid humans can acquire these fats by consuming a variety of plants or else by eating the meat of animals that have consumed these plant fats.

TABLE 2.10 Physiologically relevant fatty acids

Numerical Symbol	Common Name	Structure	Comments
14 : 0	Myristic acid	$CH_3\ (CH_2)_{12}COOH$	Often found attached to the N term. of plasma membrane -associated cytoplasmic proteins
16 : 0	Palmitic acid	$CH_3(CH_2)_{14}COOH$	End product of mammalian fatty acid synthesis
16 : $1^{\Delta 9}$	Palmitoleic acid	$CH_3(CH_2)_5C{=}C(CH_2)_7COOH$	

18 : 0	Stearic acid	$CH_3(CH_2)_{16}COOH$	
18 : $1^{\Delta 9}$	Oleic acid	$CH_3(CH_2)_7C{=}C(CH_2)_7COOH$	
18 : $2^{\Delta 9,12}$	Linoleic acid	$CH_3(CH_2)_4C{=}CCH_2$ $C{=}C(CH_2)_7COOH$	Essential fatty acid
18 : $3^{\Delta 9,12,15}$	Linolenic acid	$CH_3CH_2C{=}CCH_2C{=}CCH_2$ $C{=}C(CH_2)_7COOH$	Essential fatty acid
20 : $4^{\Delta 5,8,11,14}$	Arachidonic acid	$CH_3(CH_2)_3(CH_2C{=}C)_4$ $(CH_2)_3COOH$	Precursor for eicosanoid synthesis

Glycerol

Glycerol is a poly hydroxy alcohol that forms a basic component of triglycerides, a major form of storage lipids. Triacylglycerides are composed of a glycerol backbone to which 3 fatty acids are esterified.

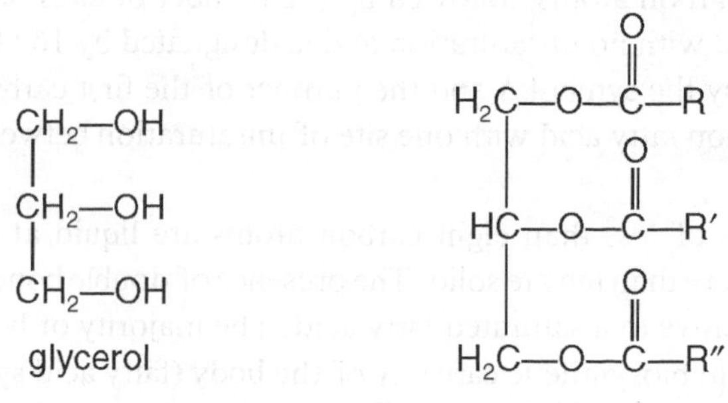

Figure 2.51 The structure of Glycerol and the Basic composition of a Triacylglycerides. R, R′, R″ are the fatty acids attached to the glycerol backbone.

Basic Structure of Phospholipids

The basic structure of Phospholipids is very similar to that of the triacylglycerides except that C-3 (*sn*3)of the glycerol backbone is esterified to phosphoric acid.

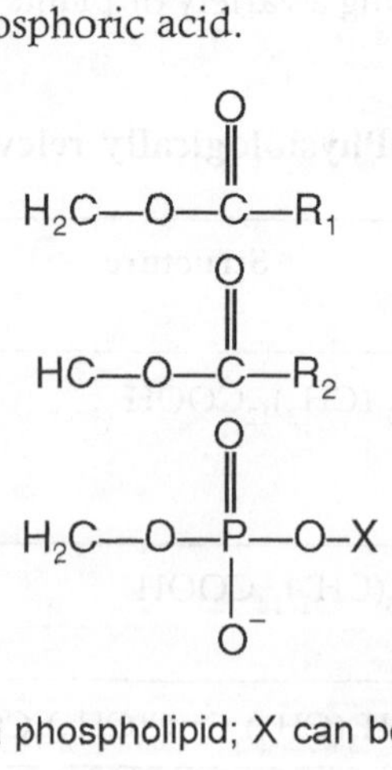

Figure 2.52 Basic structure of a phospholipid; X can be a number of different substituents.

The building block of the phospholipids is **phosphatidic acid**, which results when the X substitution in the basic structure shown in the Figure below is a hydrogen atom. Substitutions include ethanolamine (**phosphatidylethanolamine**), choline (**phosphatidylcholine**, also called **lecithins**), serine (**phosphatidylserine**), glycerol (**phosphatidylglycerol**), *myo*-inositol (**phosphatidylinositol**, these compounds can have a variety in the numbers of inositol alcohols that are phosphorylated generating **polyphosphatidylinositols**), and phosphatidylglycerol (**diphosphatidylglycerol** more commonly known as **cardiolipins**).

Steroids

These are special type of lipids, derived from the building block cholesterol. Cholesterol is the most abundant steroid present in animals. A number of steroids particularly the steroid hormones like sex hormones–estrogens, testosterone and other growth hormones and factors such as immuno suppressive factors and vitamins like Vitamin-D are derived from the basic molecule of cholesterols. Plants also contain a verity of steroids and its derivatives carrying out various biological functions.

Chemical Properties of Lipids

There are a number of chemical methods to detect and to quantify fats and lipids. Glycerol, one of the main components of triglycerides and phospholipids when heated with potassium hydrogen sulphate, it is dehydrated to an unsaturated aldehyde called acrolein. It can be very well identified by its characteristic pungent smell.

Unsaturated lipids can declourise the coloured solutions of halogens like bromine water. Unsaturated fatty acids add halogens (*e.g.* Bromine) across their double bonds, and that leads to the discolouration of the solution.

Lipids and Biological Membranes

Lipids are a class of hydrophobic molecules but very much miscible with organic solvents like hexane, chloroform, ethyl acetate etc. There are a group of biologically important lipids that show amphipathic character. An amphipathic molecule is that one, in which one end of the molecule exhibits hydrophobic (water hating) nature and the other end is having hydrophilic (water loving) nature. They are the phospholipids, the main components of biological membranes. The amphipathic nature of phospholipids makes them ideal candidates for the formation of biomembranes, which are composed of two layers of phospholipids. In aqueous medium phospholipids or any other amphipathic molecules will arrange themselves in a such a way that the hydrophobic ends are kept away from water and the hydrophilic ends are in contact with water molecules. Depending on the molecular weight and quantity of the amphipathic molecules and quantity of water, they can form different structures like vesicles, bilayers or single layered structures.

Phospholipids

Phospholipids are structural molecules forming the major component of all membranes of cells. In addition to lipids there are different types of proteins distributed along the bilayer of membranes depending on the special function of the membranes with tin the cells.

Structure of phospholipid consists of a glycerol molecule with two fatty acids esterified at positions R1 and R2 positions and a phosphoric acid esterified at the position X.

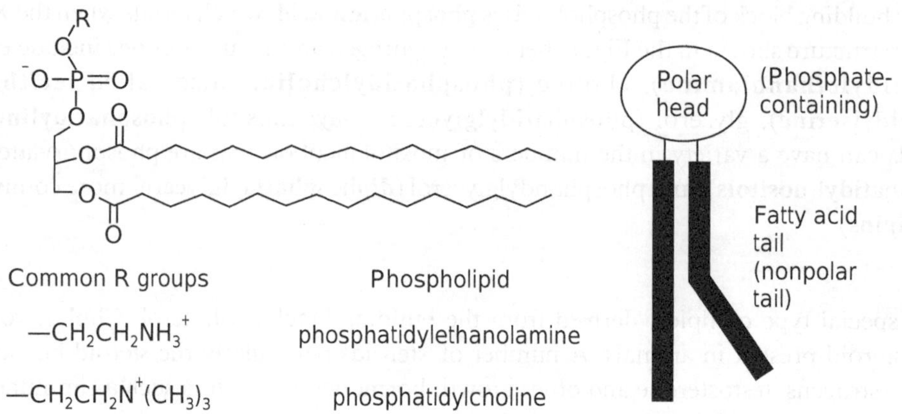

Figure 2.53 Diagrammatic sketch of phospholipid and its derivatives.

The benefit of the phospholipid structure is that the phosphate region makes the molecule highly amphipathic, ideal for the cell membrane structure :

- Hydrophilic portion in the phosphate region.
- Hydrophobic portion in the fatty acid tails.

The most common phospholipid is **lecithin**. Phospholipids also make excellent emulsifiers and are used in a number of food and household products. Cholesterol and its derivatives are also form another component of the cell membranes, but only in animal cells. It is not found in plant cell membranes. Different derivatives of phospholipids like Sphingolipids, sphingomyelin, glycosphingolipids, Cerebrosides etc. are present in certain specialised cells like neurons.

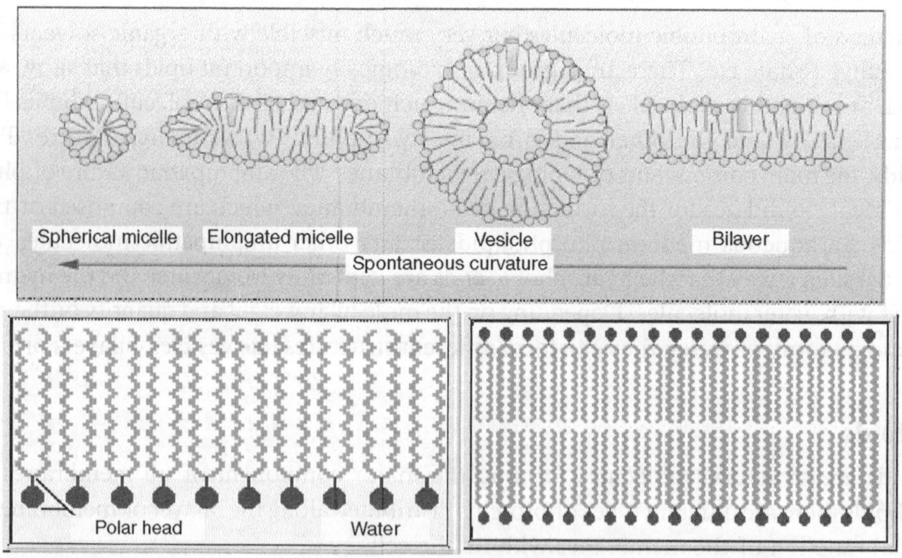

Figure 2.54 Various structures that can be formed when hydrophobic phospholipids are free in the aqueous medium.

All types of cells including both eukaryotes and prokaryotes and the cell organelles present in side the eukaryotic cells are enveloped by lipid bilayer membranes, which are strong and selectively permeable protecting the cytoplasmic content from the osmotic changes and other disturbances.

Properties of the Lipid Bilayer

As we have already mentioned, the most important property of the lipid bilayer is that it is a highly impermeable structure. Impermeable simply means that it does not allow molecules to freely pass across it. Only water and gases can easily pass through the bilayer. This property means that large molecules and small polar molecules cannot cross the bilayer, and thus the cell membrane, without the assistance of other structures.

Another important property of the lipid bilayer is its fluidity. The lipid bilayer contains lipid molecules, and, as we will discuss later, it also contains proteins. The bilayer's fluidity allows these structures mobility within the lipid bilayer. This fluidity is biologically important, influencing membrane transport. Fluidity is dependent on both the specific structure of the fatty acid chains and temperature (fluidity increases at lower temperatures). Structurally, the lipid bilayer is asymmetrical: the lipid and protein composition in each of the two layers is different.

Architecture of Membranes

Subcellular fractionation techniques can partially separate and purify several important biological membranes, including the plasma and mitochondrial membranes, from many kinds of cells. These preparations are often contaminated with membranes from other organelles. However, the plasma membranes from human erythrocytes can be isolated in near purity because these cells contain no internal membranes.

All membranes contain phospholipids and proteins. The protein: lipid ratio varies greatly :

- The inner mitochondria membrane is 76% protein.
- The myelin membrane is only 18% protein.

Because of its high phospholipid content, myelin can electrically insulate the nerve cell from its environment. The lipid composition varies among different membranes. All membranes contain a substantial proportion of phospholipids, predominantly phosphoglycerides, which have a glycerol backbone. All membrane phospholipids are amphipathic, having both hydrophilic and hydrophobic portions. Sphingomyelin, phospholipid that lacks a glycerol backbone, also is commonly found in plasma membranes. Instead of a glycerol backbone it contains sphingosine, an amino alcohol with a long unsaturated hydrocarbon chain. A fatty acyl side chain is linked to the amino group of sphingosine by an amide bond to for a ceramide. The terminal hydroxyl group of sphingosine is esterified to phosphocholine, thus the hydrophilic head of sphingomyelin is similar to that of phosphatidylcholine.

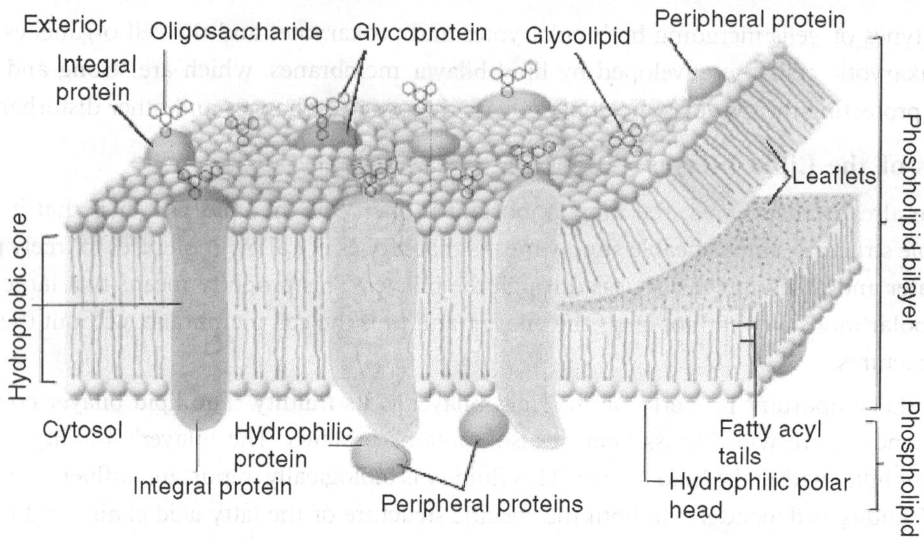

Figure 2.55 Fluid Mosaic model of a typical Plasma membrane.

Functions of Biomembranes

The biomembranes act as a very good barrier, which protects the contents from spilling out or extracellular contents entering into the cells or cell organelles. Cell membranes are selectively permeable. Water-soluble compounds like sugars, organic acids and salts like NaCl, KCl cannot freely pass through lipid membranes. There are special protein molecules positioned in the lipid bilayer to the facilitate transport of selected molecules.

The functions of biomembranes are determined by the quality of the proteins, which are distributed along with the lipids in the membranes. This makes a biomembrane distinct from another in its function. For example, the cellular envelop, mitochondrial membranes, membranes of chloroplasts are specialised for carrying out a specific function which is closely related to the types of protein present in the respective membranes. Certain cell membranes have certain informational molecules like glycoproteins, which can act as the receptors of signal molecules like hormones.

• REVIEW QUESTIONS •

1. What does anomer mean? What are the possible types of anomers in glucose ?
2. Give the total number of possible stereoisomers in aldohexoses.
3. What is the meaning of L-prefix in the naming of amino acid signify ?
4. Differentiate anomer and epimer.
5. What is a reducing sugar ?
6. Is sucrose a reducing sugar ? What is the reason ?
7. Differentiate between geometrical and optical isomerism.

8. What is the type of reactions that responsible for the polymerization of monomers ?
9. What is a glycosidic linkage ?
10. Ethanol fermentation is a waste of energy. Why ?
11. Differentiate between fermentation and aerobic respiration.
12. What are the types of photosynthetic pigments present in different organisms ?
13. What are the various functional groups present in various biomolecules ?
14. Differentiate a nuclcosidc from nucleotide.
15. Which is the site of oxidative Phosphorylation ?
16. What are the differences between starch and cellulose ?
17. Differentiate between glycogen and starch.
18. What are the differences between DNA and RNA ?
19. What are the special features of DNA that made it the genetic material ?
20. What are the various forms of DNA? Explain their structural difference and significance.
21. What is meant by primary structure of protein ?
22. Describe the various types of secondary structures of protein ?
23. What is the implication of the presence of tertiary structure and secondary structure in proteins ?
24. Enzymes are biological catalysts. What is the significance of catalysts in biological system ?
25. How an enzyme differ from an inorganic catalysts.
26. What are the factors that influence the velocity of an enzyme catalysed reaction ?
27. What does one unit of enzyme mean ?
28. Explain the reasons for the high catalytic power and specificity of enzymes.
29. What is meant by the prosthetic group of an enzyme ?
30. Give the Michaelis-Menton equation and explain V_{max} and K_M.
31. What is the structure of a biological membrane ?
32. Even though biological membranes have the same basic lipid bilayer structure there is a great difference of its biological functions. Why ?
33. What are the functions of plasma membranes ?

• SUGGESTED READING •

1. Principles of Biochemistry, Albert L. Lehninger. D.L. Nelson and M.M. Cox; Macmillan Worth Publisher, USA, 2000.
2. Biochemistry, by Mary K. Campbell, Saunders College Publishing, USA, 1999.
3. Biochemistry, Lubert Stryer, J.M. Berg and J.L.Tymoczko, W.H. Freeman and Company, USA.
4. Biochemistry and Molecular Biology, by William H. Elliott and Daphne C. Elliott, Oxford University Press, 2001.

3 CELL DIVISION

3.1 INTRODUCTION

Cell division is an integral part of the process of reproduction growth and repair of the body. During cell division, the genetic material is distributed among the daughter cells and this is an important phenomenon taking place during all types of cell divisions. There are two types of cell division observed in the eukaryotic cells. They are mitosis and meiosis. The distribution of genetic materials among the daughter cells are so important because their future development and generation of cells developing from that will be entirely depended on the genetic material that they receive from the parent cell. Mitosis is the cell division that occurs in the somatic cells (body cells) and meiosis takes place in the sex organs for the production of gametes. The main features and the process of mitosis and meiosis are discussed below.

3.2 MITOSIS

Mitosis is the process that facilitates the equal partitioning of replicated chromosomes into two identical groups. Two new daughter cells arise from one original cell. All the cells created through mitosis are genetically identical to one another and to the cell from which they came. The main purpose of Mitosis in Eukaryotic Cells is :

(*a*) Growth of the individual,

(*b*) When tissues have been damaged and need to be repaired (as in healing of a cut) and

(*c*) To reproduce asexually.

Mitosis is a nuclear division in which the daughter cells receive the same number of chromosomes as that of the parent cell. The nuclear division is some time referred to as karyokinesis, which is followed by the cytoplasmic division known as cytokinesis. The daughter cells resulting from mitosis are identical to each other and also to the parent cell in the quantity of and quality of genetic material. The genetic information, which the cell is copying and distributing during mitosis, is contained in the form of chromosomes. The period between two successive cell divisions is referred to as the **interphase**. It is not a part of mitosis, but forms the preparatory stage for cell division. The main metabolic activities during this stage of cell are respiration, protein synthesis, and the duplication of genetic material (DNA replication). Cells are usually small with no large vacuoles.

The interphase stage is subdivided into 3 parts: **G1,S and G2**. In the G1 stage the cell carries out its "housekeeping" functions while it collects the materials it will need to divide. S stands for DNA synthesis and this is the stage in which the DNA makes a copy of itself. An enzyme (DNA polymerase) assists each DNA double helix in making a copy of it, known as the **DNA replication**. The two identical DNA double helices are represented in the chromosomes as the **sister chromatids** and they are held together by the **centromere**, which is visible as a constricted area somewhere along the length of the chromosome. The G2 stage of mitosis comes next and in this stage the cell checks to see that everything is ready to begin mitosis. When the cell has duplicated its DNA and checked it to make sure that there are no errors, it is ready to begin the distribution of the DNA to two separate cells.

Mitosis is divided into four stages – prophase, Metaphase, Anaphase and Telophase. The important changes that the cells undergo in each phase are described below.

Prophase

During prophase the nuclear membrane and nucleolus disappear. The chromosomes themselves condense from long thin filaments into compact rods so that they can be more easily moved. The apparatus needed to move the chromosomes around is set up. In animal cells this is done by the two centrioles, which move toward opposite poles of the cell and begin to form the mitotic **spindle**. Plant cells do not have centrioles; they use a different mechanism to form the spindle fibers. Each chromosome has a spindle fiber attached to it at the centromere and extending off to either side of the cell where the centrioles are located. These spindle fibers are exerting tension on the chromosomes and when the centromere splits later on, this will allow the chromosomes to be pulled to either end of the cell.

Metaphase

Metaphase is an easy stage to recognize because all of the chromosomes are lined up at the equator of the cell. Depending on where the viewer is "standing", it can look as though the chromosomes are lined up east to west or it can look as though the chromosomes are aligned north to south.

Anaphase

In anaphase the centromere on each chromosome splits. Because the spindle fibers are exerting tension on the chromosomes, when the centromere splits each chromatid is pulled toward the spindle pole that it faces. Once a chromatid has its own centromere it can be called a chromosome.

Telophase

In telophase the chromosomes reach the opposite poles of the cell and their attached spindle fibers disappear. A new nuclear envelope forms, the nucleolus reappears, and the condensed chromosomes expand once more. The cytoplasm divides in a process called **cytokinesis**, and this forms the two daughter cells. In animal cells the cell membrane pinches in from sides to form a constriction between the daughter nuclei that cleaves the cell into two. In a plant, a new cell wall begins to form in the middle of the cell and gradually grows longer from each end as it works its way toward the edges of the cell. Mitosis is finished and you now have two cells, which are identical to each other, and identical to the cell from which they came.

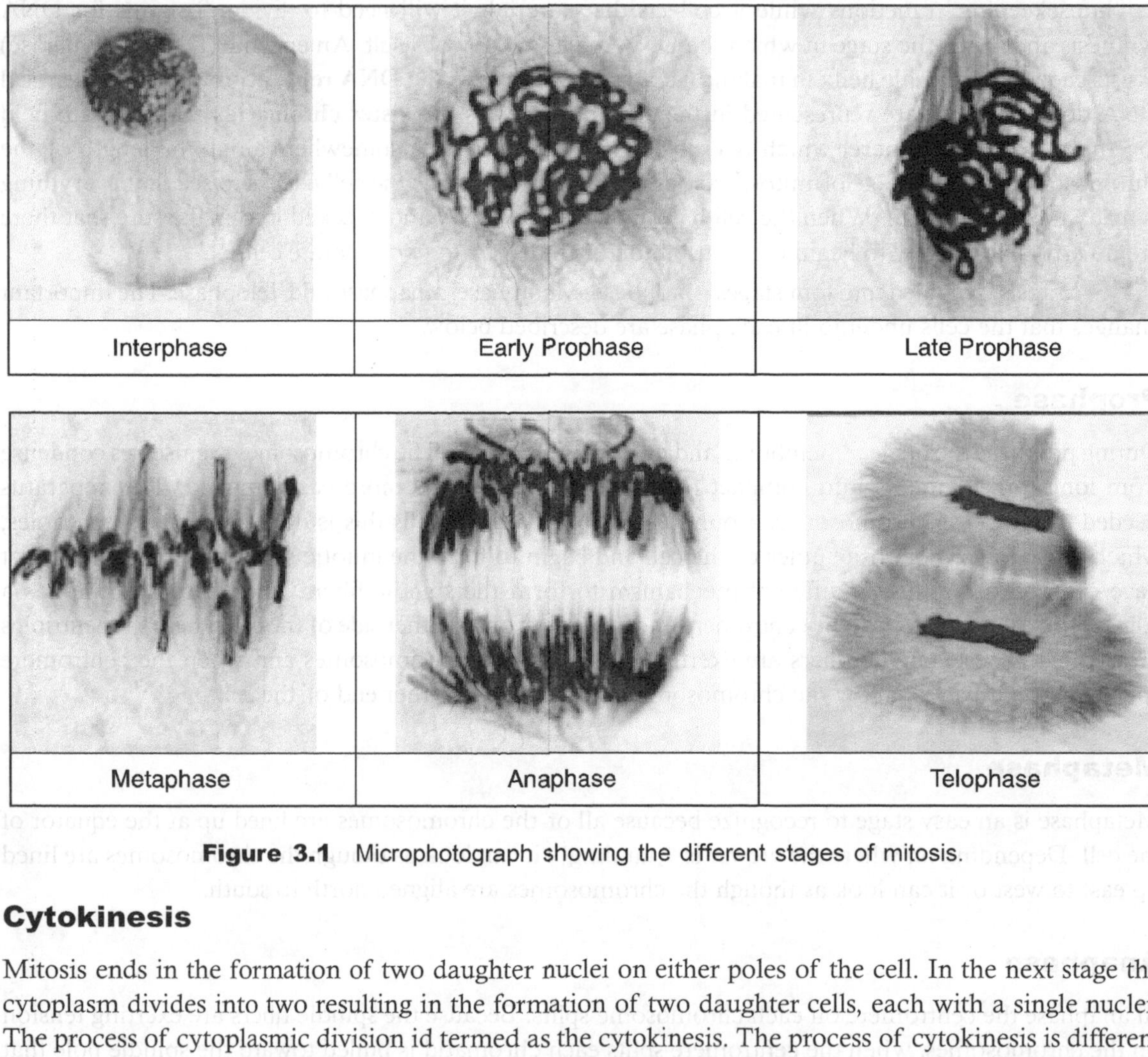

Figure 3.1 Microphotograph showing the different stages of mitosis.

Cytokinesis

Mitosis ends in the formation of two daughter nuclei on either poles of the cell. In the next stage the cytoplasm divides into two resulting in the formation of two daughter cells, each with a single nuclei. The process of cytoplasmic division id termed as the cytokinesis. The process of cytokinesis is different in animal cells and plant cells. In animal cells, following the separation of chromatids and the reappearance of nuclei at the end of anaphase, a cleavage furrow cuts between the two daughter nuclei or the separating chromatids resulting in two daughter cells. Thus the two cells receive the identical genetic materials. In higher plants after the nuclear division, a cell plate is formed between the two nuclei by the deposition of actin filaments of the microtubules. The cell wall materials are deposited in either side of the cell plate resulting in the formation of two separate cells.

3.3 MEIOSIS–THE CELL DIVISION OF SEX CELLS

Meiosis is the reduction division. It involves a replication of the genetic material followed by two successive nuclear divisions resulting in four daughter nuclei with haploid set of genome or chromosomes. This process of cell division occurs in reproductive organs and is essential for the production of gametes. The formation of haploid gametes and the fusion of male and female gametes resulting in the diploid zygote are involved in the process of sexual reproduction. The absence of meiotic cell division could result in diploid gametes, which in turn would result in the doubling of genomes with each new generation. This reduction division can occur at different stages of the life cycle depending on the type of organism. For convenience meiosis is divided into two—Meiosis I and Meiosis II.

Meiosis I

The first meiotic division is the reduction division because at the end of the first division the chromosome number is reduced to half because of the segregation of homologous chromosomes. It is divided into Prophase I, Metaphase I, Anaphase I, and Telophase I and the important events in each stage is summarised below.

Prophase I

The prophase of the Meiosis I is the longest of all stages and is further divided in to five stages for convenience. They are Leptotene, Zygotene, pachytene, Diplotene, and diakinesis. During Leptotene chromosomes start condensing. Chromosomes are visible as long thread like structures. During Zygotene the homologous chromosomes starts pairing. The pairing of homologous chromosomes is called synapsis and the paired chromosomes are known as bivalent chromosomes or tetrads of chromatids. The stage pachytene starts when the process of synapsis ends. Pachytene is very important because the phenomenon of crossing over between the homologous chromosomes occurs during this stage. The exchange of parts between homologous chromosomes is called the crossing over. Crossing over has got very important genetic significance because it is responsible for generating genetic variation among the progenies. The nuclear membrane starts disappearing. The next stage, Diplotene is marked by the tendency of the bivalent chromosomes to separate from each other. As the homologous chromosomes tries to pull away from each other, they seem to remain attached to certain specific points known as chiasmata. Chiasmata are the points on the bivalent chromosomes at which cross over have taken place. During these processes the chromosomes are undergoing progressive condensation and become more and more shorter. The last stage of prophase I is the Diakinesis. During this phase, the meiotic spindle is assembled and the chromosomes become very short. Nuclear membrane and nucleolus disappear completely. The movement of chromosomes to the equitorial plate marks the end of prophase-I and diakinesis.

Metaphase I

The tetrads line up along the equatorial plate of the cell by their kinetochore. The chromosomes are arranged in such a way that the sister chromatids of each chromosomes face the same pole. Because of this the spindle fiber from the same pole is connected to both the chromatids of a single chromosome.

Anaphase I

Homologous chromosomes separate and move toward opposite poles of the cell. It is important to note that the chromatids are still attached to each other and they **DO NOT** separate at this time.

Telophase I

The homologous chromosomes are segregated and are dragged to the poles of the cells during this phase. Nuclear envelop reappear around the chromosomes and nucleolus also appears in the nucleus. The cytoplasm splits in to two producing two daughter cells. In some cases the cytoplasmic division takes place after the meiosis I and II.

Prophase II

Prior to Prophase II cells may enter a kind of interphase, however, there is no S-phase here and the DNA amount remains the same. Chromosomes condense. Note that there is no pairing up of homologous chromosomes because there are only single chromosomes comprised of two chromatids.

Metaphase II

The chromosomes line up along the equator of the cell. Note that unlike metaphase I, each chromosome is held in place by spindle fibres from each pole of the cell (as in mitosis).

Anaphase II

As in mitosis anaphase II results in sister chromatids being separated and moving toward opposite poles of the cell. However, here there is half the number of chromosomes because DNA was not replicated following Telophase I.

Telophase II/Haploid Cell Formation

The nuclei form at the opposite poles of the cells. Following cytokinesis, four cells are formed, each with half the DNA of the original cell at the start of meiosis (*i.e.* haploid cells if the original cell was diploid). These cells then differentiate into the gametes or spores.

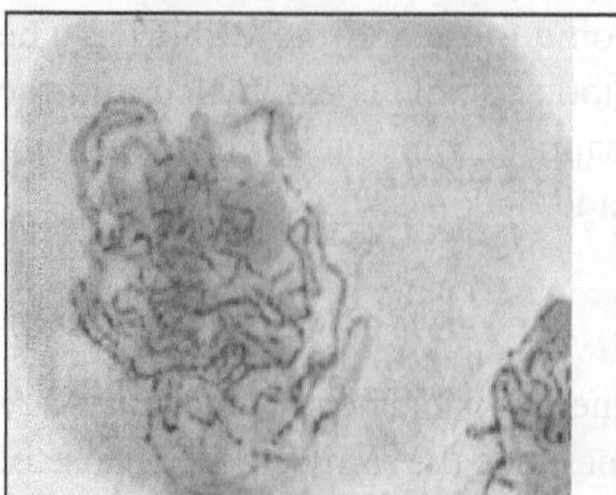

Mid Prophase I

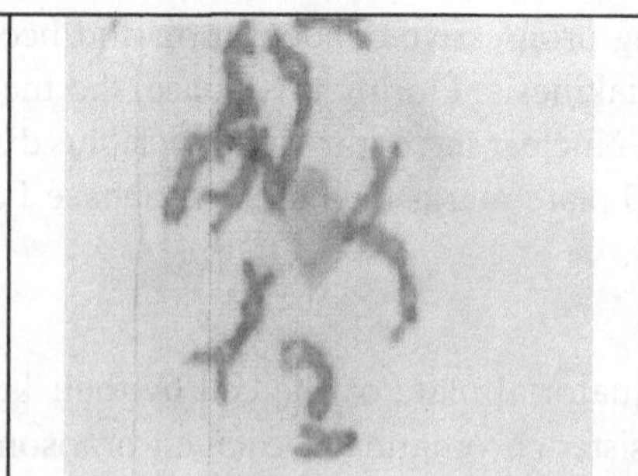

Late Prophase I

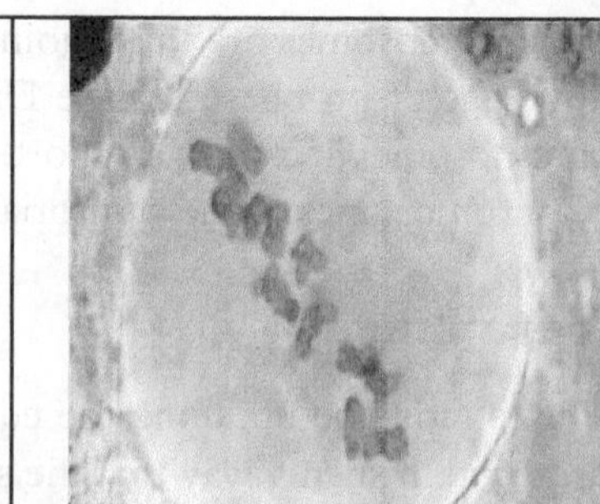

Metaphase I

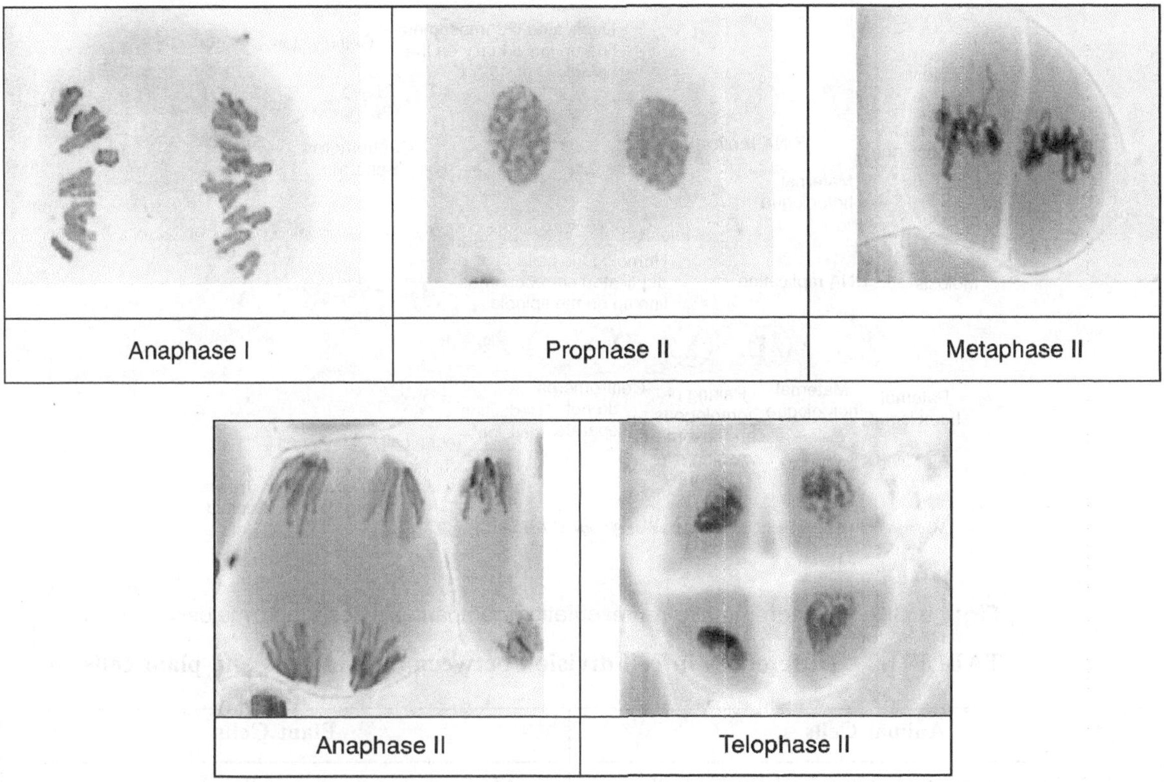

Figure 3.2 The Stages of Meiosis (The Microphotographs show meiosis in *Lilium longiformum* (lilly). Note that as in mitosis the process is dynamic and that the terms do not refer to static stages.)

TABLE 3.1 Mitosis versus Meiosis

	Mitosis	**Meiosis**
Purpose	Produces somatic cells (Body, growth)	Produces reproductive cells
Process	Cell duplication (Diploid → diploid)	Reduction division (diploid . haploid)
Number of Divisions	1 cell division	2 cell divisions
Product	1 → 2 daughter cells Daughter cells identical (To each other & mother cell)	1 → 4 cells (gametes) daughter cells different

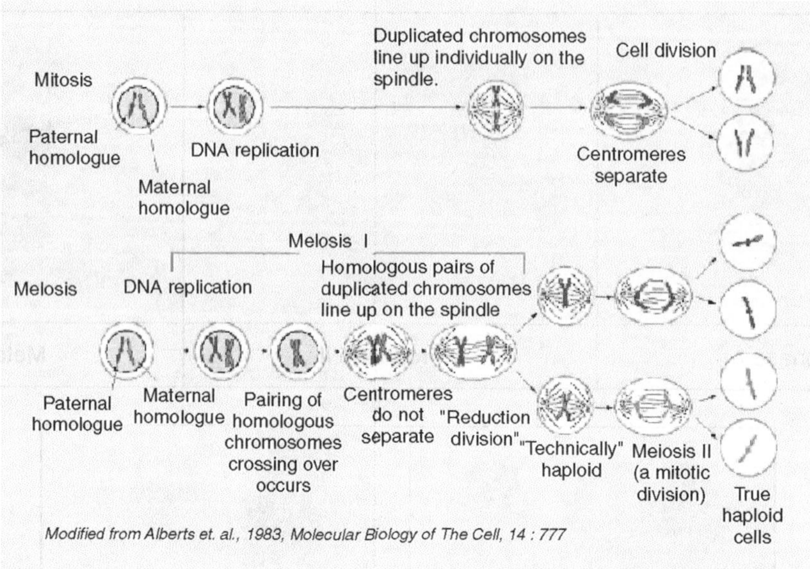

Figure 3.3 A diagrammatic representation comparing mitosis and meiosis.

TABLE 3.2 Differences in cell division between animal cells and plant cells

Animal Cells	Plant Cells
Late in anaphase, the plasma membrane pulls inward forming a cleavage furrow, which divides the cell in two during telophase.	After anaphase, a new cell wall forms between the two new nuclei to create two cells.
A centriole is found at each pole during mitosis.	No centrioles are found in plant cells.

3.4 CELL CYCLE

A eukaryotic cell undergoes alternate phases of division and non-division. The sequences of events occurring from the completion of one division to the next division form one cell cycle. The stage or the interval between two consecutive mitotic divisions is the interphase. Therefore a cell cycle can be considered as interphase + Mitosis. The interphase is metabolically very dynamic and is further divided in to S phase, G_1 and G_2 phase. The period between M and S is called $\mathbf{G_1}$; that between S and M is $\mathbf{G_2}$. A cell cannot divide into two, the two into four, etc. unless two processes alternate: **doubling** of its genome (**DNA**) in **S phase** (synthesis phase) of the cell cycle and **halving** of that genome during mitosis (**M phase**).

Stages of Cell Growth

1. **G_1 phase** : primary growth phase. Cell does its House keeping activities.
2. **S phase** : DNA replication.

3. **G_2 phase** : Chromosome condensation, cell organelle replication.
4. **M phase** : mitosis (nuclear division) (Prophase, metaphase, anaphase, & telophase).
5. **C phase** : cytokinesis (cytoplasmic division), **formation of daughter cells.**

Interphase is the longest stage of cell cycle. (G_1, S, G_2). Human cells contain 46 chromosomes during the G_1 stage of interphase. This is doubled to 92 during the S stage of interphase.

A typical cell takes about 16 hours to complete cell cycle. The actual process of cell division or mitosis occupies only a small part of this cycle, approximately one hour. The length of S and G_2 are almost equal in all cell types, but the length of G_1 phase varies considerably between cells.

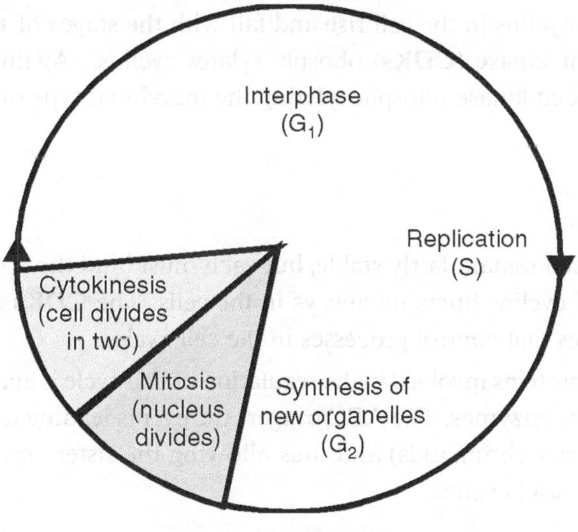

Figure 3.4 Cell Cycle : Cell Life Cycle. The repeating sequence of growth and division through which eukaryotic cells pass each generation.

Normally the cells at G_1 phase can follow either of the two paths. The cell after the cell division may withdraw from cell cycle and enter in to a resting phase termed as **G_0 Phase**, or it can enter into the G_1Phase of the cell cycle. Cell in the G_0 phase are viable and metabolically active but it can be stimulated to enter in to the G_1 phase at any time and start cell cycle again. Often G_0 cells are terminally differentiated: they will never re-enter the cell cycle but instead will carry out their function in the organism until they die.

For other cells, G_0 can be followed by re-entry into the cell cycle. Most of the lymphocytes in human blood are in G_0. However, with proper stimulation, such as encountering the appropriate antigen, they can be stimulated to re-enter the cell cycle (at G_1) and proceed on to new rounds of alternating **S phases** and **mitosis**. G_0 represents not simply the absence of signals for mitosis but an active repression of the genes needed for mitosis. Cancer cells cannot enter G_0 and are destined to repeat the cell cycle indefinitely.

Regulation of the Cell Cycle

The cell cycles in almost all eukaryotic cells are essentially the same with minor variations in the duration of each phase. There are some Proteins in the cytoplasm that control and coordinates the passage of a cell in the correct order through the cell cycle. The major proteins involved in the regulation are termed as **Cyclins.** There are three groups :

- G_1 cyclins
- S-phase cyclins
- M-phase cyclins

The levels of these cyclins in the cell rise and fall with the stages of the cell cycle. The group of enzyme**, Cyclin-dependent kinase (CDKs)** phosphorylates cyclins. Again, there are three groups of CDKs or the cyclin depended kinase phosphorylating the individual type of cyclins.

- G_1 CDKs
- S-phase CDKs
- M-phase CDKs

Their levels in the cell remain fairly stable, but each must bind the appropriate cyclin in order to be activated. The levels of cyclins fluctuate always in the cells. The **CDKs** add phosphate groups to a variety of protein substrates that control processes in the cell cycle.

The third group of proteins involved in the regulation of cell cycle is anaphase**-promoting complex (APC)** and other proteolytic enzymes. The APC triggers the events leading to destruction of the cohesin (a protein that joins the sister chromatids) and thus allowing the sister chromatids to separate. It also degrades the mitotic (M-phase) cyclins.

The important Steps in the Cell Cycle

- A rising level of **G_1 cyclins** signals the cell to prepare the chromosomes for replication.
- A rising level of **S-phase promoting factor (SPF)** prepares the cell to enter S phase and duplicate its DNA (and its centrioles).
- As DNA replication continues, one of the cyclins shared by G_1 and S-phase CDKs (cyclin E) is destroyed and the level of mitotic cyclins begins to rise (in G_2).
- **M-phase promoting factor** (the complex of mitotic cyclins with M-phase CDK) initiates the assembly of the mitotic spindle, breakdown of the nuclear envelope and condensation of the chromosomes.
- These events take the cell to **metaphase** of mitosis.
- At this point, the M-phase promoting factor activates the **anaphase promoting complex (APC)** which allows the sister chromatids at the metaphase plate to separate and move to the poles (= anaphase), completing mitosis. It destroys the M-phase cyclins. It also turns on synthesis of G_1 cyclins for the next turn of the cycle and degrades **geminin**, a protein that has kept the freshly-synthesized DNA in S phase from being re-replicated before mitosis.

Checkpoints : Quality Control of the Cell Cycle

The cell has several systems for interrupting the cell cycle if something goes wrong. There are some points in the cell cycle where the cell can check the sequence of activities related to its replication. These points are known as **checkpoints**. The following are the important checkpoints in the cell cycle to monitor the healthy progress of cell division.

- **DNA damage checkpoints.** These sense DNA damage before the cell enters S phase (a G_1 checkpoint), during S phase, and after DNA replication (a G_2 checkpoint). A check on completion of S phase. The cell seems to monitor the presence of the Okazaki fragments on the lagging strand during DNA replication. The cell is not permitted to proceed in the cell cycle until these have disappeared.
- **Spindle checkpoints.** These are the final checkpoints also known as M Phase checkpoints. They detect any failure of spindle fibers to attach to **kinetochores** and arrest the cell in **metaphase** (M checkpoint). They monitor the alignment and detect improper alignment of the spindle itself and block cytokinesis. If the damage caused to the DNA is irreparable, it triggers the process of apoptosis or cell death.

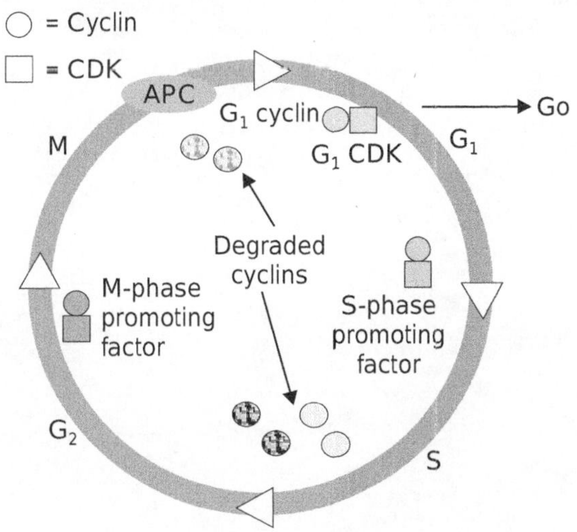

Figure 3.5 Regulation of cell cycle.

All the checkpoints examined require the services of a complex of proteins. Mutations in the genes encoding some of these have been associated with cancer; that is, they are oncogenes. This should not be surprising since checkpoint failures allow the cell to continue dividing despite damage to its integrity.

• REVIEW QUESTIONS •

1. Explain and compare Mitosis and meiosis.
2. Explain the methods of genetic recombination in bacteria.
3. Explain the regulation of cell cycle.
4. What is the need of meiosis ?
5. Explain the various stages in the prophase of meiosis.

• SUGGESTED READING •

1. Concepts of Genetics, W.S. Klug and R.A. Cummings; Prentice Hall, New Jersey, 2000.
2. Biological Science, N.P.O. Green, G.W. Stout and D.J. Taylor, Cambridge University Press, Cambridge, 1995.
3. Principle of Genetics, E.J. Gardener, M.J. Simmons and D.P. Snusted; John Wiley & Sons. Inc., New York.

4 GENES

4.1 INTRODUCTION

The science of heredity or genetics deals with the mechanism by which traits or characters are transmitted from generation to generations. There are similarities and differences among the members of a species. The members of a particular species are basically similar with respect to a set of characters. At the same time they are different from other species in these basic characteristics. For example all human have a set of characteristics that define us as a species. We belong to vertebrates, because we have a backbone and a spinal cord like any other vertebrates. Among the vertebrates we belong to the group mammals because we are hot blooded animals, feeding the young ones with milk from mammary glands, like any other mammal. These characters define us as a mammal. Then, among the mammals we belong to a type of primates in finer details. We habitually stand upright with long legs, relatively little body hair, a large brain, distinct lips, small teeth and a jutting chin. These characters separate us from other primates like chimpanzees, Gorillas and other monkeys. These biological characters, which define us as a species, inherit with out any change from generation to generation. These characters will not differ from individual to individual. Thus there are some basic traits, which are common for all its members. By these characters one species identified from the other.

At the same time there are a lot variations or differences among the members of a species, by which an individual can be identified from the other. Within the human species there are lot of variations for a large number of traits. Characters such as eye colour, hair colour, skin colour, height, weight, blood glucose level, mental depressions, predisposition to high blood pressure, diabetics, and a number of other personality traits are highly variable among individuals. Most of these characters are inherited biologically and others are inherited culturally and influenced by environment. For example, traits such as skin colour, eye colour etc are inherited biologically, but the native language that we speak is inherited culturally and influenced by the environment. Many traits are influenced jointly by biological inheritance and environmental factors. For example, skin colour, even though inherited biologically is tremendously influenced by the environment. The body weight is determined in part by biological inheritance, but partly by eating habits and the level of physical exercise. Thus there are similarities as well as differences among the members of a species.

The tendency of the young ones to resemble the parents is called the heredity and the variations that appear between the parents and the young ones and also among the members of the young ones or a species is called the variations. The science that deals with the study of the mechanism of heredity and variations is known as genetics. The fundamental concept of genetics is that the traits that are inherited are determined by certain factors or elements of heredity known as **genes**. These factors or genes are transmitted from parents to offspring during reproduction. The elements of heredity and some principles governing their transmission from generation to generation were discovered by Gregor Mendel, a little known Central European monk in his experiments with garden peas. His experimental results and ideas had been published in 1866, but largely went unrecognized until after his death. While Mendel's research was with plants, the basic underlying principles of heredity that he discovered also apply to people and other animals because the mechanisms of heredity are essentially the same for all complex life forms.

The modern science of genetics started from the rediscovery of the laws of inheritance in 1900, which were originally postulated by Gregor John Mendel in 1866. Three scientists—de Vries, Tshermark and Correns rediscovered the principles of genetics independently. Sutton and Boveri observed the parallelism between the Mendelian 'factors' and the chromosomes during meiosis. They established the chromosomal basis of inheritance. The term 'genetics' was proposed by Bateson in 1906 to denote all the subjects concerned with heredity and variation. He defined it as the science of heredity and variation. Bateson along with Punnet observed the variations of Mendelian inheritance and explained the phenomenon of linkage. In 1910 another scientist W. Johannsen introduced the term gene mainly to represent that factors or units which are responsible for the inheritance of a trait of an organism. Three years later the publications of Mendel's experimental results and ideas, in 1869 Friedrich Miescher discovered a new types chemical with slight acidity isolated from the nuclei of salmon sperm cells. At that time nobody thought that it would turn out to be the chemical substance by which the hereditary factors or genes are made. Miescher's weakly acidic chemical, now known as the deoxyribonucleic acid (DNA) is the chemical substance of the gene or hereditary factors. But the relation between gene and DNA was not demonstrated until about the middle of 20th century. Now it is genetically and biochemically proved fact that a gene is a stretch of DNA that codes for a specific polypeptide.

4.2 MENDELIAN GENETICS

Through the selective growing of common pea plants (*Pisum sativum*) over many generations, Mendel discovered that certain traits show up in offspring plants without any blending of parent characteristics. For instance, the pea flowers are either purple or white—intermediate colour do not appear in the offspring of cross -pollinated pea plants. Mendel observed seven traits that are easily recognized and apparently only occur in one of two forms :

1. **Flower colour** is purple or white
2. **Flower position** is axil or terminal
3. **Stem length** is long or short
4. **Seed shape** is round or wrinkled.
5. **Seed colour** is yellow or green
6. **Pod shape** is inflated or constricted
7. **Pod colour** is yellow or green

This observation that there are traits that do not show up in offspring plants with intermediate forms was critically important because the leading theory in biology at the time was that inherited traits blend from generation to generation. Most of the leading scientists in the 19th century accepted this "blending theory." Charles Darwin proposed another equally wrong theory known as 'pangenesis'. This held that hereditary 'particles' in our bodies are affected by the things we do during our lifetime. These modified particles were thought to migrate via blood to the reproductive cells and subsequently could be inherited by the next generation. This was essentially a variation of Lamarck's incorrect idea of the "inheritance of acquired characteristics."

Figure 4.1 Gregor John Mendel, the "Father of Genetics".

Mendel picked common garden pea plants for the focus of his research because they can be grown easily in large numbers and their reproduction can be manipulated. Pea plants have both male and female reproductive organs. As a result, they can either self-pollinate themselves or cross-pollinate with another plant. In his experiments, Mendel was able to selectively cross-pollinate purebred plants with particular traits and observe the outcome over many generations. This was the basis for his conclusions about the nature of genetic inheritance.

In cross-pollinating plants that either produces yellow or green peas exclusively, Mendel found that the first offspring generation (f1) always has yellow peas. However, the following generation (f2) consistently has a 3:1 ratio of yellow to green.

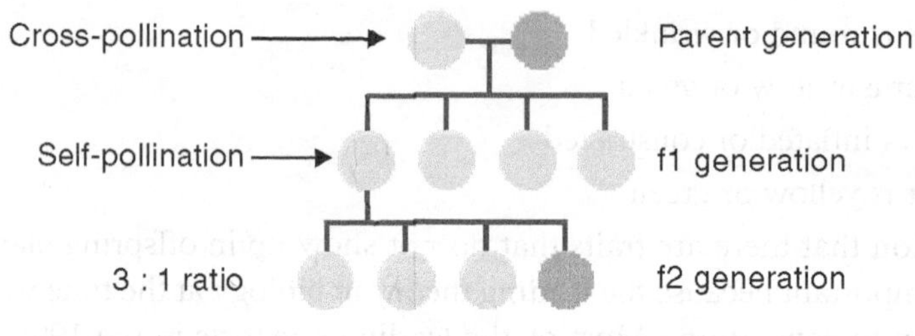

Figure 4.2 Monohybrid cross and Monohybrid ratio.

This 3 : 1 ratio occurs in later generations as well. Mendel realized that this is the key to understanding the basic mechanisms of inheritance.

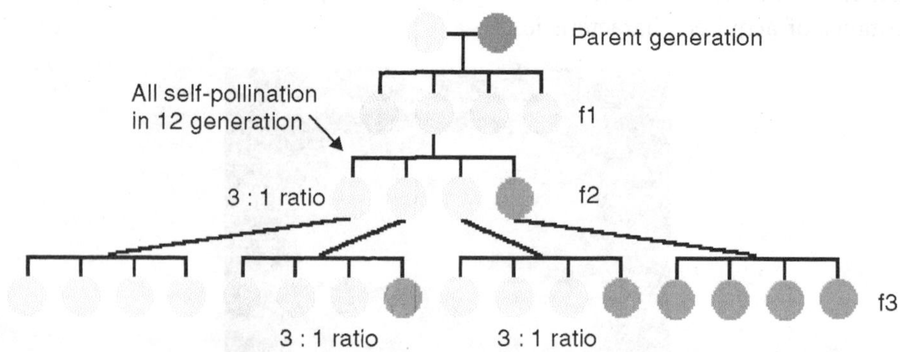

Figure 4.3 Dihybrid cross and Monohybrid ratio.

He came to three important conclusions from these experimental results :

1. That the inheritance of each trait is determined by 'units' or 'factors' (now called genes) that are passed on to descendents unchanged.
2. That an individual inherits one such unit from each parent for each trait.
3. That a trait may not show up in an individual but can still be passed on to the next generation.

It is important to realize that in this experiment the starting parent plants were homozygous for pea colour. That is to say, they each had two identical forms (or alleles) of the gene for this trait—2 yellows or 2 greens. The plants in the f1 generation were all heterozygous. In other words, they each had inherited two different alleles—one from each parent plant. It becomes clearer when we look at the actual genetic makeup, or genotype, of the pea plants instead of only the phenotype, or observable physical characteristics.

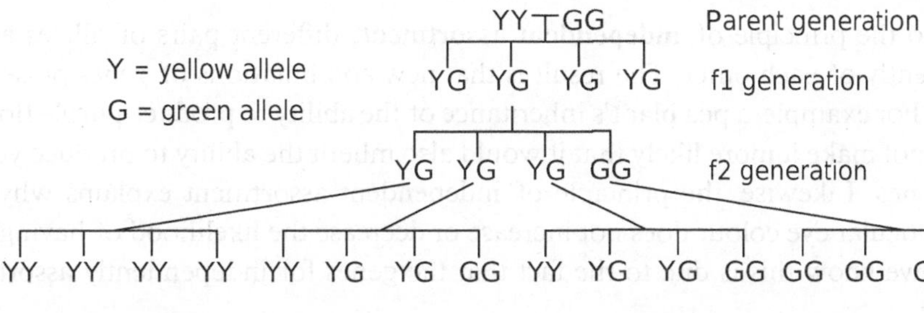

Figure 4.4 A dihybrid cross showing f1 and f2 generations.

Note that each of the f1 generation plants (shown above) inherited a Y allele from one parent and a G allele from the other. When the f1 plants breed, each has an equal chance of passing on either Y or G alleles to each offspring. With all of the seven pea plant traits that Mendel examined, one form appeared **dominant** over the other. Which is to say, it masked the presence of the other allele. For example, when the genotype for pea color is YG (heterozygous), the phenotype is yellow. However, the dominant yellow allele does not alter the recessive green one in any way. Both alleles can be passed on to the next generation unchanged.

Mendel's observations from these experiments can be summarized in three principles :

1. Law of Dominance.
2. Law of Segregation.
3. Law of Independent assortment.

The **law of dominance** states that there are at least two alleles for a gene and one of the allele can be dominant and the other is recessive. The dominant allele, if it is present, will always determine the trait. For example, in pea plant tall is dominant over dwarf.

According to the **principle of segregation,** for any particular trait, the pair of alleles of each parent separate and only one allele passes from each parent on to an offspring. Which allele in a parent's pair of alleles is inherited is a mater of chance. We now know that this segregation of alleles occurs during the process of sex cell formation (*i.e.*, meiosis).

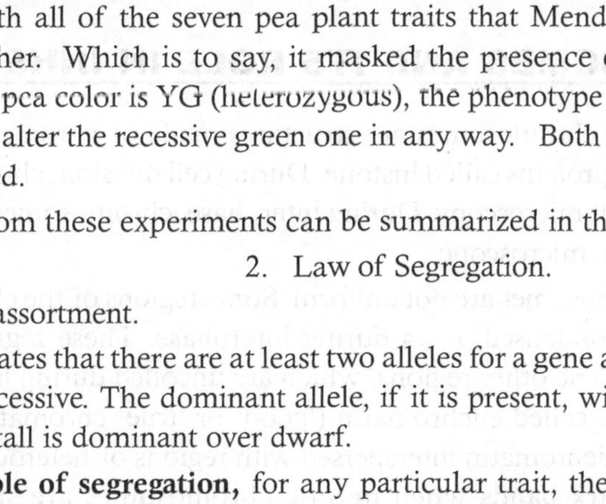

Figure 4.5 Segregation of alleles in the production of sex cells.

According to the principle of **independent assortment**, different pairs of alleles are passed to offspring independently of each other. The result is that new combinations of genes present in neither parent are possible. For example, a pea plant's inheritance of the ability to produce purple flowers instead of white ones does not make it more likely that it would also inherit the ability to produce yellow peas in contrast to green ones. Likewise, the principle of independent assortment explains why the human inheritance of a particular eye colour does not increase or decrease the likelihood of having 6 fingers on each hand. Today, we know this is due to the fact that the genes for independently assorted traits are located on different.

These two principles of inheritance, along with the understanding of unit inheritance and dominance, were the beginnings of our modern science of genetics. However, Mendel did not realize that there are exceptions to these rules.

4.3 CHROMOSOMES AND ITS ROLE IN INHERITANCE

As described elsewhere chromosomes are structures made of chromatin, which is a mixture of DNA and a specific family of proteins called histone. During cell division, chromosomes are highly condensed, becoming visible by light microscopy. During interphase, chromosomes are relatively decondensed, and are not visible under the microscope.

However, Chromosomes are not uniform. Some regions of the chromosome consist of chromatin that is always highly condensed, even during interphase. These regions are called heterochromatin ('different' chromatin). The other regions, which are uncoiled during interphase, and highly condensed during cell division, are called euchromatin ('good' or 'true' chromatin). Each chromosome of a cell consists of regions of euchromatin interspersed with regions of heterochromatin. These various regions appear as light and dark bands when mitotic chromosomes are stained with various dyes. Each chromosome has a different banding pattern, so chromosomes can easily be identified when stained in this way. Photographs of stained mitotic chromosomes for the purposes of chromosome identification shown in figure 4.6 are known as karyotypes.

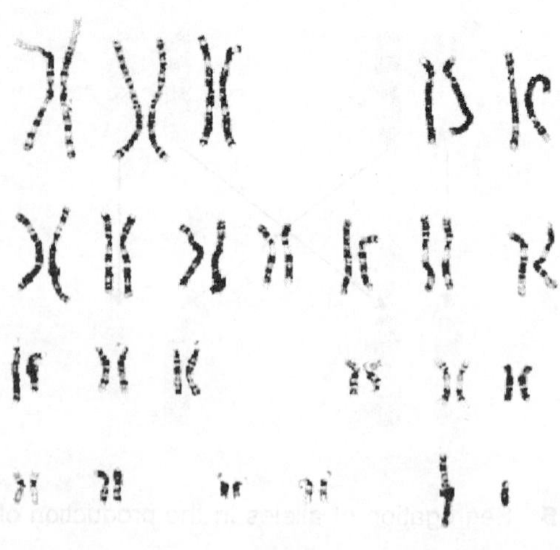

Figure 4.6 Karyotype in Humane.

4.4 SEX CHROMOSOMES

There is one example where a somatic cell may not have two copies of a particular chromosome under normal circumstances. This involves the **sex chromosomes**. As we shall see, these chromosomes are involved in determining the gender of an individual. Our consideration of sex chromosomes will focus on the sex chromosomes of *Drosophila* and humans.

Both fruit flies and humans have two different sex chromosomes: X and Y. Females have two X chromosomes in every somatic cell; males have one X and one Y. The X and Y-chromosomes are quite different, so males have two chromosomes that do not appear as a natural pair. The Y chromosome is much smaller than the X, and its centromere is closer to one end. There is also little similarity in the DNA sequences found on the two chromosomes. **However, the X and Y do behave as members of a homologous pair**. They pair up during meiosis, just as other homologues do. They separate during meiosis I, producing two types of sperm: those containing an X chromosome (female-producing sperm) and those containing a Y chromosome (male-producing sperm). Because the sex chromosomes are somewhat unique in their properties, they are classified separately from the other chromosomes. This is indicated by their name: the sex chromosomes. The remaining chromosomes are called autosomes.

4.5 THE CHROMOSOME THEORY OF HEREDITY

We have seen how chromosomes are segregated during cell division, but what exactly does this have to do with the inheritance of physical traits (*i.e.* genes) ? Fairly soon after the rediscovery of Mendel's work, many biologists believed that genes were situated on the chromosomes, but this idea required proof. Evidence came in from an experiment conducted by T. H. Morgan on an eye color mutation (white eyes) in *Drosophila*. By comparing karyotypes of specific flies with their phenotype, Morgan demonstrated that the white eye mutation was inherited along with the X chromosome. Further work led to the identification of other genes that were transmitted along with the X chromosome, and genes that were transmitted along with particular autosomes. Each chromosome appeared to contain a particular set of genes. Mapping studies showed that each gene resided at a particular locus (plural: loci) on a chromosome, indicating that chromosomes are linear arrays of genes.

The Chromosomal Basis of Mendel's Postulates

The realization that chromosomes are linear arrays of genes made it possible to explain Mendel's postulates in terms of chromosome behavior during meiosis.

Principle of Segregation

To explain Mendel's principle of segregation, let us consider a premeiotic cell from an individual who is heterozygous for a particular gene (Aa). Each member of the homologous pair of chromosomes will have a different allele of that gene.

After replication, each chromosome of the homologous pair consists of two sibling chromatids. During meiosis I, the two members of the homologous pair segregate into separate cells.

After meiosis II, two of the gametes contain a chromosome with the dominant allele, and two gametes contain a chromosome with the recessive allele. Each of these gametes has an equal chance of participating in fertilization. Therefore, if two heterozygotes mate, the offspring will have an equal chance of inheriting a dominant or recessive allele from the male parent, and an equal chance of inheriting a dominant or recessive allele from the female parent. Therefore there is an equal chance of having each of the following genotypes : AA, Aa, aA, and aa.

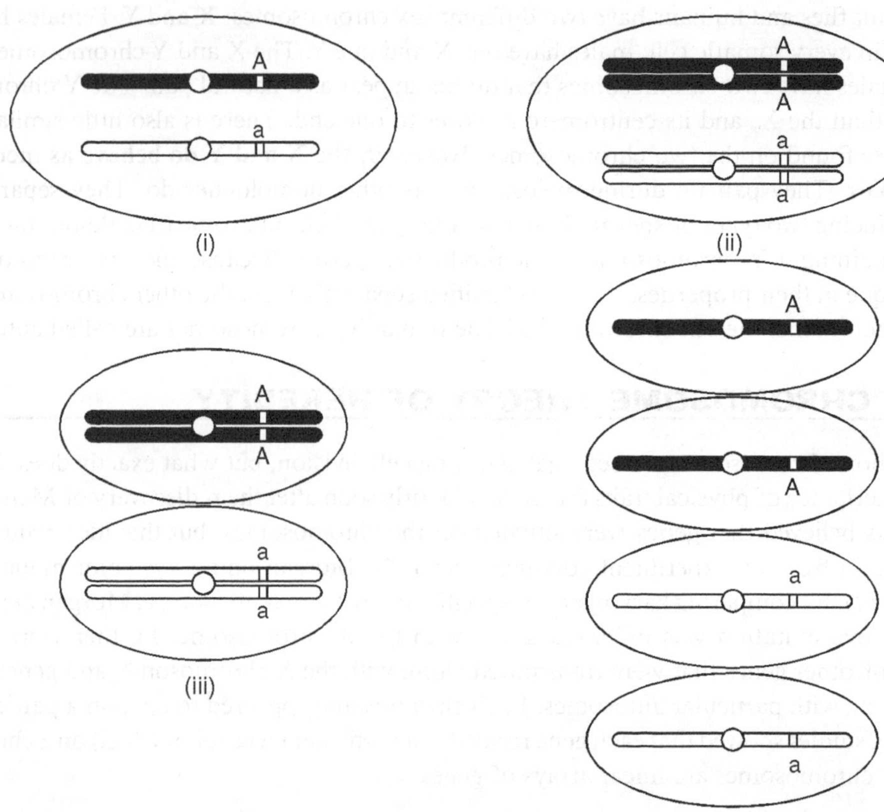

Figure 4.7 Meiosis and law of segregation.

Independent Assortment

To illustrate Mendel's principle of independent assortment, we will consider a premeiotic cell from an individual heterozygous at two gene loci (AaBb). Each of the larger chromosomes has a different allele for the A gene, and each of the smaller chromosomes has a different allele for the B gene.

After replication, each of the chromosomes consists of two sibling chromatids joined at the centromere.

The first meiotic division produces two possible combinations of chromosomes in the daughter cells, because the large and small chromosomes assort independently. One of the possible combinations (let us call it 'combination 1') is shown to the right, where the large chromosome with the dominant allele of the A gene has segregated along with the small chromosome that has the dominant allele of the B gene. Likewise, the chromosomes containing the recessive allele of each gene have segregated together [Fig. 4.8 (*iii*)].

Here is the other possible combination of chromosomes (we will call it 'combination 2') after meiosis I. In this case, the chromosome with the dominant allele of the A gene has segregated with the chromsome that has the recessive allele of the B gene, and vice versa [Fig. 4.8 (*iv*)]. Because the larger chromosome pair assorts independently of the small chromosome pair (*i.e.* segregation of the homologues of the large pair does not affect segregation of the homologues of the small pair), both of the possible combinations will occur in a large group of cells undergoing meiosis.

Here is the chromosome complement of each gamete after meiosis II resulting from combination 1 shown in figure 4.8 (*v*) and the chromosomal complement of the four gametes after meiosis II, resulting from combination 2 is given in Fig. 4.8 (*vi*). You can see that the allele combinations in the gametes are different between combination 1 and combination 2. But with that two possible chromosome combinations, gametes with all possible combinations of allclcs are obtained. In an individual, many cells would be undergoing meiosis simultaneously, and statistical probability dictates that all of the chromosome/allele combinations represented here would be present in the gamete population.

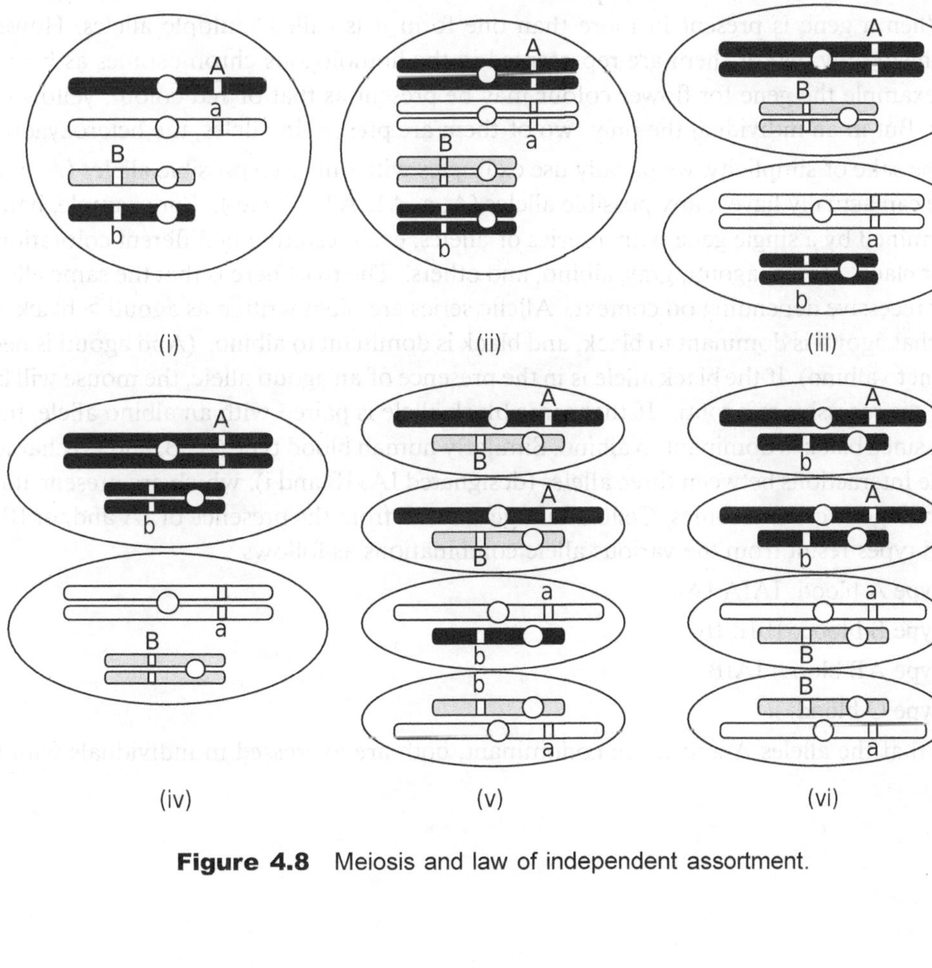

Figure 4.8 Meiosis and law of independent assortment.

But it was shown in some insects, that male insects have one X chromosomes (XO) per cell and the female cells have two X-chromosomes (XX) . There for male insects produced two types of gametes, those with X and without X in equal proportions. Example the insect *Protenor* produced two types of spermatids (male gametes) cells, with X chromosomes (X) and without X chromosomes (O) in equal numbers.

4.6 MULTIPLE ALLELES

In a diploid (2n) organism every gene must exist in pairs, and each member of that pair is called an allele or allelic pair. One of the allele is supplied by the male gamete (n) and the other one is given by the female Gamete (n) during the process of fertilization. These alleles are situated in the homologous pairs of chromosomes at specific positions called loci. Over a period time random mutations can create changes in these genes resulting in altered forms of the gene. When there are two altered forms a gene present in a diploid cell or in an allelic pair, It is known as heterozygous and when the members of an allelic pair is of the same type of genes, they are known as homozygous. But when we consider a population of a species there are innumerable copies of a single gene at any given time. By random mutation a gene can be mutated to more than one form in a population. Thus more than one alternate form a gene is created naturally. When a gene is present in more than one form it is called multiple alleles. However in a diploid organism only two of them are represented in the homologous chromosomes as heterozygous alleles. For example the gene for flower colour may be present as that of red colour, yellow colour or white colour. But in an individual the only two of them are present in alleles, the heterozygous allele.

For the sake of simplicity, we usually use examples with only two possible alleles (A and a). But a single gene can actually have many possible alleles (A, a, A1, A2, A′, etc.). For example, hair color in mice is determined by a single gene with a series of alleles, each resulting in different coloration. There are alleles for black, brown, agouti, gray, albino, and others. The twist here is that the same allele can be dominant or recessive depending on context. Allelic series are often written as agouti > black > albino. This means that agouti is dominant to black, and black is dominant to albino. (And agouti is necessarily also dominant to albino). If the black allele is in the presence of an agouti allele, the mouse will be agouti because black is recessive to agouti. If that same black allele is paired with an albino allele, the mouse will be black since black is dominant to albino. Similarly human blood type (A, B, and O) characteristics are due to the interactions between three alleles (designated IA, IB and i), which are present at the same locus of homologous chromosomes. Cellular antigens result from the presence of IA and/or IB, but not i. Four blood types result from the various allele combinations as follows :

- Type A blood: IAIA IAi
- Type B blood: IBIB IBi
- Type AB blood: IAIB
- Type O blood: ii

Note that the alleles A and B are codominant, both are expressed in individuals with type AB blood.

4.7 LINKAGE AND CROSSING OVER

Sometimes, the predictions (*i.e.* for dihybrid crosses) based on our understanding of Mendel's principles are seemingly violated; it is often found that the number of offspring obtained for each phenotype is significantly different from a 9 : 3 : 3 : 1 ratio, the Dihybrid ratio. Numbers within the recombinant phenotypes are higher than predicted and numbers of the two parental phenotypes are fewer than predicted. Typically, when this occurs, it is because the genes of the trait that we considered for the study are present on the same set of chromosomes (linkage) and crossing over during prophase I affects the outcome of meiosis. When genes distribution after meiosis varies depending on whether or not crossing over has occurred, the genes are said to be linked.

Up until now we have assumed that all genes were inherited independently. However, we have also said that genes are arranged on chromosomes, which are essentially long strands of DNA residing in the nucleus of the cell. This certainly opens the possibility that two otherwise unrelated genes could reside on the same chromosome. Does independent inheritance hold for these genes ?

To start with, we need to consider the rather complex process that forms gametes (egg and sperm cells, each with only one copy of each chromosome or haploid) from normal cells or diploid with two copies of each chromosome, one derived from each parent. At one stage of this process (please see the details of meiosis), the maternally derived chromosome lines up with the corresponding paternally derived chromosome, and only one of the two goes to a specific gamete. A chromosome is not a single gene, in fact it is a group of genes arranged in lines and migrate as single unit. A dog with 39 chromosomes does not meant that it contains only 39 genes, but there are a large number of genes in a chromosome and they move together as a single unit. Thus Mendel's law of independent assortment is based on the independent orientation genes that are located on different chromosomes. Luckily all the seven characters that Mendel has studied in pea plant were located on different chromosomes and that is why he was able to get the expected 9 : 3 : 3 : 1 ratio, the dihybrid ratio in the F_2 generation. But it was observed that all the genes are not obeying the law of independent assortment and results in the modified forms of dihybrid ratios. Because those genes which are located in the same chromosomes cannot move independently but some times behave like a single gene, when it gives monohybrid ratios for dihybrid cross. But In fact, in dihybrid crosses in addition to the formation of monohybrid ratios, there is the formation of variations of dihybrid ratios also. This shows that things are a little more complicated yet, because while the paternal and maternal chromosomes are lined up, they can and do exchange segments, so that at the time they actually separate, each of the two chromosomes will most likely contain material from both parents.

At this point we need to define a couple of terms. Two genes that are located in the same chromosomes are linked if they are close together and thus tend to be inherited together. Such genes are called linked genes or linkage groups. But if those genes located on the same chromosomes are present far apart; there is the chance of exchanging their positions between the homologous chromosomes. This interchange of genes between the homologous chromosomes is termed as crossing over. The process of crossing over gives some degree or percentage of independent assortment, which is closely related to the physical distance between the linked genes in a chromosome. In true linkage, there is always the possibility that linked genes can cross over. Crossing over involves the braking and rejoining of DNA segments (exchange) of genes between the homologous chromosomes. Thus the process of crossing over breaks the original linkage between genes present in a chromosome. This happens often enough that genes far apart on long chromosomes appear to be inherited independently, but if genes are close together, a break is much less likely to form between them than at some other part of the paired chromosomes. The process of crossing over results in the genetic recombination of paternal and maternal traits and the amount of such crossing over depends on the relative distance between the genes. Modifications of dihybrid ratios depend on this physical distance, and the frequency of recombination between two genes can be used to estimate the relative distance between them on the chromosomes. The recombination frequencies can be calculated from the phenotypes of the parental and new recombinants formed in the offspring of heterozygous individuals.

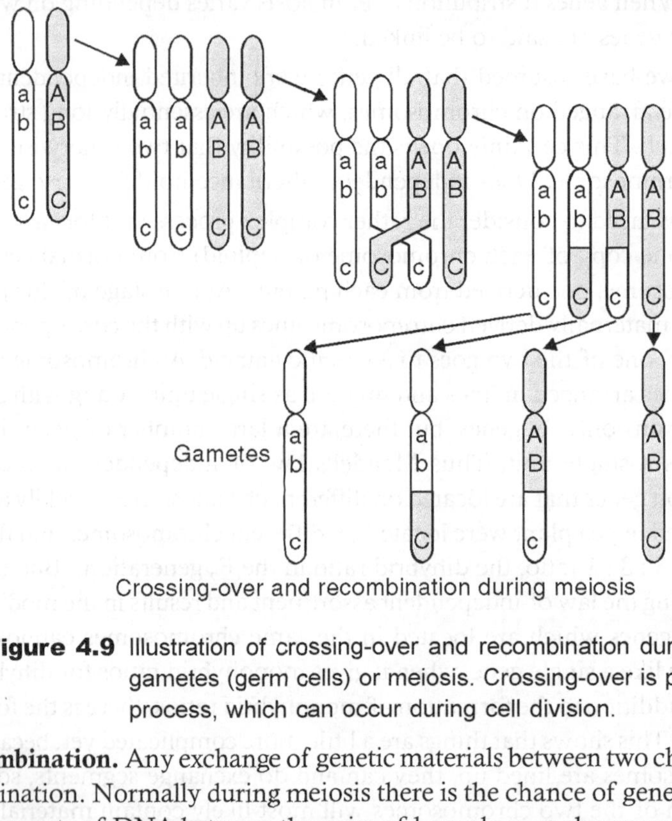

Figure 4.9 Illustration of crossing-over and recombination during the formation of gametes (germ cells) or meiosis. Crossing-over is part of a complicated process, which can occur during cell division.

Genetic Recombination. Any exchange of genetic materials between two chromosomes is defined as the genetic recombination. Normally during meiosis there is the chance of genetic recombination due to the exchange of segments of DNA between the pairs of homologous chromosomes. This creates new combinations of genes (Fig. 4.9). This process takes place during meiosis by the crossing over. It allows the mixing of paternal and maternal traits located on the same chromosomes. Process of genetic recombination is very important for the generation of variation among the progenies and in the population of sexually reproducing organisms. Mutation also produces variations, but there is difference. Recombination gives rise to different combinations of the existing genes and its alleles. The process of sexual reproduction can create innumerable combination of genes on which the natural selection has operated for millions of years to produce adaptable phenotypes and genotypes. Genetic recombination can take place only during process of meiosis and it is absent in mitosis. In mitosis the cell division is an equal division with the equal distribution of genetic materials in diploid conditions. There is no pairing of chromosomes (synapsis) and related crossing over during mitosis. This method of cell division is

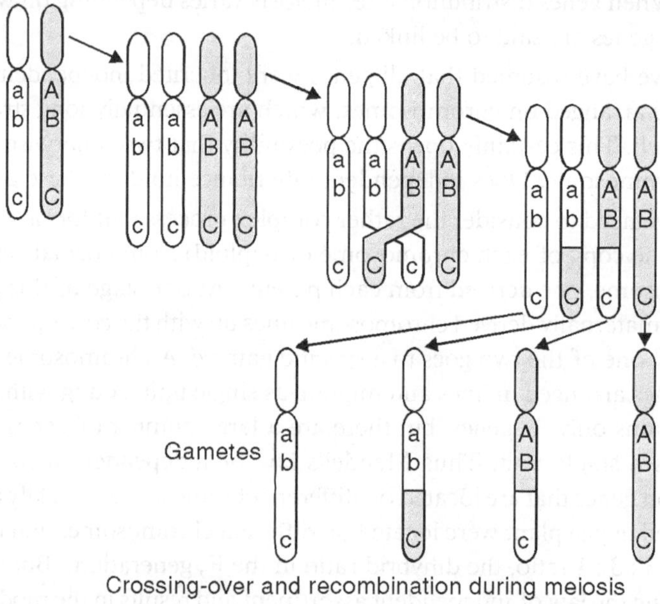

essential for the growth and repair of the body. Thus mitosis occurs in somatic cells and meiosis occurs in specialized cells of reproductive organs during the formation of gametes. The cells formed by meiosis are quantitatively and qualitatively different from the mother cells and may be dissimilar in quality among the daughter cells.

4.8 GENETIC MAPPING

Physical crossing over during meiosis I is a normal event. The effect of this event is to rearrange heterozygous homologous chromosomes into new combinations. The term used for crossing over is **recombination**. Recombination can occur between any two genes on a chromosome, the amount of crossing over is a function of how close the genes are to each other on the chromosome. If two genes are far apart, for example at opposite ends of the chromosome, crossover and non-crossover events will occur in equal frequency. Genes that are closer together undergo fewer crossing over events and non-crossover gametes will exceed than the number of crossover gametes. (Fig. 4.10).

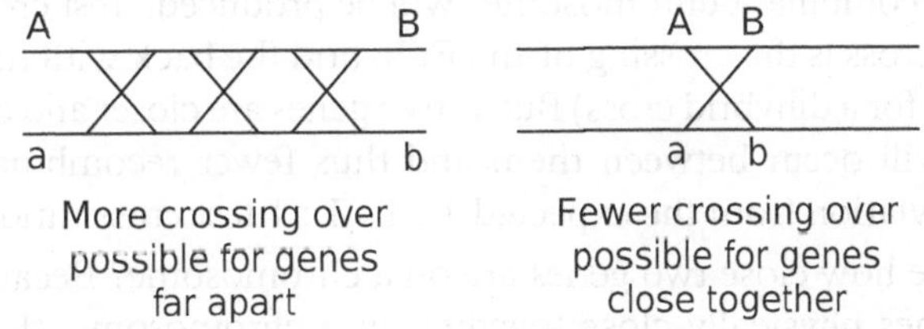

Figure 4.10 Frequency of crossing over between linked genes.

Two genes that very close to each other on the chromosome crossing over will be a very rare event. Two types of gametes are possible when we follow genes on the same chromosomes. If crossing

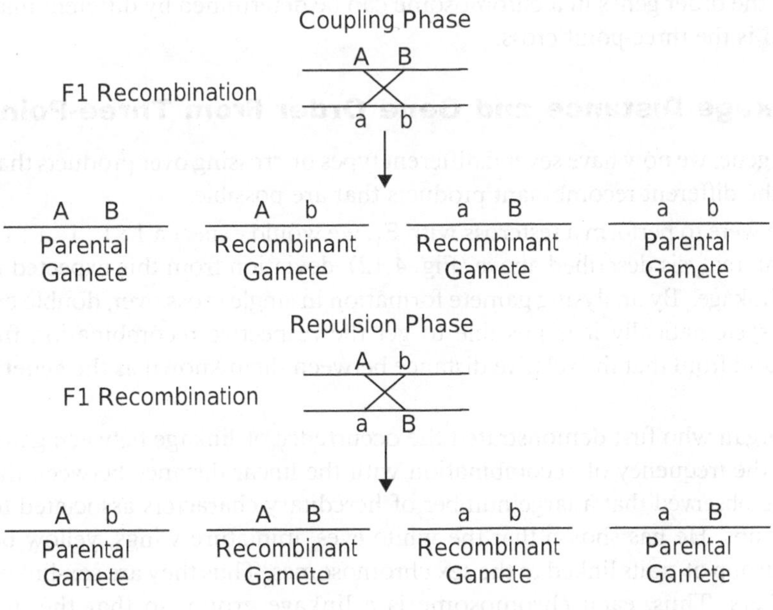

Figure 4.11 Coupling or repulsion in gamete formation.

over does not occur, the products are **parental gametes**. If crossing over occurs, the products are **recombinant gametes**. The allelic composition of parental and recombinant gametes depends upon whether the original cross-involved genes in coupling or repulsion phase. Fig. 4.11 depicts the gamete composition for linked genes from coupling and repulsion crosses.

It is usually a simple matter to determine which of the gametes are recombinants. These are the gametes that are found in the lowest frequency. This is the direct result of the reduced recombination that occurs between two genes that are located close to each other on the same chromosome. Also by looking at the gametes that are most abundant you will be able to determine if the original cross was a coupling or repulsion phase cross. For a coupling phase cross, the most prevalent gametes will be those with two dominant alleles or those with two recessive alleles. For repulsion phase crosses, gametes containing one dominant and one recessive allele will be most abundant (Fig. 4.11). Understanding this fact will be important when you actually calculate a linkage distance estimate from your data.

The important question is how many recombinant chromosomes will be produced. If the genes are far apart on the chromosome a cross over will occur every time that pairing occurs and an equal number of parental and recombinant chromosomes will be produced. Test cross data will then generate a 1 : 1 : 1 : 1 ratio. (A test cross is the crossing of the F_1 hybrid the back with the recessive parent and the test cross ratio is 1 : 1 : 1 : 1 for a dihybrid cross) But as two genes are closer and closer on the chromosome, fewer cross over events will occur between them and thus fewer recombinant chromosomes will be derived. We then see a deviation from the expected 1 : 1 : 1 : 1 test cross ratio.

How can we decide how close two genes are on a chromosome? Because fewer crossover events are seen between two genes physically close together on a chromosome, the lower the percentage of recombinant phenotypes will be seen in the testcross data. **By definition, one map unit (m.u.) is equal to one percent recombinant phenotypes.** In honor of the work performed by Morgan **one m.u. is also called one centimorgan (cM)**. To determine the linkage distances simply divide the number of recombinant gametes by the total gametes produced. The frequency of recombination between distantly placed genes and the order genes in a chromosome can be determined by different mapping techniques. One such method is the three-point cross.

Deriving Linkage Distance and Gene Order From Three-Point Crosses

By adding a third gene, we now have several different types of crossing over products that can be obtained. Fig. 4.12 shows the different recombinant products that are possible.

Now if we were to perform a testcross with F_1, we would expect a 1 : 1 : 1 : 1 : 1 : 1 : 1 : 1 ratio. As with the two-point analysis described above (Fig. 4.12), deviation from this expected ratio indicates the occurrence of at linkage. By analysing gamete formation in single cross over, double cross over and then triple cross over systematically it is possible to get the respective recombination frequencies for the individual genes and from that the relative distance between them known as the **genetic distance** can be calculated.

It was Morgan who first demonstrated the occurrence of linkage between genes in *Drosophila* in 1911 and related the frequency of recombination with the linear distance between the genes along the chromosomes. He observed that a large number of hereditary characters associated together in groups called linkage groups. He has shown that the white eyes, miniature wings, yellow body colour single bristles etc. are a group of traits linked to the sex chromosomes. Thus they are sex-linked traits associated with male members. Thus, each chromosome is a linkage group, so that the haploid number of chromosomes is an organism of its linkage groups. There are four linkage groups in *Drosophila*, 23 in man 7 linkage groups in peas and ten in maize.

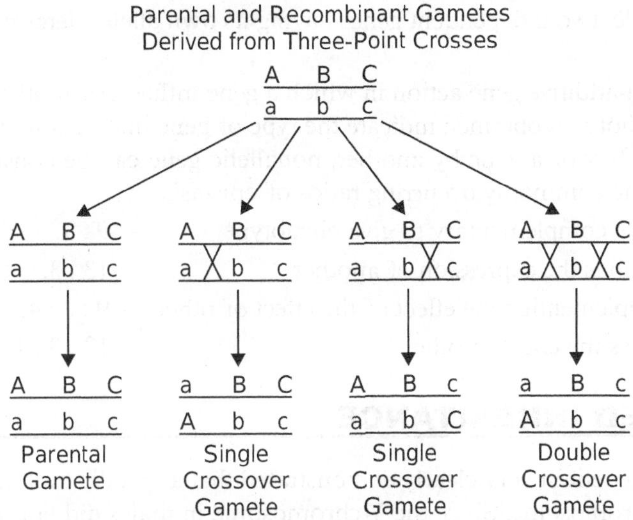

Figure 4.12 Type of gametes in a trihybrid cross.

By knowing the genetic distance and the order of genes in chromosome it is possible to make the genetic map of that chromosome.

4.9 GENE INTERACTION OR POLYGENES

In monohybrid cross, the F_2 generation will show two phenotypes in the ratio 3 : 1 known as the monohybrid ratio. This happens only when there is complete dominance. In dihybrid crosses the F_2 generation will have four phenotypes in the ratio 9 : 3 : 3 : 1. This is applicable when genes are independent (absence of linkage) and show complete dominance. In the case of incomplete dominance the heterozygous ratio of a monohybrid cross will modify into 1 : 2 : 1, which is actually the genotypic ratio. Similarly, the number of phenotypes in the F_2 generation of a dihybrid cross also under goes modification.

Multiple alleles are different forms of the same gene. That means the sequence of the bases is slightly different in the genes located on the same place of the chromosome. In polygenic characters etc. there is more than one gene involved and there may be multiple alleles for each of these genes affecting a specific character. This is far more complex involving potentially a number of chromosomes. The phenotypic characters like high blood pressure (hypertension) are not the result of a single 'blood pressure' gene with many alleles (a 120/80 allele, a 100/70 allele, a 170/95 allele, etc.). This phenotype is the result of interactions between a person's weight (one or more obesity genes), cholesterol level (one or more genes controlling metabolism), kidney function (salt transporter genes), smoking (a tendency to addiction), and probably lots of others too. Each of the contributing genes can also have multiple alleles. Similarly, eye colour in man is due to complex interactions of two genes each having pairs of incompletely dominant alleles and skin colour is determined by the interactions between at least three independent genes having two or more alleles.

These multiple gene interaction is divided into two types—additive gene interactions and non-additive gene interactions or **epistasis**. When a phenotype is influenced by two or more independent genes and each one of them contribute equally in the formation of the trait it is called the additive gene interactions. For example, two independent genes having its own alleles determine the seed coat colour in wheat.

Epistasis is a non-additive gene action in which a gene influences another gene in different ways and the ratio of the phenotypes obtained indicate the type of gene interaction or the epistasis. Any type of suppression of the effect of a gene by another, nonallelic gene can be considered as epistasis. The following are some of the commonly occurring ratios of epistasis.

1. Both genes are complementary to give phenotype —9 : 7
2. One gene inhibits the expression of another —13 : 3.
3. One gene supplementing the effect of the effect of other —9 : 3 : 4.
4. One gene hides the effect of other —12 : 3 : 1.

4.10 SEX LINKED INHERITANCE

T. H. Morgan, famous geneticist very clearly demonstrated that a specific trait namely, white-eye colour was linked to the sex chromosome. Since the Y-chromosome in males did not possess any gene for eye colour it gave a distorted segregation in the progeny. Therefore, a white eyed female will produce white eyed male progenies, but in the female progenies one may expect 50% white eyed and 50% red eyed individuals. Now it is very well known that the distorted segregation is because genes of eye colour and sex determination are tightly linked on the X- chromosome. This was the first to prove the existence of linked genes.

Sex determination in humans results from the action of a pair of chromosomes, the sex chromosomes (the other 22 pairs are called autosomes). Those chromosomes, which are responsible for the determination of the sex of an organism, are termed as sex chromosomes or X-chromosomes. The all other chromosomes of the cell are the autosomes. There are two types of sex chromosomes X-chromosomes and Y-chromosomes. X-chromosomes in homozygous conditions determine the femaleness and association of X chromosomes and Y-chromosomes produces the maleness. We designate the male XY and the female XX. An understanding of Mendelian principles allows us to predict that males and females are likely to be conceived with equal probability.

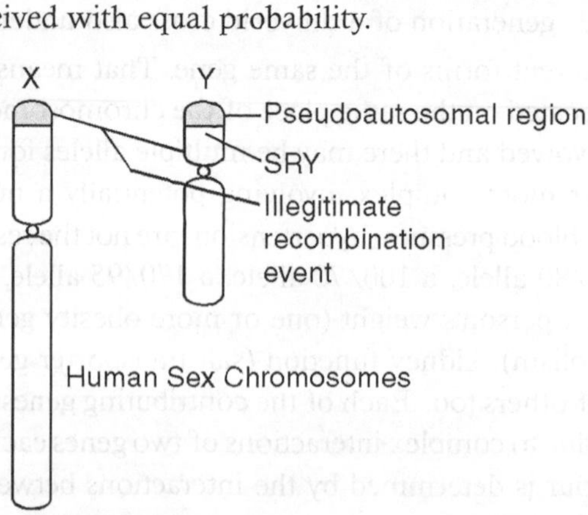

Figure 4.13 The homologous regions of X and Y chromosomes.

The X chromosome contains a significant number of genes. In contrast, the Y chromosome contains very few. Therefore, genes on the X chromosome (often called 'X-linked' genes) are in a unique situation. Females have two copies of each gene, just like the normal situation with autosomal (In humans, there are 22 pairs of autosomes) genes. Males, on the other hand, since they have only one X chromosome, have only one copy of each X-linked gene. Because of this, males cannot be homozygous or heterozygous; they are referred to as hemizygous (The condition of having only one allele of a pair). Therefore, alleles that are recessive in a female are automatically expressed in a male (because there is no second allele to overshadow the recessive one). The normal rules of dominance do not apply to males in this case. For this reason, problems associated with X-linked recessive alleles, such as hemophilia (inability of blood to clot) and color blindness are more common in males than they are in females.

Let us consider hemophilia as an example. For a female to be affected by hemophilia, she has to be homozygous for the hemophilia allele (which we will designate 'h'). If an affected female mates with an unaffected male :

P1 : ♀hh X H–♂

↓

F1 : ♀Hh, ♂h–

All of the male offspring would be affected by hemophilia. All of the female offspring would be unaffected, but would be carriers. This criss-cross pattern of inheritance of a phenotype (from mother to sons) is one of the features of X-linked inheritance. If the female parent is a carrier, and the male parent is a hemophiliac, the following offspring are obtained :

P1 : ♀Hh X h–♂

↓

F1 : ♀1/2 Hh, 1/2 hh

♂1/2 H–, 1/2 h–

Half of the offspring of each gender will, on average, be hemophiliac.

The origin of the hemophilia that plagued the crowned families of Europe is actually of English/ German origin. Queen Victoria was the original carrier. Her son the Prince of Wales did not inherit it, and so the British royal family has been free of the disease. The royal families of France, Spain, and Russia, into which were married daughters of Victoria who were carriers of the hemophilia gene, were not so lucky.

Test yourself. *What offspring would be produced from a mating between a homozygous normal female and a hemophiliac male ?*

Sex influenced traits are related to sex but not coded for by genes on the sex chromosomes (baldness, gout, allergies and cleft palate are all generally more prevalent in men than women). Typically, they are influenced by the hormone, testosterone (in both men and women).

Chromosomal Nondisjunction is the improper segregation of chromosomes during cell division. During meiosis this phenomenon can results in a gamete (and hence, possibly an offspring) having an abnormal number of chromosomes. Often, it occurs in sex chromosomes of males or females but it may occur in autogenous model as well. For example, in a female cell undergoing meiosis, if nondisjunction occurs once during the first meiotic division, the result with respect to sex chromosomes could be as shown below.

Sex chromosomes in the original cell	:	XX
After Interphase	:	XXXX
After Meiosis I	:	XXXX + O
Defective eggs	:	XX + XX + O + O

Note that it would be possible for a nucleus to be XXXX, if

Nondisjunction were to occur during both Meiosis I and II

Defective sperm, which could result from nondisjunction in a male :

O, XX, YY, XY.

Unfortunately, such defective gametes sometimes participate in fertilization, yielding zygotes. Resulting offspring could have ***Trisomy X*** (fertile females with XXX), ***Klinefelter's Syndrome*** (sterile males with XXY), ***Turner's Syndrome*** (sterile females with X), or be XYY males. Down's Syndrome results from nondisjunction involving autosomal chromosomes. All of these genetic abnormalities and others can be diagnosed using ***amniocentesis***, a procedure in which fetal cells obtained from the amniotic fluid of a pregnant woman are karyotyped. Another technique, ***chorionic villi*** sampling, may also be used to determine the genetic characteristics of a foetus.

4.11 EXTRA NUCLEAR INHERITANCE

According to the traditional concept the genome is located in the nucleus of a eukaryotic cell. All genetic loci discussed to date have been located on the chromosomes in the cell's nucleus. But there are some genes present in the cytoplasm, inside the organelles like mitochondria and chloroplasts. These organelles have a circular DNA molecule similar to that of a eukaryotic system. The characters inherited by them are known as extra chromosomal inheritance or cytoplasmic inheritance. We have studied about the process of fertilization. During fertilization only the nucleus of the male gamete enters the egg, leaving the cytoplasm outside. Thus, the cytoplasm or the cytoplasmic genes of the zygote are contributed only by the egg and not by the male gamete. Therefore, the extra chromosomal inheritance is also known as maternal inheritance. There are three types of cytoplasmic inheritance known today :

A. Maternal influence (Egg cytoplasm influences the phenotype of the offspring)

B. Organelle heredity (Mitochondria and Chloroplasts)

C. Infectious heredity (An infectious particle is transmitted during conjugation)

Cytoplasmic inheritance shows certain special features : (1) Lack of Mendelian segregation and typical Mendelian ratios, (2) Persistence of characters for many generations, (3) Controlled by mitochondrial and chloroplast DNA and (4) shows maternal inheritance as these characters are transmitted only by female gametes.

1. Variegation in Four o'clock Plant and Maternal Inheritance

The classic study of maternal inheritance was performed by Correns on the four o'clock plant. This plant can have green, variegated (white and green) or white leaves. Flower structures can develop at different locations on the plant and the flower colour corresponds to the leaf colour. When Correns crossed the different coloured flowers from different locations on the female plant with pollen obtained from flowers of the three different colours, the progeny that resulted from the cross always exhibited the colour of the leaf of the female. That is, regardless of whether the pollen was from a leaf which was green, variegated or white. If the female flower came from a region where the leaves were green, all the progenies were green. Similar results were seen when the female was from a region on the plant where the leaves were either variegated or white. In comparison to traits controlled by maternal effects, those traits controlled by maternal inheritance, the female phenotype is always expressed in its offspring.

Female	Male	Progeny Phenotype
Green	Green, Variegated or White	Green
Variegated	Green, Variegated or White	Variegated
White	Green, Variegated or White	White

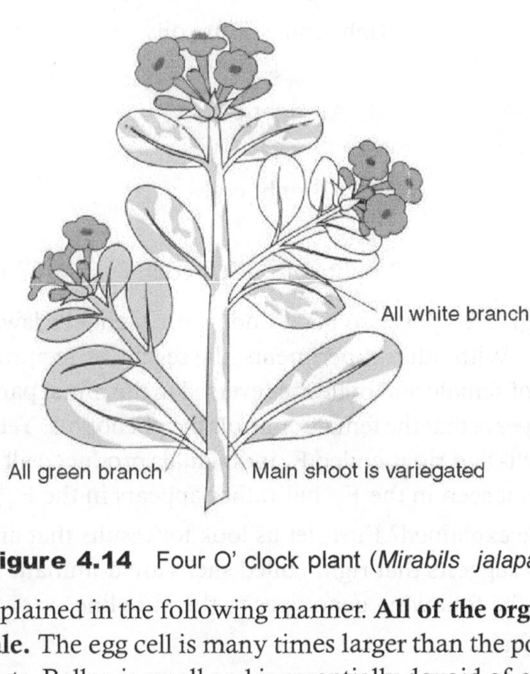

Figure 4.14 Four O' clock plant (*Mirabils jalapa*).

The results can be explained in the following manner. **All of the organelle DNA that is found in an embryo is from the female.** The egg cell is many times larger than the pollen cells, and contains both mitochondria and chloroplasts. Pollen is small and is essentially devoid of organelles, and thus organelle DNA. So any trait that is encoded by the organelle DNA will be contributed by the female. In the case of the four o'clock plant, the different colours of the leaves are a result of the presence or absence of chlorophyll in the chloroplast, a trait that can be controlled by the chloroplast DNA. Thus, green shoots contain chloroplasts that have chlorophyll, the chloroplasts in the white shoots contain no chlorophyll, and the variegated shoots contain some chloroplasts with chlorophyll and some without chlorophyll. Thus, depending upon the location in the plant where the flower comes from, the egg can have chloroplast with chlorophyll, without chlorophyll or a mixture of the two types of chloroplasts. This is the biological basis of maternal inheritance.

2. Snail Shell Coiling and Maternal Effects

The embryo is formed when a female gamete unites with a male gamete. In the vast majority of species, the female gamete is physically larger than the male gamete and provides the cytoplasm for the developing embryo. Within this cytoplasm are factors that were released by the nuclear genes of the female. Those factors may have specific effects upon the developing embryo. The female cytoplasm also contributes the mitochondria for all species as well as the chloroplast for plant species. These two organelles contain DNA and control certain traits in the offspring. Those phenotypes that are controlled by nuclear factors found in the cytoplasm of the female are said to express a **maternal effect**. Those phenotypes controlled by organelle genes exhibit **maternal inheritance**.

The classic phenotype, which exhibits maternal effects, is coiling direction of snail shells. The **coiling phenotype** that is seen in the offspring is **controlled by the genotype of the mother**. The following crosses were made between pure line snails, and the results were seen. By convention, the female is always given first.

	Cross 1	**Cross 2**
Parents	Right coil × Left coil	Left coil × Right coil
	↓	↓
F_1	All right coil	All left coil
	↓	↓
F_2	All right coil	All right coil
	↓	↓
F_3	3 right coil : 1 left coil	3 right coil : 1 left coil

These results at first glance appear to be at odds with Mendel's laws. First, the F_1 phenotype is not the same for both crosses. With other experiments, the results of reciprocal crosses (complementary crosses were the phenotypes of female and male are reversed in the initial parental cross) were equivalent, but with this experiment it appears that the female controls the phenotype. Yet, the F_2 appears to contradict this hypothesis because the left- and right-coiled F_1 individuals produced all right progeny. Furthermore, the 3 : 1 Mendelian ratio is not seen in the F_2, but rather appears in the F_3 generation.

How can this result be explained? First, let us look for results that are familiar. The F_3 ratio of 3 rights : 1 left for both crosses suggests that right-coiled shells are dominant to left-coiled shells. If this is the case, then we can assign the following genotypes to the pure lines :

- Right-coiled shell : s^+s^+
- Left-coiled shell : ss

The next observation is that the phenotype of the F_1 generation is always that of the female parent. One hypothesis would suggest that the genotype of the female controls the genotype of its offspring. Can this result be confirmed in the subsequent generations ? If the genotypes we assigned to the parents are correct, then the genotype of F_1 individuals from each cross are s^+s (from s^+s^+ × ss and ss × s^+s^+). If the female genotype does control the phenotype of its offspring, then we would predict that all the F_2 snails would have right coils. This is the exact result that is seen. But what would the genotypes of the F_2 snails be ? If we intermate snails with the genotype s^+s the genotypic ratio should be $3s^+$ to 1 ss. These genotypes would not be expressed as a phenotype until the F_3 generation. These are the results that were obtained. A general conclusion from all traits that express a maternal effect is that the normal Mendelian ratios are expressed one generation than expected.

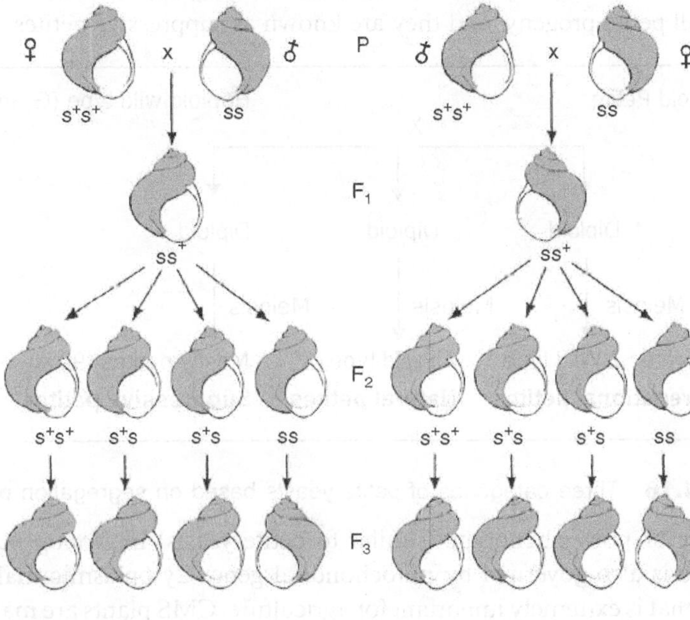

Figure 4.15 Maternal effects in shell coiling of snails.

Inheritance of Pyrenoids in Chloroplasts of *Spirogyra triformus*

Spirogyra contains two chloroplasts in each cell. In 1920 VanWisselingh found a cell in which one chloroplast had pyrenoids and the other chloroplast did not have pyrenoids. When this cell divided, both daughter cells were like the mother cell. Genes in the nucleus could not have caused this because both chloroplasts would have been the same genes in the cytoplasm outside of the chloroplast couldn't have caused this because both chloroplasts are in the same cytoplasm and would be the same. This could only be caused by a genes in the chloroplast itself. This work proved that genes are present in the chloroplasts themselves, which control the phenotype of chloroplasts. Genes in the nucleus also control the phenotype of chloroplasts. There are hundreds of known genes located on chromosomes that control the phenotype of chloroplasts. *Example*: albino gene in maize.

Petite mutations in *Saccharomyces cerevisiae* (baker's and brewer's yeast, Ephrussi et al., 1956): Yeast cells usually forms large colonies called grande on nutrient agar plates. Rarely small or minute colonies known as petite also appear among them. Petite yeast produce small colonies because they are unable to carry out aerobic respiration because they have defective mitochondria,(having defective electron transport proteins–cytochromes a, b, and c). Segregational petites are due to genes in the nucleus. Neutral and suppressive petites are due to genes in the cytoplasm. Thus, there are genes in the nucleus and also in the cytoplasm that control the phenotype of the mitochondria. There are three types of petite mutants that differ in their mode of inheritance. The inheritance of a nuclear gene of haploid yeast shows Mendelian segregation of ascospores into 2 : 2 ratio. Such a ratio is found when the first category of petite mutants is crossed with grande wild type and thus a nuclear gene controls this phenotype, the Segregational petites. In the second type the cross between the grande and petite phenotypes results in progeny with all normal phenotype known as neutral petites. In the third category the cross between grande and petite phenotypes produce all petite progeny, and they are known as suppressive petites. (Fig. 4.16)

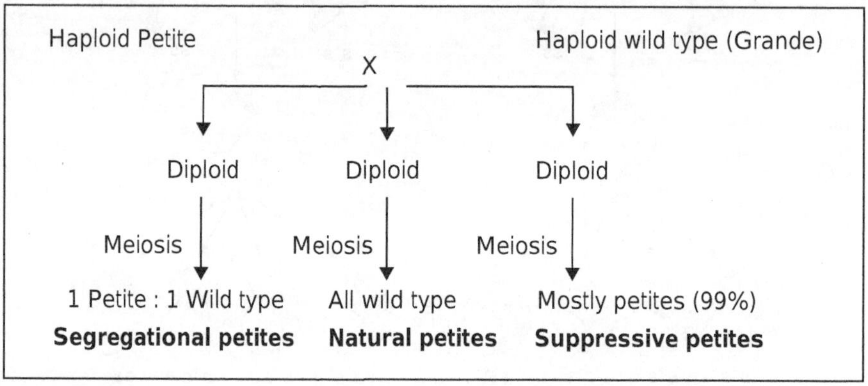

Figure 4.16 Three categories of petite yeasts based on segregation patterns.

The inheritance of poky phenotype (similar to petite yeasts) in *Neurospora crassa*, cytoplasmic male sterility in maize is also governed by mitochondrial gene. Cytoplasmic male sterility (CMS), a cytoplasmic mutation that is extremely important for agriculture. CMS plants are male-sterile and female-fertile. These mutations are in the mitochondria and are maternally inherited.

Infectious heredity, an invading microorganism causes a mutant phenotype that is then transmitted by maternal inheritance.

Genetic information is present in chloroplasts and mitochondria

DNA was shown to be present in chloroplasts (1963) and in mitochondria (1964). Chloroplasts and mitochondria have complete transcriptional and translational apparatuses. The complete nucleotide sequence of human mitochondrial DNA has been determined (16,569 nucleotides).

1. Human mitochondrial DNA codes for 13 proteins, 22 tRNAs, and 2 rRNAs.
2. The complete nucleotide sequences have been determined for mitochondria and chloroplasts of many other organisms.
3. Mitochondria have hundreds of proteins, but only a small number (13 in humans) are coded for by the mitochondrial genome.
4. Most proteins in mitochondria are coded for by the nuclear genome, translated on cytoplasmic ribosomes, and imported into the mitochondria. (Fig. 4.17).

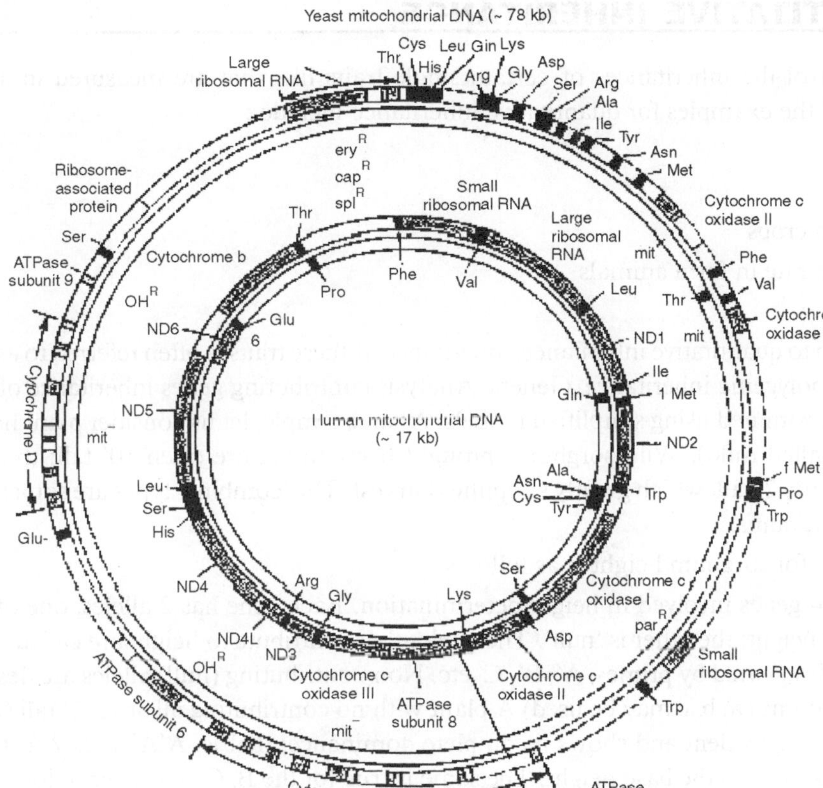

Figure 4.17 An overview of the mitochondrial genome.

Evolution of eukaryotic cells

Endosymbiont hypothesis, the hypothesis that is most widely accepted and the possible steps involved in the evolution of eukaryotic cells are as follows :

- An endosymbiont is a symbiont that is present inside a cell.
- Anaerobic prokaryotes were thought to be the first living things on earth.
- Anaerobic prokaryotes evolved into aerobic prokaryotes.
- The oxygen released by aerobic prokaryotes was toxic to the anaerobic prokaryotes, and this presumably was the driving force that caused anaerobic prokaryotes to evolve into aerobic prokaryotes.
- Anaerobic prokaryotes evolved into anaerobic eukaryotes.
- The anaerobic eukaryotes took up aerobic prokaryotes as an endosymbiont, and the aerobic prokaryotes became mitochondria.
- These cells then took up photosynthetic prokaryotes as an endosymbiont, and the photosynthetic prokaryotes became chloroplasts.
- Most of the genes of the original endosymbionts were lost because the same types of genes were present in the nucleus of the host eukaryotic cell. For this reason, the genomes of mitochondria and chloroplasts are much smaller than prokaryotic genomes.

4.12 QUANTITATIVE INHERITANCE

Many genes control the inheritances of certain single traits like that are measured in a quantitative manner. Some of the examples for quantitative inheritance include :

A. Height

B. Weight

C. Yield in crops

D. Growth rate in farm animals

E. IQ etc.

In addition to quantitative inheritance, inheritance of these traits is often referred to as "cumulative gene action" or "polygenic-inheritance. Genetic Analysis contributing genes inheritance of quantitative traits can best be examined using simplified models. As an example, let us consider plant height in grain sorghum (often called milo). Wild sorghums brought from Africa are often 10′ tall; those grown for grain are more nearly 2′ tall, which greatly simplifies harvest. The 'combine' types are sometimes referred to as '4-dwarf' sorghums.

The model for sorghum height is as follows :

There are 4 genes involved in height determination. Each gene has 2 alleles, one of which adds about one foot of height; the other is 'null'. The alleles that contribute to height are called "contributing alleles", and are designated by primes, A′, B′, C′ etc. Non-contributing (null) alleles are designated with A, B, C etc. (Sometimes a, b, c may be used) A plant with no contributing alleles is 2′ tall (AA, BB, CC, DD). Each gene is equivalent and shows incomplete dominance; that is, A′A′ adds 2 feet, A′A adds 1 foot and AA adds 0 feet to the base height. The same is true for the B, C, and D gene loci. A plant with the genotypes A′A′, BB, CC, DD or AA, BB, C′C′, DD will be 4′ tall.

A′A′, BB, CC, DD (4′) × AA, BB, CC, DD (2′)

↓

A′A, BB, CC, DD (3′)

If we allow the 3′ tall F_1 progeny to self pollinate, which is normal in sorghum, we expect to see a 1 : 2 : 1 ratio of F_2 plants that are 4′ : 3′ : 2′ tall, respectively. This is true because only one gene with incomplete dominance is 'segregating' (heterozygous) in the F_1 hybrid. (A′A × A′A gives 1/4th A′A′ : 2/4th A′A : 1/4th AA). The same would be true if AA, B′B′, CC, DD or any other 4′ tall true breeding plant (homozygous at all 4 gene loci) was crossed to the AA, BB, CC, DD parent. What if we cross A′A′ BB, CC, DD to AA, B′B′, CC, DD ? Both parents are 4′ tall as is the F_1, but the F_1 is heterozygous for 2 genes(A′A, B′B, CC, DD). Now in the F_2, we can predict that 1/4th of the progeny will be A′A′ and that 1/4th of these will also be B′B′; thus 1/16th of the progeny should be 6′ tall! Likewise, 1/16th should be AA, BB, CC, DD or only 2′ tall having no primes. When the range in the F_2 progeny goes beyond the original parents, we see "**transgressive segregation**". Transgressive segregation implies that the parents donate contributing alleles from different genes to the hybrid.

There should also be plants in the F_2 generation that have 1 contributing allele (A′A, BB, CC, DD or AA, B′B, CC, DD). These should represent 1/4th (1/8th + 1/8th) of the progeny, and they will be 3′ tall. The same is true for plants with 3 contributing alleles (A′A′, B′B, CC, DD and A′A, B′B′, CC, DD).

Genes that contribute to quantitative traits are referred to as polygenes, or QTLs, which stands for Quantitative Trait Loci. When this model applies, the hybrid will always show a phenotype that is the average of the parents, and there will be more variation among the F_2 progeny than in either P_1 or F_1 progeny. It would not be difficult to imagine cases where some genes made larger or smaller contributions than others, or where one or more genes may be dominant. Other than skewing the expected frequencies somewhat, neither change from the model significantly alters the analysis of polygenic traits.

Environmental Effects

Most or all quantitative traits are also influenced by environmental factors. In the case of sorghum height, plants will not reach their genetic potential without water, nutrients (fertilizer) and sunlight. Environmental effects means that some differences can be seen within a purebred parent or in an F_1 population, where all of the plants have the exact samc gcnotypc. Environmental effects will also "smooth out" the stair-step effect expected for genotypes in the F_2 that differ in the number of contributing alleles. Since it is not possible to 'count' the number of classes in an F_2 population when environmental effects smooth away the genotypic differences, or to identify individuals in the extremes, the number of genes that contribute to the trait cannot be simply estimated.

Normal statistical methods can be used for studying the inheritance of quantitative traits and to analyze the contributions of genes and environment to a trait. The concept is to partition the sources of variation that lead to differences among individuals in the sample, and to identify the portion of variation that results from segregating genes. Total variation (V_t), which is also often called phenotypic variation (V_p) arises from differences in genotype (V_g), the environment (V_e) and may also result from interactions ($V_{g \times e}$) where some genotypes do better in one environment and others in another.

$$\mathbf{V_t = V_p = (V_g + V_e + V_{g \times e})}$$

In many cases, the interaction component cannot be measured, so is ignored or handled by working within a specific environment or only working with a specific 'breed' or 'cross'.

Heritability (H^2 or broad-sense heritability) is the fraction of variation due to genetic differences, *i.e.* V_g. $H^2 = V_g/V_t$. It is relatively simple to make H^2 estimates in plants, since pure-breeding, completely homozygous parents can be maintained. Any variation within a purebreeding homozygous parent, or in the F_1 progeny of a cross between 2 purebreeding parents must result from V_e, since all plants within each of the populations have the exact same genotype. Thus, these plants can be used to estimate V_e. Variation among F_2 plants arises both from differences in genotype and from local environments, so the variation in the F_2 is a measure of V_t. In the example below, two true breeding corn parents, one with an average row number of 6 is crossed to another with an average of 14 rows. The F_1, as expected, is right between the parents, having an average of 10 rows. Variance (V) for row number in the parents and F_1 is low (arbitrarily measured as 1) Variance in the F_2 is 8.

$V_t(8) = V_g + V_e\,(1)$; ignoring $V_{g \times e}$ since all plants were grown in the same environment, V_g must be 7. Therefore, $H^2 = 7/8$.

It is critical to realize that :

1. Heritability measures are only valid for the population that was measured.
2. Genetic differences will not be measured unless the parents have different alleles.
3. Environment by gene interactions may be important, but are generally ignored.

Plant and animal breeders are interested in heritability, because it allows them to predict if selective breeding can be used to 'improve' a trait. For example, if rate of weight gain in nursing Hampshire pigs is highly heritable, saving those that grow the fastest for breeding purpose will lead to improved weight gain in future generations. If H^2 is low, changes in the diet may be more important. Selection will make progress so long as genetic differences can be combined to make an improved genotype; Selection for one trait may be balanced by loss in another; for example selection for increased egg size in Leghorns is successful, but the hens lay fewer eggs.

4.13 GENE AT THE POPULATION LEVEL

Most of the species live in groups or population and each individual of the population has a specific genome mix up or allele combination at the individual level. In individual cases we have seen various types of allelic combinations and its interaction leading to a specific type of phenotype. In a diploid organism every gene exists in pair, the allelic pair or allele. When both the members of the allele are identical the allelic pair is homozygous and when they are different forms, the allele or the allelic pair is heterozygous. Then the phenotype of the organism then depends on the nature of the genes, whether they have complete dominance, incomplete dominance or co-dominance. Various types of interactions between the members of multiple alleles and the nature of dominance may generate several alternate phenotypes. There may be the effect of polygenes in which a trait may be governed by more than one gene and its alleles. Similarly there can be the effect of cumulative gene or interactions of various types similar to that of epistasis. But all these are at the individual level.

Population genetics is concerned with changes in the **gene frequencies** of sexually reproducing populations over time. Major changes in the **gene pool** create a potential for genetic differences, which may result ultimately in speciation and adaptive radiation.

The Gene Pool

The term gene pool refers to the total number and variety of genes and alleles present in a sexually reproducing population, which is available to be passed on to the next generation.

A Model of a Gene Pool

Suppose we have 100 homozygous wild type (++) female Drosophila and 100 homozygous vestigial (vv) male Drosophila in a population. In the gene pool, half the genes for this character would be '+' and the other half would be 'v'. To put it another way, we could say that the **Gene Frequency** of both + and v is 0.5 (50%). In the next generation, the phenotypes of the flies produced by crossing ++ females with vv males will be wild and their genotypes will be +v. The proportions of the genes present have not changed and therefore their gene frequencies will remain at 0.5. By using coloured counters, we can make a model to represent this situation and then trace the relative frequencies of genes in the population from generation to generation.

Now consider a mathematical model of a gene in a population (gene pool) : Consider a population of Drosophila, which started, with a mixture of wild and vestigial winged flies (Fig. 4.18). Assume that during the period of the investigation there has been :

- A large population (say 250+).
- Random mating between all genotypes.

- No mutation (or mutation is in equilibrium).
- No selective advantage for any genotype (all genotypes survive equally well).
- No gene flow between this population and others (no immigration/emigration or balanced migration).
- Normal meiosis (*i.e.* chance is the only factor affecting gamete formation).

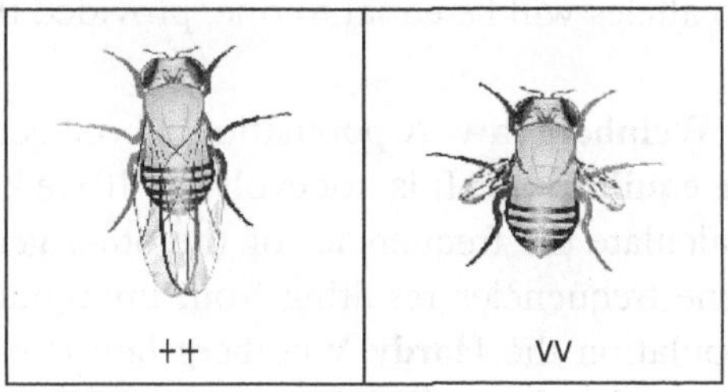

Figure 4.18 (++) Female Drosophila and (vv) male Drosophila.

In this population there will be three genotypes, ++, +v, and vv.

Let the frequency of allele + in the gene pool = **p**

Let the frequency of allele v in the gene pool = **q**

Then **p + q = 1** (100% of the gene pool)

Flies of genotype ++ can only be produced by the union of two gametes each carrying a + allele. The probability of one + bearing gamete uniting with another + bearing gamete is the product of their frequency in the population. The frequency of genotype ++ is, therefore

$$\mathbf{p \times p = p^2}$$

Similarly, flies of genotype vv can only be produced by the union of gametes each carrying a v allele. The probability of this occurring is the product of their frequencies in the population.

The frequency of genotype vv is, therefore

$$\mathbf{q \times q = q^2}$$

Flies of genotype +v can be produced by the union of a + carrying sperm and a v carrying egg. The probability of the genotype being formed in this way is, therefore

$$\mathbf{p \times q = pq}$$

Flies of genotype +v can also be formed by the union of a v sperm with a + egg. The probability of this occurring is, therefore

$$\mathbf{q \times p = qp}$$

The total frequency of genotype +v is, therefore

$$\mathbf{pq + qp = 2pq}$$

Since the frequencies of the three genotypes together **must add up to 1** (*i.e.* 100% of the population), we can write

$$\mathbf{p^2 + 2pq + q^2 = 1}$$

This basic equation of population genetics was worked out in 1908 by the English mathematician G. H. Hardy and a German physician W. Weinberg and is usually referred to as the **Hardy Weinberg**

Equation. They realized that provided certain conditions were met (the assumptions given above) the gene frequencies in a population will not change from one generation to the next. In the same way we can make a mathematical model for the distribution of a gene in human with its two alleles T and t. The allele T is for tasting a chemical PTC and t for the absence of the ability to taste the chemical. T is dominant over the recessive t. In exactly the same way we can find out the frequency (p and q) of distribution of these genes in the population. According to the Hardy Weinberg law at any point of time the total frequency of these two alleles will be equal to one, provided the above mentioned criteria are fullfilled.

This is called the **Hardy Weinberg law.** A population whose gene frequencies are not changing is said to be **in Hardy Weinberg equilibrium.** It is not evolving. If we know the value of p or q or q^2 in any population, we can then calculate the frequencies of the other genotypes. This may enable us to detect long term changes in gene frequencies resulting from immigration, emigration, mutation and selection. But for a normal population the **Hardy Weinberg law** is not applicable because it cannot satisfy some or all of the conditions for Hardy-Weinberg equilibrium and therefore the population will not be static and will be under the process of evolution.

Hardy and Weinberg established the conditions under which populations remain in genetic equilibrium. (*i.e.* their gene frequencies remain constant from generation to generation).

It follows that the opposites of those conditions are the basic reasons for changes in a population's gene frequencies, and are therefore the basic cause of evolutionary change.

- The population is small (*e.g.* < 100).
- Mating is not random (*e.g.* how do wingless Drosophila catch a mate ?).
- Mutation is not in equilibrium.
- Some genotypes have a selective advantage (*e.g.* wingless Drosophila on a windy island).
- Gene flow is not balanced.
- Meiosis is not normal.

Note that the Hardy Weinberg Law does not apply to nonsexually reproducing populations or monoploid populations.

4.14 NATURE OF GENETIC MATERIAL

Mendel helped to establish that heredity was controlled by 'factors' and chromosomes were soon suspected of carrying the factors or genes. This followed the discovery of the scientific proof that DNA is the genetic material or the material of the gene and chromosomes. DNA was first identified in 1868 by Friedrich Miescher, a Swiss biologist, in the nuclei of pus cells obtained from discarded surgical bandages. He called the substance **nuclein,** and in 1914 Feulgen perfected a specific DNA stain (Feulgen stain); however, the connection between DNA and heredity was not made until many years later. It was only in the middle of the twentieth century that the role of DNA as the genetic material was established. The following are two of the classic experiments that helped to demonstrate the role of DNA as the carrier of genetic information.

In 1928 the British scientist Frederick Griffith was working with the bacterium *Streptococcus pneumoniae,* a causing agent of pneumonia. He worked with two types of bacteria, the strain S, which is highly infectious and grows in smooth and shiny colonies, and the strain R, which is harmless and grows in rough colonies. The S strain is surrounded by a polysaccharide coat, which is responsible for its virulence. The R strain lacks that coat. Griffith injected mice with different strains of the bacterium. Initially, he used R strains and the mice were unaffected. Then, he used S strains and the mice died of pneumonia. Living bacteria could be isolated from the dead mice. Next, heat-killed S bacteria and injected them into the mice, and the mice survived, indicating that the bacteria has to be alive and to have the coat to be infective. Finally, Griffith injected mice with a mix of living R and heat-killed S strains of the pathogen. To his surprise, the mice died and living S bacteria was recovered from the mice's blood. Griffith concluded that some R bacteria had been transformed into S by interaction with the dead S bacteria. He thought that there was a **transforming principle** responsible for this change in genetic material, and that it could be a protein (Fig. 4.19).

Avery's transformation experiments confirmed the chemical nature of Transforming Principle

Oswald Avery, together with his co-workers Colin MacLeod and Maclyn McCarty continued Griffith's experiments in the 1930's and 1940's to identify what was the transforming principle. They cultured S

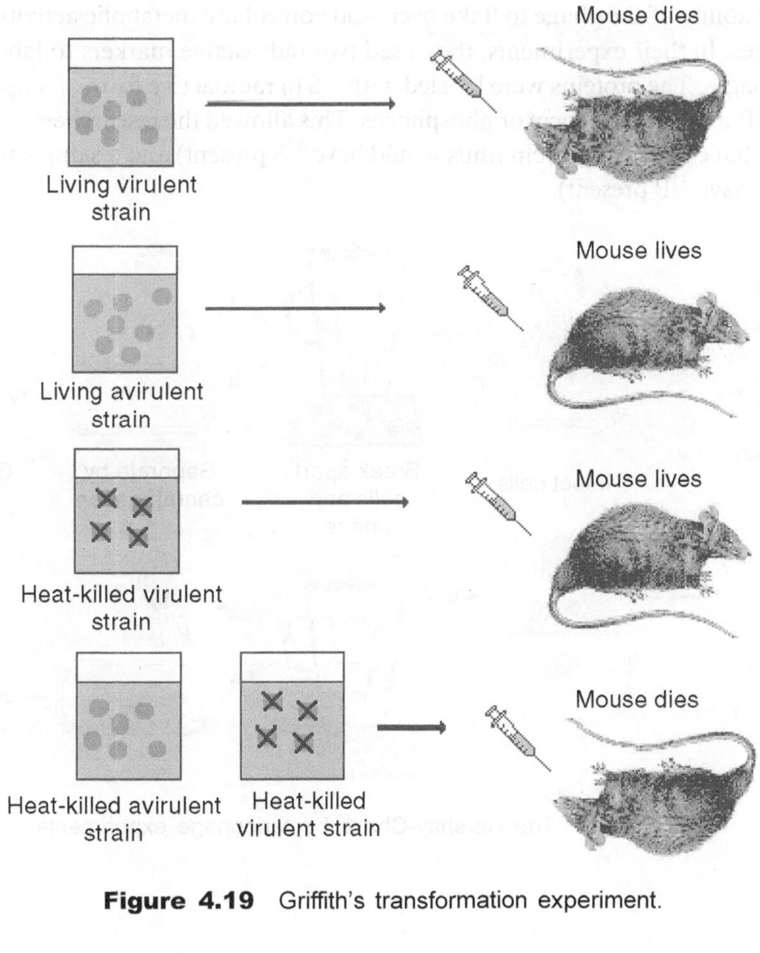

Figure 4.19 Griffith's transformation experiment.

cells and broke them open, and separated the different cellular components of that cell extract (*i.e.*, lipids, polysaccharides, proteins and nucleic acids). Then, they tested each of these macromolecular components to check which of them could transform R cells into S bacteria. They found that the only components capable of transformation were the nucleic acids. At this point RNA was not separated from DNA so they conceived another experiment to identify which of these two nucleic acids was the transforming principle. They took the nucleic acids and treated them with the enzyme **ribonuclease**, which degrades RNA and not DNA, but the transforming principle was still present. Then, they treated the nucleic acids with **deoxyribonuclease,** which degrades DNA but not RNA, and no transformation resulted. So, they suggested that the genetic material was DNA.

The Hershay-Chase Bacteriophage Experiments

In 1952, the research team of Hershey and Chase published a report that concluded that DNA is the genetic material of the bacteriophage (a virus that infects bacteria) T2 (strange name!). They knew that when a bacterium was infected with this phage, the bacteria soon became a machine that produced new phages. Phages have a very simple structure—they are composed of a strand of DNA surrounded by a protein coat. Hershey and Chase set out to determine which component (the DNA or the protein) was responsible for the ability of the phage to 'take over' and control the metabolic activity of the bacteria to produce new phages. In their experiments, they used two radioactive markers to label the proteins and the DNA of the phages. The proteins were labeled with ^{35}S (a radioactive form of sulphur) and the DNA was labeled with ^{32}P, a radioactive form of phosphorus. This allowed the researchers to easily differentiate between a sample that contained protein (thus would have ^{35}S present) and a sample that only contained DNA (thus would have ^{32}P present).

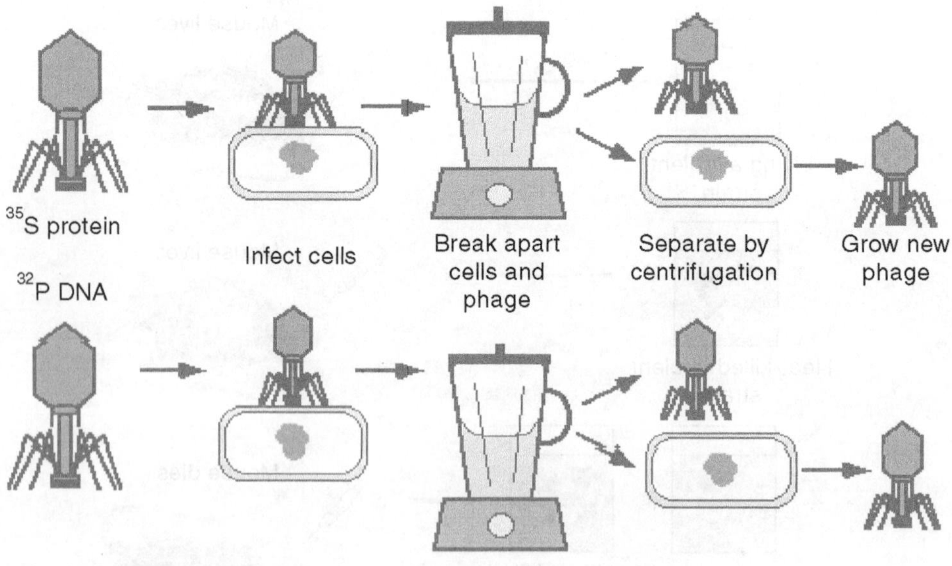

Figure 4.20 The Hershay-Chase bacteriophage experiments.

During a phage infection, it was hypothesized that some part of the phage was injected into the bacterium and it was this injected material that conveyed the genetic material necessary to produce new phages. Hershey and Chase determined that the phage injected only the DNA into the bacterium and concluded that DNA must be the genetic material in phages. This was done by a series of two experiments in which different sets of non-radioactive bacteria were incubated with phages that had either their protein or their DNA labeled (as described above). They allowed the phages to infect the bacteria for a short time; they agitated the incubations to dislodge any loose parts of the phages. The bacteria cells were then pelleted in a centrifuge and the location of the radioactivity (in the pellet with the bacteria or in the supernatant). They found that the radioactive DNA was always found with the bacteria cells and that the radioactive protein was always in the supernatant. This suggested that the DNA was injected into the bacteria but the protein coat was not. Thus all of the information needed to produce new viruses was contained in the DNA and not the protein.

4.15 GENOME ORGANIZATION

The complete Genetic or DNA complement of an organism is termed as the genome. It includes the genetic material of the nucleus and cytoplasm. All organisms have genome made up of DNA, containing genes. Genomes may vary in their size, number of genes, number of chromosomes, and how genes are organized within chromosomes(s), and the DNA may be circular or linear. Generally, genome size increases with complexity of the organisms. There is considerable variation in the size of the genome among the organisms. As the size of the genome increases the number of genes also increases correspondingly. But there are exceptions. Among the various groups of organisms, viral genomes are usually relatively small and come in sizes ranging from 5 kb (SV40) to about 250 kb (Vaccinia-virus and Cytomegalovirus). (1 kb = 1000 base pairs). Bacterial genomes on the other hand range from 600 kb (Mycoplasm) to more than 7000 kb (Streptomycetes), but smaller than eukaryotic genomes. Genome size and its organization vary greatly even among the eukaryotic organisms.

1. Viral Genomes

Viruses are nonliving particles that must infect and subvert cellular machinery of host to reproduce. Viruses composed of protein coat and a core that contains the genome. Even though genome of viruses is the most simplest and smallest, they are the most diverse due to diverse evolutionary histories as well as diversity in strategies for packing genome in coat. It can be made of single stranded DNA (ssDNA), double stranded DNA (dsDNA), single stranded RNA (ssRNA), or even double stranded RNA (dsRNA). These molecules of nucleic acids may be linear or circular. The complete nucleotide sequences of many viral genomes have been determined. Viral genomes are very small, have little intergenic space, and few genes (herpes virus contains ~200 genes, and many others have far fewer genes). There are various types of viruses. They can be morphologically very variable and even complex. Their morphology is one way of classifying viruses. They contain DNA or RNA but never both. Although they have a protein coat around their hereditary material, they lack properties of cells such as membranes, ribosomes, enzymes and ATP synthesis ability. Thus, they can only proliferate by using a host cell's translational machinery. In other words, they are obligate intracellular parasites. Their replication cycle consists of attachment and entry into the cell; replication of viral nucleic acid; synthesis of viral proteins; and finally, assembly of viral components and escape from host cells. Bacteriophages have received greatest attention from molecular geneticists. A bacteriophage is a virus that infects bacteria only.

Bacteriophages have a simple structure, which consists of their double-stranded DNA (in some cases single stranded, *e.g.* M13) surrounded by protein coat. Only the DNA enters into the bacteria. Some phages supply their own enzyme components (own s factors) to instruct the bacteria to transcribe phage genes preferentially. Bacteriophages usually infect only one species of bacteria but there are some who can infect several species even in different genera. Their life cycle may be either lytic (virulent phage) or lysogenic (temperate phage). Their DNA may be integrated into the host chromosome and remain as a prophage. Integration is achieved by recombination between a 15 bp sequence called *att* (for attachment) in the host chromosome and an identical sequence in the phage chromosome. This recombination requires an integrase (Int) enzyme encoded by the phage. Bacteriophages are used for gene cloning in molecular biology. DNA fragments can be inserted into a phage and following transfection of competent bacteria; many copies of the desired DNA fragment can be obtained.

The size of the bacteriophage genome also varies in different strains. For example, M13 and T7 contain genomes that are 6.4 kb and 39 kb respectively. (kb = Kilo base) . The complete nucleotide sequences of many phage genomes have been determined and it reveals that they contain relatively few genes. For example, M13 phage have only 10 genes, that include genes of coat protein and enzymes needed for phage DNA replication. These genomes in most cases have overlapping genes *i.e.* genes within gene, coding for different proteins. The eukaryotic viral genome is more or less similar to that of bacteriophages, but have introns (non coding regions between a gene), which is a unique feature of eukaryotic genome.

2. Prokaryotic Genome

In prokaryotes genome size is relatively small, but bigger than that of virus. There is considerable variation in the size among prokaryotes and in some cases genome is bigger than certain eukaryotic cells. For example the genome size of *Bacillus megasterium* is 30 Mb and that of Yeast is 12.1 Mb, which is an eukaryote. The gnome sizes in other bacteria are smaller than this. For example, *Mycoplasma genitalium* has a genome size of 580 kb, *Methanococcous jannaschii* has a genome of 1739kb and the genome size in *E. coli* is 4639 kb. In most cases, the genome consists of a single circular chromosome, but there are exceptions. They also posses small independently replicating, circular double stranded DNA molecules known as plasmids. Plasmids can be 1 to 300 kb long and may exist as multiple, free copies. They also carry genes, which are absent in bacterial chromosome. *Vibrio Cholera* has two different circular chromosomes and *Borrelia burgdoferi* (Lyme disease) has a linear chromosome, and 17 plasmids.

Generally the bacterial genome has a high gene density, *i.e.* genes are close together with very small intergenic regions. Unlike eukaryotic systems, major part of the genome is functional. About 85-90% of total DNA is involved in protein coding, about 20–10 percentage is the intergenic DNA and less than 2% of DNA is made up of transposable elements. Functionally related genes located together as a group called operons and introns are extremely rare. Overlapping genes are also common in bacterium as in the case of viruses. Bacterial genomes have associated proteins, used to pack it into a **nucleoid**, but the proteins not similar to the histones of eukaryotes.

In *E. coli*, 4639 kb DNA contains about 2400 genes, which make about 80% of the total DNA. The remaining 20 per cent is made up of intergenic region. There are three cases of overlapping genes and about 27% of the transcriptional units are organized into 75 polycistronic groups or operons. The genes in an operon are functionally related. (In *Aquifex aeolicus*, operons genes are not functionally related as per our ideas about operons).

Another example, *Methanococcus jannaschii*, extreme thermophilic bacteria has three circular chromosomes—one 166 MB, plus two smaller ones of 58.4 and 16.5 kb. The entire genome has sequenced in 1996; its 58% of genes do not match any known genes. Majority of genes involved in energy production, cell division, and general metabolism more closely resemble eubacterial genes. It has some similarities with that of eukaryotic genes involved in RNA synthesis, protein synthesis, and DNA synthesis. It contains histone genes and DNA organized into chromatin, and has introns in tRNA genes.

3. Eukaryotic Genomes

Eukaryotic genomes are much more complex than bacterial genomes. They cover an enormous size range (more than three orders of magnitude) and display different levels of structural complexity. Initially it is important to obtain an idea of the relative sizes of various eukaryotic genomes and to put them into perspective with bacterial and viral genomes. In principle the rule holds that an increase in structural complexity of an organism goes hand in hand with a more complex genome (since it takes obviously more DNA to encode more proteins); but there are some exceptions to this rule such as the extraordinary large sizes of some amphibian and plant genomes. There are two types of genome in a eukaryotic system. They are nuclear genome and cytoplasmic genome. The cytoplasmic genome includes mitochondrial genome in the case of animals and both mitochondrial and chloroplast genome in the case of plants. For example, the human genome, the term used to describe the total genetic information (DNA content) in human cells consists of two genomes: a complex nuclear genome which accounts for 99.9995% of the total genetic information, and a simple mitochondrial genome which accounts for the remaining 0.0005%.

Nuclear Genome

The size of genomes in eukaryotes vary very significantly. The genomes of the structurally simplest unicellular or filamentous eukaryotes such as the fungi, *Saccharomyces cerevisiae* or *Schizosaccharomyces pombe* are around 12 and 18 Mb respectively *i.e.* only about four times than the size of the average bacterial genome. Multicellular eukaryotes with differentiated cells (tissues) again contain more DNA in their somatic/germ line cells: Nematodes (*Caenorhabditis elegans*) and small dicot plants (such as *Arabidopsis thaliana*) around 100 Mb; insects (dipters) have genomes nearly twice as large again (165 Mb for *Drosophila melanogaster*). But then there is a large jump in genome size when one looks at the animal phylum of the vertebrates: Bird genomes have around 1100 Mb to 2000 Mbp, reptiles between 1600 and 5100 Mbp and mammals between 2300 and 5600 Mb); fish genomes range from 2600 to 7000 Mbp. The genomes of many monocot plants (cereals) also lie in the lower Mb range but the maize genome has 15 Mbp. Few genomes tend to be even more complex (Salamander genome: 90 Gbp, some flowering plants 120 Gbp) but this increased size is mostly not linked to more structural complexity.

In prokaryotic genomes, a large portion of the total DNA is functional, codes for different proteins and RNAs. In eukaryotes a great portion of the genome is non functional and the coding region is relatively small. According to the draft human genome project the total number of genes present in human genome is only about 33,000. It comes around 2-5% of the total genome.

Most of the genomic DNA are intergenic region and are non-coding. It includes the highly repetitive sequences and introns. The repetitive sequences are individual, short or small sequences, which repeat either in tandem arrays or interspersed throughout the genome. Simple Sequence Repeats (SSR) or microsatellites are the regions of DNA where one to few bases are tandemly repeated for few to hundreds of times. Some of the repetitive DNA sequences may be present several million times in the genome and others only as ten to several hundred copies.

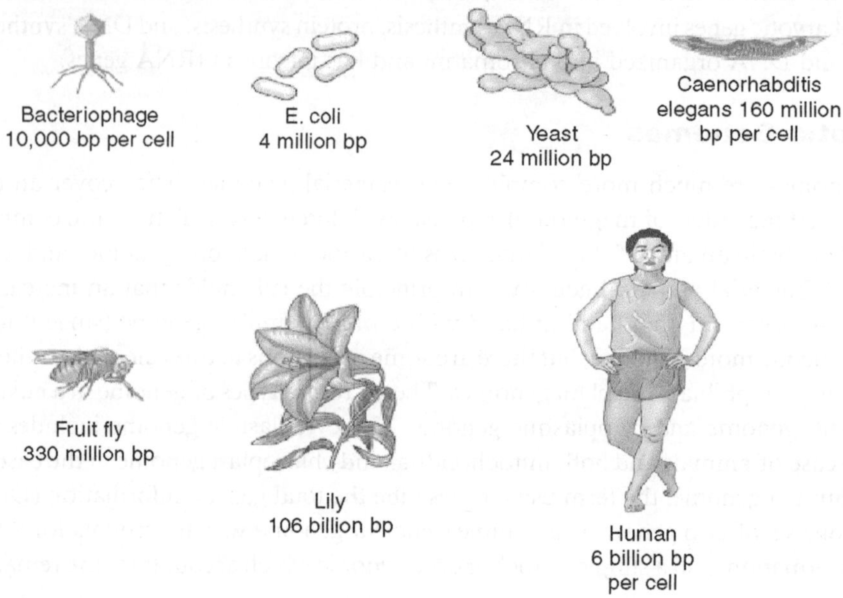

Figure 4.21 Relative size of various eukaryotic Genomes (DNA in base pairs).

The length of DNA in the nucleus is far greater than the size of the compartment in which it is contained. To fit into this compartment the DNA has to be condensed in some manner. The degree to which DNA is condensed is expressed as its packing ratio. It may be worth recapitulating that DNA comes as double stranded right handed helix with three isoforms, two with right handed turns (A and B conformation) with a major and minor groove and one left handed one (Z-DNA) with only one groove. The B-conformation with 10.5 bp per helical turn is the in vivo conformation. In order to accommodate the immense length of eukaryotic DNA inside the minute size of the cell, the DNA strand is wound up into so-called superstructures.

The first level of such a superstructure is the *nucleosome* in which the DNA is wound in two full turns as left-handed super helix around a protein disc consisting of histone proteins forming the so called **histone octamer**. These nucleosomes are separated by short linker DNA regions stabilized by histone H1 giving the structure some resemblance to "beads on a string". The next higher structural order is the 30 nm chromatin structure, which is clearly visible under the electron microscope. This structure arises by a further supercoiling of the string of nucleosomes around a fixed axis with about 6-7 nucleosomes per turn. The next more complex structural feature is called the 100 nm chromosomal domain in which 20-100 kb of DNA are held together by so called matrix proteins; this structure is visible under the light microsocope in **lampbrush chromosomes** of *Chironomus tentans* oocytes for example.

Further (but transient) condensation of chromosomes occurs during cell division in late G_2 phase and during mitosis (metaphase chromosomes) in which the DNA assumes a clear cytological structure which can be visualized for example in **karyotyping**.

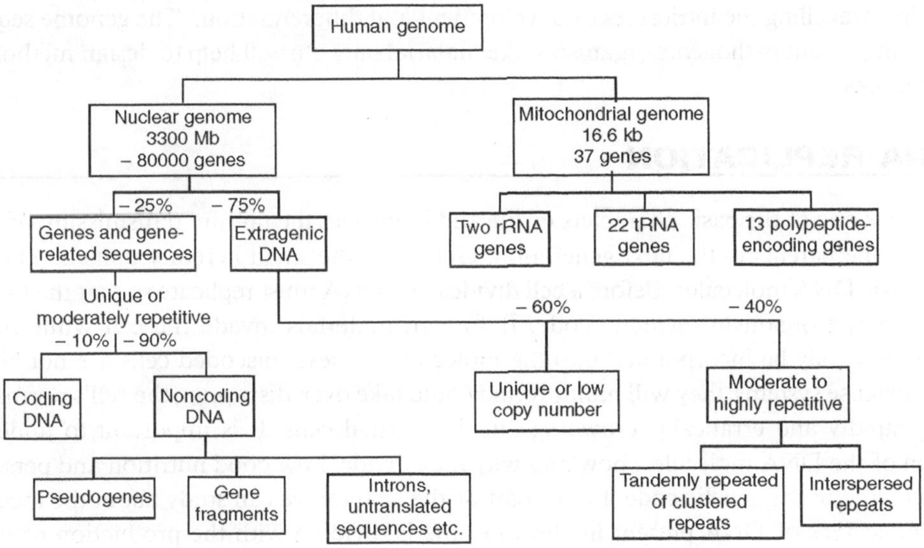

Figure 4.22 Organization of Human genome.

Organellar Genomes

The eukaryotic organelles—mitochondria and chloroplast also contain its own genome in the form of a circular double stranded DNA that carry limited number of genes which are essential for the function and maintenance of the organelles. They have the small ribosomes similar to that of prokaryotes—bacteria and cynobacteria. There is a great variation in the size of mitochondrial genome (Fig. 4.17) from organism to organisms, but in the case of chloroplast genome the variation is very low among different types of plants. The genome organization of genes and its expression hold a lot of similarities with that of prokaryotic system. It supports the theory of prokaryotic origin of chloroplasts and mitochondria.

4.16 GENOME SEQUENCING PROJECTS

The genomes of a large number of organisms were sequenced and a number of them are in the final stages by different genome project groups. The genome projects are having a profound impact on health care discovery. The main purpose of genome projects are to access the entire genome sequences and that can be used to find out the probable genes and their functions of the various organisms. Some important genomes that are partially or completely sequenced include that of *E. coli*, Mycobacterium tuberculosis, yeast, Arabidopsis, Drosophila, mouse humans, rice, wheat etc. Complete sequences are available for Arabidopsis, yeast, Human, Mycobacterium etc. By 2005 most of the important genomes would be sequenced and the sequence will be available for non-commercial purposes from the databases. The availability of these databases for researchers will help them to better understanding of the gene and its expression and thereby the better understanding of the fundamental biological processes. For example, the human genome project would greatly help the development of gene therapy. The final and the complete sequence of the human genome would unravel the mystery of heredity and reveal the sources of many diseases like diabetes, heart diseases, cancers, hemophilia, Alzheimer's diseases and cystic fibrosis. It can also help unravelling the intricacies of development and differentiation. The genome sequencing of a number of important pathogenic organisms like malarial parasite will help to design methodologies to control the disease.

4.17 DNA REPLICATION

Reproduction is one of the basic characters of life and it involves the faithful transmission of the genetic material from the parents to the next generation. Genes are the small heredity units on chromosomes and are made of DNA molecules. Before a cell divides, the DNA must replicate so that the two new cells will each have the organism's genetic code. If foreign materials invade the cell while the DNA is duplicating, they may be incorporated into the molecule. If these miscoded cells are not killed by the body's own defense systems, they will multiply and could take over, disrupting the cell's normal activities or dividing rapidly and erratically, crowding out the normal cells. It is important to understand the construction of the DNA molecules; how and why they divide; how good nutrition and personal habits can help to maintain the genetic code. In this part of the chapter we can study about the mechanism by which the molecules of DNA present in chromosome is dividing with the production of two exactly similar copies.

The Principle of DNA Replication

The structure of DNA suggested to Watson and Crick the mechanism by which DNA-hence genes-could be copied faithfully. They proposed that when the time came for DNA to be replicated, the two strands of the molecule will be separated from each other but remained intact as each served as the template for the synthesis of a complementary strand. When the replication process is complete, two DNA molecules—identical to each other and identical to the original have been produced. This mode of replication is described as **semi-conservative replication**—one-half of each new molecule of DNA is old and the other half new.

In DNA replication, the two chains separate from each other like a zip unfastening as the hydrogen bonds, which link the bases of one chain with the bases of the other are not very strong. Any free nucleotides then come along and form hydrogen bonds with each of the two chains. These nucleotides then join together through their sugar and phosphate groups; and two DNA molecules result. The complementary relationship between the bases ensures that each of the DNA molecules is identical to the original one. Because the sequence of the bases in the two daughter molecules is exactly the same with that of the parent molecule (*i.e.* A to T and C to G), accurate replication occurs. The enzyme that carries out this reaction is called DNA polymerase. DNA replication occurs only at a specific step in the cell cycle. The following table describes the cell cycle for a hypothetical cell with a 24 hr cycle.

TABLE 4.1 Different phases of cell cycle

Stage	Activity	Duration
G_1	Growth and increase in cell size	10 hr
S	DNA synthesis	8 hr
G_2	Post-DNA synthesis	5 hr
M	Mitosis	1 hr

Messelson and Stahl Experiment

When Watson and Crick published their famous short paper on the structure of DNA, they suggested a mechanism by which the molecule would replicate: that each strand would be copied into a new strand with a complementary sequence. However, it fell to Matthew Meselson and Franklin Stahl at Caltech to demonstrate that this was true. Meselson and Stahl in 1957 gave experimental evidence that each DNA strand served as a template for new synthesis, a process called **semi-conservative replication.**

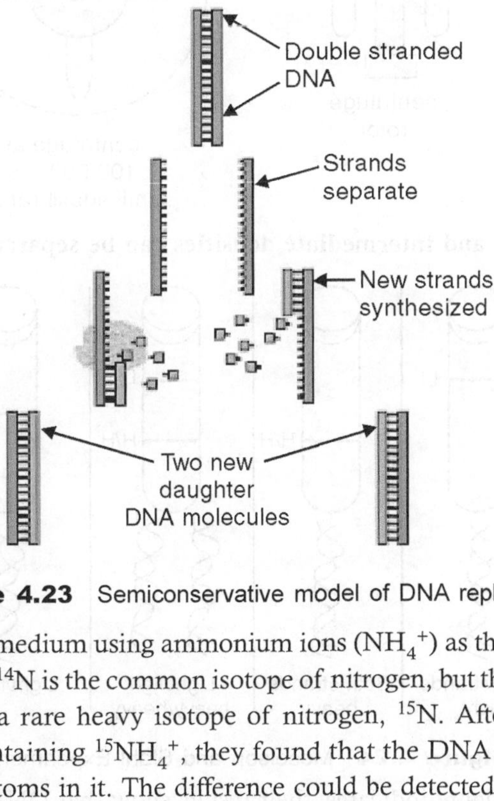

Figure 4.23 Semiconservative model of DNA replication.

They grew *E. coli* in a medium using ammonium ions (NH_4^+) as the source of nitrogen for DNA (as well as protein) synthesis. ^{14}N is the common isotope of nitrogen, but they could also use ammonium ions that were enriched for a rare heavy isotope of nitrogen, ^{15}N. After growing *E. coli* for several generations in a medium containing $^{15}NH_4^+$, they found that the DNA of the cells was heavier than normal because of the ^{15}N atoms in it. The difference could be detected by extracting DNA from the *E. coli* cells and spinning it in an ultracentrifuge (Fig. 4.24). The density of the DNA determines where it accumulates in the tube. Then they transferred more living cells that had been growing in $^{15}NH_4^+$ to a medium containing ordinary ammonium ions ($^{14}NH_4^+$) and allowed them to divide just **once**. The DNA in this new generation of cells was exactly intermediate in density between that of the previous generation and the normal. It tells us that half of the nitrogen atoms in the new DNA are ^{14}N and half are ^{15}N. But it tells us nothing about their arrangement in the molecules. However, when the bacteria were allowed to divide **again** in normal ammonium ions ($^{14}NH_4^+$), two distinct densities of DNA were formed: half the DNA was normal and half was intermediate.

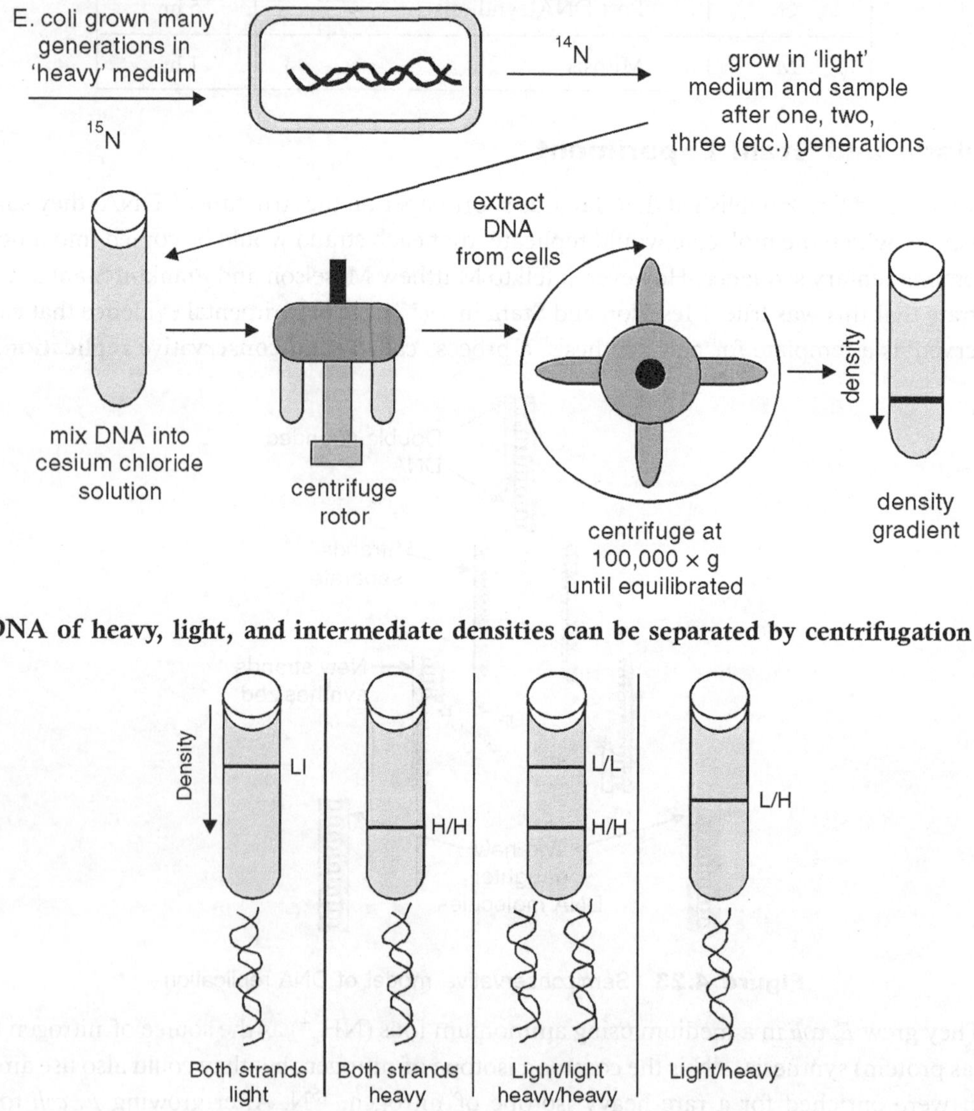

Figure 4.24 Meselson and Stahl Experiment.

As this interpretative figure indicates, their results show that DNA molecules are not degraded and reformed from free nucleotides between cell divisions, but instead, each original strand remains intact as it builds a complementary strand from the nucleotides available to it. This is called **semiconservative replication** because each daughter DNA molecule is one-half 'old' and one-half 'new'.

The different steps in the experiment that demonstrated the semiconservative replication of DNA in *E. coli* is illustrated below.

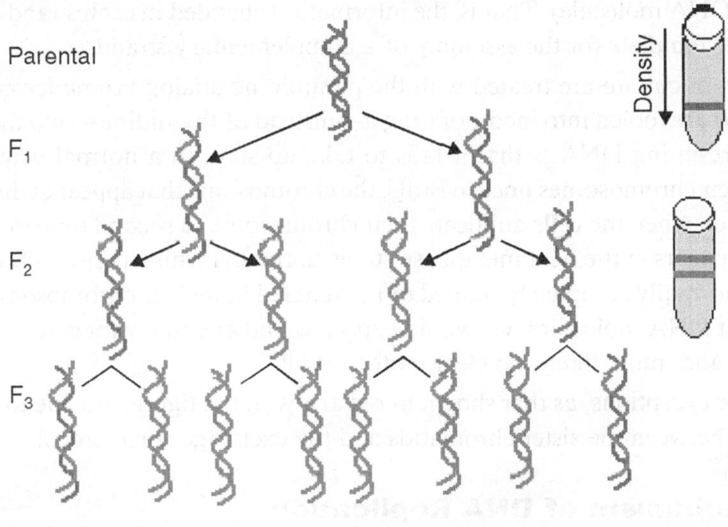

Figure 4.25 Meselson and Stahl Experiment–DNA of three generations.

E. coli is a prokaryote, but semi conservative replication of DNA also occurs in eukaryotes. And because each DNA molecule in a eukaryote is incorporated in **one chromosome**, the replication of entire chromosomes is semiconservative as well.

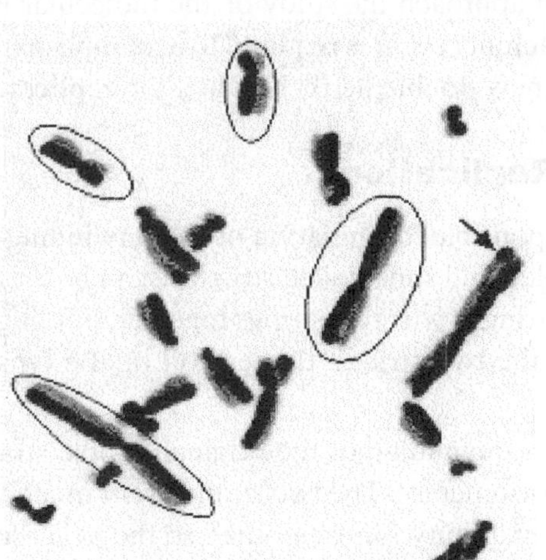

Figure 4.26 Semiconservative replication of chromosomal DNA.

The semiconservative replication can be demonstrated in the case of chromosomes also. Actually each of the single chromosomes that you observe under 440X magnification of your laboratory microscope contains a single molecule of DNA. For some of our chromosomes, this molecule—if stretched out—would extend 5 cm (2 inches). The following image demonstrates that each chromosome contains but a single molecule of DNA and that the replication of a chromosome is **semi-conservative** (just as is the replication of a single DNA molecule). That is, the information encoded in each strand of DNA remains intact as it serves as the template for the assembly of a complementary strand.

If cells dividing in culture are treated with the pyramidine analog **bromodeoxyuridine (BrdU)**, during S phase, the cells are fooled into incorporating it—instead of thymidine—into their DNA. One of the properties of the resulting DNA is that it fails to take up stain in a normal way. When cells are allowed to duplicate their chromosomes **once** in BrdU, the chromosome that appear at the next metaphase stain normally. However, when the cells duplicate their chromosomes a second time in BrdU, one of the sister chromatid that appears at the next metaphase stains normally, while its sister chromatid does not. Each of chromatid is normally completely stained or not (circled in red). If chromosomes were made up of numbers of different DNA molecules, we would expect to find at best a variegated pattern of stained and unstained regions and, more likely, no clear pattern at all.

But there can be exceptions, as that shown in red arrow in the figure, are due to the spontaneous exchange of segments between the sister chromatids and the exchange is reciprocal.

Biochemical Mechanism of DNA Replication

It is very important to know that DNA replication is not a passive and spontaneous process. Many enzymes are required to unwind the double helix and to synthesize a new strand of DNA. DNA replication has two important requirements in addition to the enzymes and the nucleotides: DNA template and free 3′-OH group. The parent strands provide the template and the free 3′-OH group is provided by a primer, a small oligonucleotide. We will approach the study of the molecular mechanism of DNA replication from the point of view of the machinery that is required to accomplish it. The unwound helix, with each strand being synthesized into a new double helix, is called the **replication fork**.

The Enzymes of DNA Replication

1. **Topoisomerase** is responsible for initiation of the unwinding of the DNA. The tension holding the helix in its coiled and supercoiled structure can be broken by nicking a single strand of DNA. Try this with string. Twist two strings together, holding both the top and the bottom. If you cut only one of the two strings, the tension of the twisting is released and the strings untwist.
2. **Helicase** accomplishes unwinding of the original double strand, once supercoiling has been eliminated by the topoisomerase. The two strands very much want to bind together because of their hydrogen bonding affinity for each other, so the helicase activity requires energy (in the form of ATP) to break the strands apart.
3. **DNA polymerase** proceeds along a single-stranded molecule of DNA, recruiting free dNTPs (deoxy-nucleotide-triphosphates) to hydrogen bond with their appropriate complementary dNTP on the single strand (A with T and G with C), and to form a covalent phosphodiester

bond with the previous nucleotide of the same strand. The energy stored in the triphosphate is used to covalently bind each new nucleotide to the growing second strand. There are different forms of DNA polymerase—DNA Polymerase I (Pol I) was the first enzyme discovered with polymerase activity, and it is the best characterized enzyme. Although this was the first enzyme to be discovered that had the required polymerase activities, it is not the primary enzyme involved with bacterial DNA replication. That enzyme is DNA Polymerase III (Pol III). Three activities are associated with DNA polymerase I;

- 5′ to 3′ elongation (polymerase activity)
- 3′ to 5′ exonuclease (proofreading ativity)
- 5′ to 3′ exonuclease (repair activity)

The second two activities of DNA Pol I are important for replication, but DNA Polymerase III (Pol III) is the enzyme that performs thc 5′-3′ polymerase function.

DNA polymerase cannot start synthesizing de novo on a bare single strand. It needs a **primer** with a **3′OH group** onto which it can attach a dNTP. DNA polymerase is actually an aggregate of several different protein subunits, so it is often called a **holoenzyme**. The holoenzyme also has proofreading activities, so that it can make sure that it inserted the right base, and nuclease (excision of nucleotides) activities so that it can cut away any mistakes it might have made.

4. **Primase** is actually part of an aggregate of proteins called the **primeosome**. This enzyme attaches a small RNA primer to the single-stranded DNA to act as a substitute 3′OH for DNA polymerase to begin synthesizing from. This RNA primer is eventually removed by **RNase H** and the gap is filled in by **DNA polymerase I**.
5. **Ligase** can catalyze the formation of a phosphodiester bond given an unattached but adjacent 3′OH and 5′phosphate. This can fill in the unattached gap left when the RNA primer is removed and filled in. The DNA polymerase can organize the bond on the 5′ end of the primer, but ligase is needed to make the bond on the 3′ end.
6. **Single-stranded binding proteins** are important to maintain the stability of the replication fork. Single-stranded DNA is very labile, or unstable, so these proteins bind to it while it remains single stranded and keep it from being degraded.

Semidiscontinuous Replication

The two strands in a DNA molecule are antiparallel. One strand runs in 3′-5′ direction and the other strand runs in the opposite direction *i.e.* 5′-3′ direction. This suggests that during DNA replication one of the newly synthesized strand is polymerized in the 5′-3′ direction while the other strand is polymerized in the 3′-5′ direction. But it is known that all known DNA polymerases can carry out the polymerization only in the 5′-3′ direction. That is it can add fresh nucleotides only to the 3′OH groups of sugar and not to the 5′ terminal of the growing nucleic acid chain. This presents a problem, that how does the synthesis of 3′-5′ direction take place ? The answer is that the strand which grows in the 3′-5′ direction (copying of the 5′-3′ strand) , known as the lagging strand is synthesized in discontinuously as a series of short DNA fragments known as Okazaki fragments after its discoverer Reiji Okazaki. Each of these fragments is synthesized by DNA polymerase III, with usual 5′-3′ polarity, as shown in Fig. 4.27.

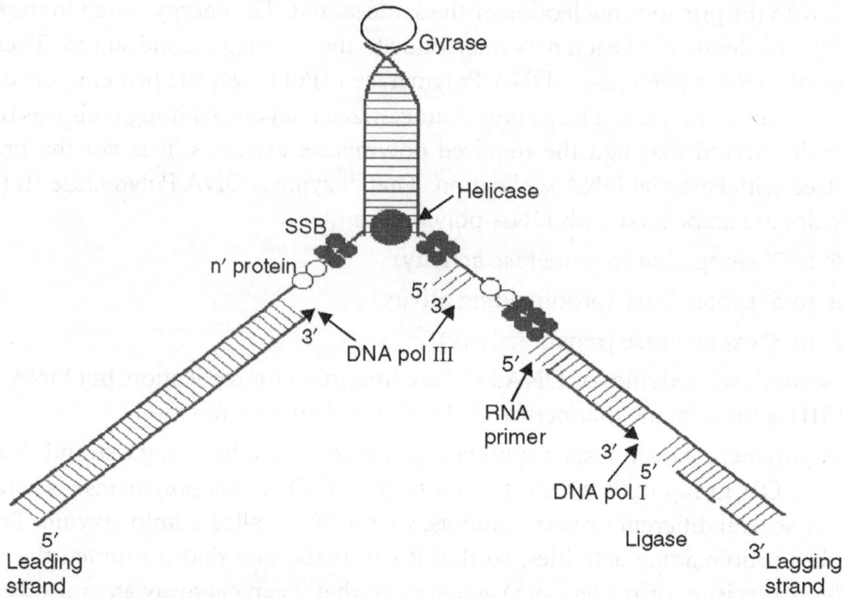

Figure 4.27 DNA replication fork.

In bacterial systems the Okazaki fragments are 1000 to 2000 nucleotides in length, but in eukaryotes these fragments are relatively small and in most cases it is less than 200 nucleotides long. Finally as the replication proceeds, these Okazaki fragments are joined together to produce intact daughter strand. Thus during DNA replication one strand is synthesized continuously and the other strand is synthesized discontinuously, and therefore, DNA replication is called semidiscontinuous.

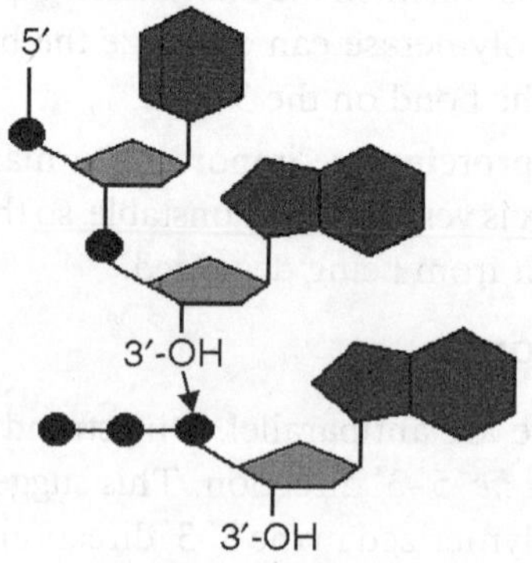

Figure 4.28 Addition of a dNTP to a growing oligonucleotide.

We already have observed that DNA polymerase does not have the de novo synthesis of DNA strands. It can only carry out the addition of new nucleotides to a growing nucleic acid strand. Thus, DNA synthesis is primed by short segments of RNA made by the enzyme 'primase' or DNA polymerase a (in the case of eukaryotes). The leading strand requires only single priming, which is in the beginning or initiation step and it does not need any more priming. But for the lagging strand, the priming is a routine process and it repeats at every time a new Okazaki fragment is initiated. Finally the RNA primer fragments are excised by the 5′ to 3′ exonuclease action of Polymerase I, and the gap so created will be filled by the DNA polymerase I by its 5′-3′ polymerase activity. The adjacent Okazaki fragments are finally joined by DNA ligase enzyme resulting in an intact continuous lagging strand. The movement of the replication fork (Fig. 4.29) may be unidirectional or bi-directional. The final product is a pair of identical DNA molecule or chromosome conserving half of the strand of the parent molecule and it does not have any RNA stretches in it.

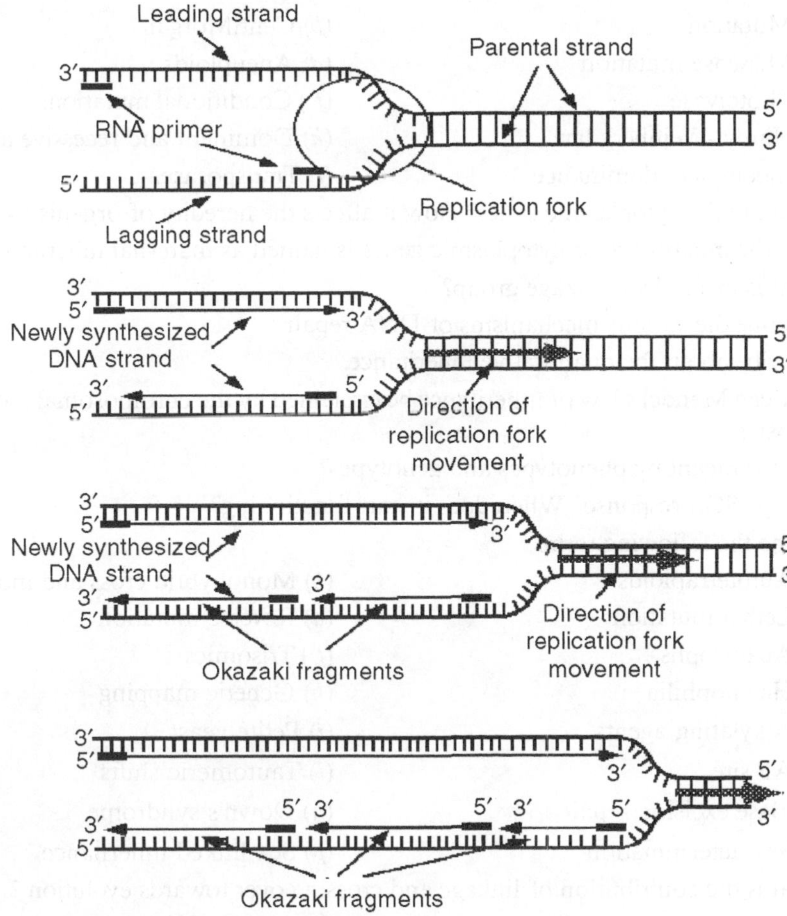

Figure 4.29 Replication fork with leading strand and lagging strand showing Okazaki fragments.

• REVIEW QUESTIONS •

1. Describe the basic laws of heredity.
2. What is meant by gene interaction?
3. What is meant by DNA repair mechanism? What is the need of repair mechanism?

4. Differentiate between multiple allele and multiple genes.
5. What is meant photodimers? Describe two mechanisms by which it can be repaired.
6. If a point mutation occurs in the coding region of a gene what will be the nature of its protein ?
7. What is meant by a frame shift mutation ?
8. Explain what is Xeroderma pigmentosum.
9. Give definitions for the following :
   (*a*) Mutation
   (*b*) CentiMorgan
   (*c*) Missense mutation
   (*d*) Aneuploids
   (*e*) Photolyase
   (*f*) Conditional mutations
   (*g*) Hardy -Weinberg law
   (*h*) Dominant and recessive allele
   (*i*) Incomplete dominance
   (*j*) Transposase.
10. What are the cytoplasmic genes? How it affects the heredity of organisms?
11. Why the inheritance of cytoplasmic genes is named as maternal inheritance?
12. What is meant by a linkage group?
13. Describe the various mechanisms of DNA repair.
14. Describe about the quantitative inheritance.
15. How can Mendel's law of inheritance be explained by the chromosomal movements during meiosis?
16. What is meant by phenotypes and genotypes?
17. What is SOS response? When does it start functioning?
18. Define the following terms :
   (*a*) Autotetraploids
   (*b*) Monohybrid cross and monohybrid ratio
   (*c*) Lethal mutation
   (*d*) Reverse mutation
   (*e*) Auxotrophs
   (*f*) Trisomics
   (*g*) Haemophilia
   (*h*) Genetic mapping
   (*i*) Alkylating agents
   (*j*) Petite yeast
   (*k*) AP site
   (*l*) Tautomeric shifts
   (*m*) Base excision repair
   (*n*) Down's syndrome
   (*o*) Sex determination
   (*p*) Sex linked inheritance.
19. What is the contribution of linkage and crossing over towards evolution ?
20. What is meant by spontaneous mutation and induced mutations ?
21. What are the various types of gene mutations ?
22. Explain the biochemical mutations.
23. Differentiate between euploids and aneuploids.
24. What are the various types of chromosome mutations ?
25. Explain the cytoplasmic inheritance in *Mirabilis jalapa*.
26. 5-bromouracil can reverse the mutation effect created by EMS. Explain.
27. Explain the photoreactivation mechanism of DNA repair.

• SUGGESTED READING •

1. Genes VII, Benjamin Lewin. Oxford University Press, New York, 2000.
2. The Metabolic and Molecular Basis of Inherited Disease, Scriver, C.R., Beaudet, A.L., Sly, W.S. and Valle, D. (Eds.) Seventh Edition, McGraw Hill, New York.
3. DNA Repair and Mutagenesis, Friedberg, E.C., Walker, G.C., and Siede, W., ASM Press, Washington, D.C. 1995, pp. 1-698.
4. Molecular Biology of the Gene, J. D. Watson, N. H. Hopkins, J. W. Roberts, J. A. Steitz and A. M. Weiner, The Benjamin/Cummings Publishing Company, Inc. Menlo Park, California.
5. Principles of Genetics, E. J. Gardener, M. J. Simmons and D. P. Snusted, John Wiley & Sons, Inc. New York.
6. Molecular Cell Biology, H. Lodish, A. Berk, S. L. Zipursky, P. Matsudaira, D. Baltimore and J. Darnell, W. H. Freeman and Company, New York.
7. Genetics: A Molecular Approach, T. A. Brown, Nelson Thrones Ltd., Cheltenham, London.

UNIT 2

5 GENE EXPRESSION

5.1 FINE STRUCTURE OF THE GENE

Gene is the unit of heredity—it is the unit of recombination, mutation and cystron. Based on the data obtained from the experiments on pea plants, Mendel has very correctly predicted the presence and nature of genes, (which he interpreted as particulate factors) including the alternate forms of a gene called alleles. Today we know both its biochemical as well as its genetical aspects. A gene is defined as a segment of DNA molecule (A DNA molecule = a chromosome), which carries all the biochemical information of a polypeptide or RNA molecule, along with its regulatory elements. A complete gene should have the coding region of the protein or RNA along with its control elements. The size of the genes varies greatly. The minimum number or the average number of nucleotides in a gene can be considered as 1000 bp. The part of the gene that holds the sequences for the polypeptide chain or the RNA molecule (in the case of non protein genes like that of rRNA and tRNA etc.) is called the cistron or the coding region. Some genes as in the case of prokaryotes have more than one cistron in a gene controlled by single regulatory elements and such genes are known as polycistronic genes. The controlling elements are known as promoters and are located upstream to the coding region of a gene. The end of the cistron or the coding region is marked by specific signals called termination codons. In some cases the regulatory elements also includes certain sequences in addition to the promoters known as enhancers near or way from the promoters.

The biochemical information that present in the genes are in the form of specific nucleotide sequences or instructions known as codons for the synthesis of RNA molecules that ultimately direct the synthesis of the corresponding polypeptides or proteins. This process of DNA directed RNA synthesis and RNA directed protein synthesis are known as gene expression and forms the Central dogma of life. According to what has been called the central dogma of molecular genetics, the function of DNA is to store information and pass it on to RNA, while the function of RNA is to read, decode and use the information received from DNA to make proteins.

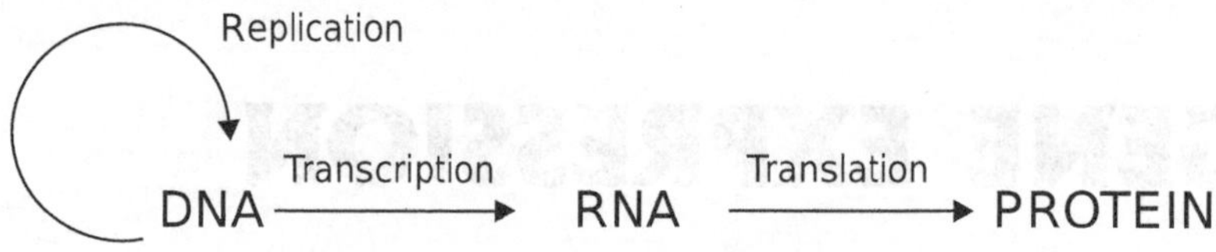

Figure 5.1 Central dogma of life.

The three fundamental processes that take place in the transfer and use of genetic information are the following.

1. Replication is the process by which a replica, or identical copy, of DNA is made. Replication occurs every time a cell divides so that information can be preserved and handed down to offspring. This is similar to making a copy of a file onto a disk so you can take that file to a different computer.

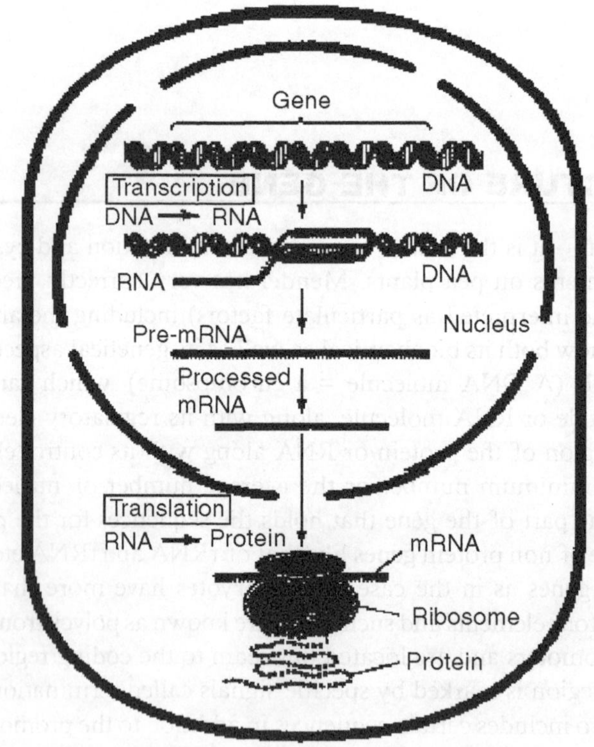

Figure 5.2 Pathway of gene expression.

2. Transcription is the process by which the genetic messages contained in DNA are 'read' or transcribed. The product of transcription, known as messenger RNA (mRNA), leaves the cell nucleus and carries the message to the sites of protein synthesis. This tutorial explains later why this step is necessary in organisms with a nucleus.

3. Translation is the process by which the genetic messages carried by mRNA are decoded and used to build proteins.

Intergenic DNA separates the genes from one another. In the case of virus genome the genes are packed very closely and therefore this intergenic regions are very low or almost nil. In higher organisms the genes are more spread out and are usually separated by very long intergenic regions. Intergenic regions are comparatively less in the case of prokaryotes. The number of genes varies greatly from organisms to organisms. *E. coli* carries approximately 2400 genes, yeasts with sixteen chromosomes have about 6600 genes. It was estimated that the human genome contains about 30,000 to 35,000 genes. But it forms only about 2-3% of the total DNA. The remaining major parts of the DNA are non-coding and non-functional consisting of various types of repeated intergenic regions. In addition to these intergenic repeated sequences there are non-coding regions within the coding regions of the genes known as introns. The coding regions of the genes, which are intervened by introns, are called exons. Thus, the coding regions of the genes in eukaryotes are splited apart by the introns and such genes are called the split genes. They are also known as mosaic genes or discontinuous genes. Split genes and introns are very rare in the case of prokaryotes. In higher organisms where the introns are very common there is mechanism known as RNA splicing to remove the introns from the transcribed RNA molecule. Since prokaryotes do not have the introns, splicing mechanism is absent in them.

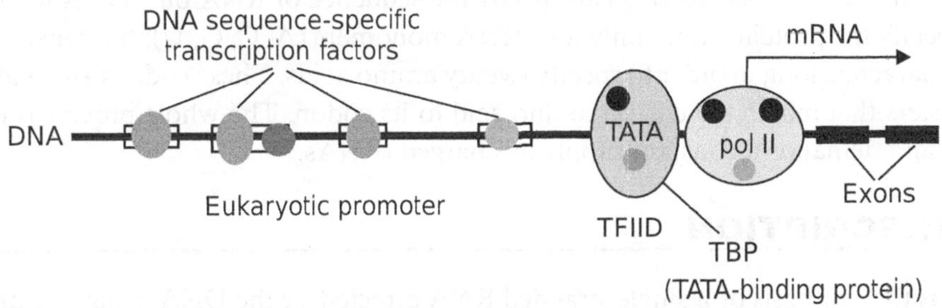

Figure 5.3 A diagrammatic sketch of a eukaryotic gene.

Genes may occur individually or in clusters. Clusters of genes are very common in bacteria and virus, where the DNA—gene ratio is low. In bacteria the genes, which are functionally related, are grouped together under a single regulatory element and such gene clusters or polycistronic genes are known as operons. In plants and animals such grouping of functionally related genes are present but they are regulated individually by its own regulatory elements. Such groups are called multigene family. In eukaryotes there are some copies of genes with coding regions or cistrons alone, without any regulatory elements. Such genes are not functional due to the lack of promoters and other accessories. Such genes are known as silent genes or pseudogenes.

5.2 FROM GENE TO PROTEIN

Now the question is how is the information stored in DNA translated into biological action? Remember, DNA by itself is inert. It is akin to having a book that tells you the recipe for dynamite. By itself, the book is useless. In the hands of someone who knows how to read the instructions, collect the raw materials, and has the requisite technical skill, though, the information is potentially quite explosive.

One Gene-One Enzyme

The action of enzymes on their substrates is the basis of most of the metabolic and biochemical reactions that occur in biological systems. Each metabolic reaction is carried out by an enzyme that has been specified by a unique gene. The synthesis of proteins and enzymes directed by the corresponding genes is known as process of gene expression. Now it is almost clear that a gene specifies an enzyme or a polypeptide and that determines the fate of metabolic reaction.

5.3 GENE EXPRESSION

Think about the genetic information as a recipe. If you had only one copy of your recipe, and any damage would be irreplaceable you would like to safeguard that information. If someone approached you for that recipe, you would not want to let him walk off with the book. You might instead give him a xerox that you did not need back. In the same way, the first step in gene expression is to copy the genetic information into mRNA (**transcription**). Then, the protein synthetic machinery (**ribosomes, tRNA**, amino acyl tRNA synthetases, amino acid biosynthetic pathways) can use it to construct the appropriate protein. The mRNA serve as the template for the synthesis of proteins in the cytoplasm along with other components and this process is known as translation.

It is easy to see how base pairing can specify the sequence of RNA on a DNA template and how mRNA can specify the protein. With only four RNA monomers (A, C, G, U), the genetic word must be at least three characters long in order to specify twenty amino acids. These codons are read one at a time by tRNA adapters that match the correct amino acid to its codon. The whole process is carried out on the ribosome, and demands a constant supply of charged tRNAs.

5.4 TRANSCRIPTION

Transcription is the synthesis of a single stranded RNA directed by the DNA using one of its strands as the template. During this process the genetic information present in the DNA or the gene is transfer to the RNA molecule. A gene that is meant for a protein or enzyme will be transcribed in to RNA called messenger RNA or mRNA, because it carries the message of genetic information for protein synthesis to the site of protein synthesis in the cytoplasm. The two strands of the DNA molecule are separated from one another, exposing the nitrogenous bases. Only one strand is actively used as a template in the transcription process, this is known as the **sense** strand, or **template** strand. The complementary DNA strand, the one that is not used, is called the **nonsense** or **antisense** strand.

The two strands of the DNA molecule are separated from one another, exposing the nitrogenous bases. Only one strand is actively used as a template in the transcription process, this is known as the **template** strand or the **antisense** strand. The complementary DNA strand, the one that is not used, is called the **sense** or the **coding** strand. Because the RNA molecule transcribed is a direct **complementary copy** of the nitrogenous bases in the template or antisense strand and it is exactly identical to the non-template strand of the DNA except for the presence of Uracil instead of thymine. In short the RNA transcribed will be an exact copy of the sense strand or the coding strand and is complementary to the template strand or the antisense strand.

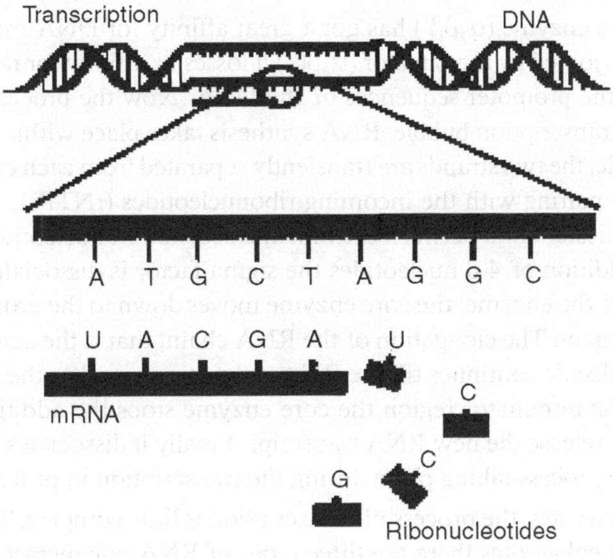

Figure 5.4 The process of transcription.

A multisubunit enzyme called RNA polymerase or more specifically DNA dependant RNA polymerase catalyzes the process of transcription. It needs the ribonucleoside triphosphates (rNTPs)—ATP, UTP, CTP and GTP, double stranded DNA for template. One at a time, this enzyme adds ribonucleotides to a growing RNA strand by joining incoming ribonucleoside triphosphates to the ribose sugar molecule of the last nucleotide of the growing RNA strand. Two of the phosphate groups are removed from the triphosphate and a covalent bond is formed between the remaining phosphate and the third carbon atom of the ribose sugar at the end of the RNA strand. There is an important difference between RNA polymerase and DNA polymerase. RNA polymerase does not need a primer and it can start the synthesis of a new chain. Whereas DNA polymerase cannot synthesize a new strand of DNA, but it can only add new nucleotides to an existing strand. That is DNA polymerase is primer dependent.

Initiation of the transcription process begins with binding of the RNA polymerase enzyme to the DNA molecule at a region known as the **promoter** site. This site is right in front of a gene where transcription will begin. In bacteria, this region contains two short sequences of bases that appear to be the same or similar in all, or most, promoters. The RNA polymerase enzyme does not copy the promoter into the RNA, but begins the synthesis of the RNA at a specific nucleotide sequence called the **start signal** or **initiation site** which is often the bases GTA on the DNA (which then become the bases CAU on the RNA molecule). The first nucleotide corresponding to the position at which transcription begins is marked as +1 and is the start site.

RNA polymerase of *E. coli* has been studied extensively and is considered as a model to explain the activity of the enzyme in other bacteria. In bacteria a single enzyme is responsible for the transcription of all types of RNA molecules including mRNA, rRNAs and tRNAs. This enzyme is a multimeric molecule consisting of five subunits. The enzyme molecule can be divided into two components—a core enzyme and a sigma factor. The core enzyme consists of four subunits—two α subunits , one copy of β and β′ subunits. The core enzyme ($\alpha_2\beta\beta'$) has got a great affinity for DNA molecule and its binding is random. But when it was joined by the σ factor ($\alpha_2\beta\beta'$) it losses its affinity for random DNA but acquires specificity and binds to the promoter sequences of the DNA. Now the process of transcription begins with the formation of a transcription bubble. RNA synthesis takes place within the transcription bubble. In the transcription bubble, the two strands are transiently separated from each other to make the template strand available for base pairing with the incoming ribonucleotides (rNTPs). The holoenzyme ($\alpha_2\beta\beta'$) makes the first phosphodiester bond between the first and second ribonucleotides and thus the initiation is achieved. After the addition of 4-8 nucleotides the sigma factor is dissociated from the holoenzyme and the remaining part of the enzyme, the core enzyme moves down to the promoter and carries out the elongation of the RNA chain. The elongation of the RNA chain, that is the addition of new nucleotides to the growing RNA molecule continues till the RNA polymerase reaches the terminator sequences on the DNA template. In the terminator region the core enzyme stops the addition of nucleotides to the growing RNA chain and release the new RNA transcript. Finally it dissociates from the DNA template. This is the outline of the process-taking place during the transcription in prokaryotes.

In the case of eukaryotes, the process of transcription is little complex. The difference starts from the RNA polymerase. In eukaryotes there are three types of RNA polymerase—RNA polymerase I, II and III, each type transcribing different class of RNA molecules. RNA polymerase transcribes rRNAs, RNA polymerase II transcribes mRNA and RNA polymerase III transcribes tRNA and other RNAs like small nuclear RNAs (snRNAs). Each of these RNA polymerases contain more than subunits (8-14 subunits). There is no difference in the basic mechanism of transcription between the eukaryotic and prokaryotic RNA polymerases. In eukaryotes also the RNA is synthesized in the 5′-3′ direction using the four rNTPs under the direction of the DNA template. Another important difference between the two types of RNA polymerases are that non of the eukaryotic RNA polymerases can directly recognize the promoters of the eukaryotic genes. Therefore, the recognition of the promoter and initiation of RNA synthesis a large number of small protein molecules known as transcription factors (TF) which are responsible for the recognition of promoters by the RNA polymerase. TF molecules and its interaction with different types of promoters will differ from tissues to tissues or cell to cell and are controlled by many other factors. It is the basis of differential expression of genes or tissue specific or tome bound expression of certain specific genes in a specific cells or tissues.

In eukaryotes all the types of RNAs—mRNA rRNA and tRNAs are transcribed in the form of precursor RNA molecule known as primary transcript. The primary transcript is processed to create the actual functional form of RNA. This mechanism of maturation of the primary transcript leading to the formation of functional form RNA molecule is called the posttranscriptional modification. For example, the primary transcript of mRNA molecule is called the heterogenous RNA or hnRNA, and are processed in the nucleus before they are transported to the cytoplasm as mRNA for translation.

The following are the major posttranscriptional modifications for the formation of a mature mRNA from hnRNA.

- *Capping at the 5′ end*: An extra guanosine residue is first adds to the 5′ terminal nucleotide of the primary transcript. This guanosine further modified by the addition of methyl groups. This is the 5′ cap, which prevents the 5′ end of the mRNA from exonuclease enzymes present in the cytoplasm. It also helps in the transport of this mRNA out of nucleus and helps in the initiation of translation processes.

- *Addition of Poly A tail at 3′ end*: At the 3′ end of the mRNA a string of adenine residues are added to form a poly A tail. This also helps the mRNA from the attack of exonuclease.
- *Splicing*: Removal of intervening non-coding sequences or introns, which are also transcribed along with the coding sequences. This process of cutting and removing the introns and joining the parts of exons is known as mRNA splicing. This is very important for the formation of a functional enzyme or proteins.

The mature RNA thus formed leaves the nucleus and gets transported to the cytoplasm. The mRNA in prokaryotes does not need the process of splicing because the genes are without any introns and therefore the RNA produced are directly can be used for the process of translation.

5.5 GENETIC CODE

DNA is a two-stranded molecule. Each strand is a polynucleotide composed of **A** (adenosine), **T** (thymidine), **C** (cytidine), and **G** (guanosine) residues polymerized by 'dehydration' synthesis in linear chains with specific sequences. Each strand has polarity, such that the 5′-hydroxyl (or 5′-phospho) group of the first nucleotide begins the strand and the 3′-hydroxyl group of the final nucleotide ends the strand; accordingly, we say that this strand runs 5′ to 3′ ("*Five prime to three prime*") . It is also essential to know that the two strands of DNA run *antiparallel* such that one strand runs 5′ → 3′ while the other one runs 3′ → 5′. At each nucleotide residuc along the double-stranded DNA molecule, the nucleotides are complementary. That is, **A** forms two hydrogen bonds with **T**; **C** forms three hydrogen bonds with **G**. In most cases the two-stranded, antiparallel, complementary DNA molecule folds to form a helical structure, which resembles a spiral staircase. This is the reason why DNA has been referred to as the "Double Helix".

One strand of DNA holds the information that codes for various genes; this strand is often called the template strand or antisense strand (containing anticodons). The other, and complementary, strand is called the coding strand or sense strand (containing codons). Since mRNA is made from the template strand, it has the same information as the coding strand. Figure 5.5 refers to triplet nucleotide codons along the sequence of the coding or sense strand of DNA as it runs 5′ → 3′ ; the code for the mRNA would be identical but for the fact that RNA contains **U** (uridine) rather than **T**.

An example of two complementary strands of DNA would be :

(5′ → 3′) **ATGGAATTCTCGCTC** (Coding, sense strand)
(3′ ← 5′) **TACCTTAAGAGCGAG** (Template, antisense strand)
(5′ → 3′) **AUGGAAUUCUCGCUC** (mRNA made from Template strand)

Since amino acid residues of proteins are specified as triplet codons, the protein sequence made from the above example would be Met-Glu-Phe-Ser-Leu... (MEFSL...).

Practically, codons are 'decoded' by transfer RNAs (tRNA), which interact with a ribosome-bound messenger RNA (mRNA) containing the coding sequence. There are 64 different tRNAs, each of which has an anticodon loop (used to recognize codons in the mRNA). 61 of these have a bound amino acyl residue; the appropriate 'charged' tRNA binds to the respective next codon in the mRNA and the ribosome catalyzes the transfer of the amino acid from the tRNA to the growing (nascent) protein/ polypeptide chain. The remaining 3 codons are used for 'punctuation'; that is, they signal the termination (the end) of the growing polypeptide chain.

First Position	Second Position of Codon: T	C	A	G	Third Position
T	TTT Phe [F]	TCT Ser [S]	TAT Tyr [Y]	TGT Cys [C]	T
	TTC Phe [F]	TCC Ser [S]	TAC Tyr [Y]	TGC Cys [C]	C
	TTA Leu [L]	TCA Ser [S]	TAA Ter [end]	TGA Ter [end]	A
	TTG Leu [L]	TCG Ser [S]	TAG Ter [end]	TGG Trp [W]	G
C	CTT Leu [L]	CCT Pro [P]	CAT His [H]	CGT Arg [R]	T
	CTC Leu [L]	CCC Pro [P]	CAC His [H]	CGC Arg [R]	C
	CTA Leu [L]	CCA Pro [P]	CAA Gln [Q]	CGA Arg [R]	A
	CTG Leu [L]	CCG Pro [P]	CAG Gln [Q]	CGG Arg [R]	G
A	ATT Ile [I]	ACT Thr [T]	AAT Asn [N]	AGT Ser [S]	T
	ATC Ile [I]	ACC Thr [T]	AAC Asn [N]	AGC Ser [S]	C
	ATA Ile [I]	ACA Thr [T]	AAA Lys [K]	AGA Arg [R]	A
	ATG Met [M]	ACG Thr [T]	AAG Lys [K]	AGG Arg [R]	G

	G	GTT Val [V]	GCT Ala [A]	GAT Asp [D]	GGT Gly [G]	T	
		GTC Val [V]	GCC Ala [A]	GAC Asp [D]	GGC Gly [G]	C	
		GTA Val [V]	GCA Ala [A]	GAA Glu [E]	GGA Gly [G]	A	
		GTG Val [V]	GCG Ala [A]	GAG Glu [E]	GGG Gly [G]	G	

Figure 5.5 The Universal Genetic Code—Triplet codons and its corresponding amino acids.

Lastly, the Genetic Code in the figure above has also been called "The Universal Genetic Code". It is known as 'universal', because all known organisms use it as a code for DNA, mRNA, and tRNA. The universality of the genetic code encompasses animals (including humans), plants, fungi, archaea, bacteria, and viruses. However, all rules have their exceptions, and such is the case with the Genetic Code; small variations in the code exist in mitochondria and certain microbes. Nonetheless, it should be emphasized that these variances represent only a small fraction of known cases, and that the Genetic Code applies quite broadly, certainly to all known nuclear genes.

5.6 TRANSLATION

The process of protein synthesis is called translation because there is a change in the language. The language of DNA and RNA are the same, the alphabets being the four letters of nucleotides A, T (U), G and C. The language of protein is different. It consists of the alphabets of 20 amino acids. The story written in the language of nucleotides (four letters) are Xeroxed into another paper (mRNA) and is taken to the cytoplasm, where it is translated into another language, the language of amino acids (20 letters). That is why the process of protein synthesis is called translation and synthesis of RNA is named as transcription.

The 'tools' needed for translation is a single strand of RNA copied from DNA molecule. This is the **messenger RNA (mRNA)** molecule that carries the codes; ribosomes, which are the sites of assembly of the polypeptide; all 20 Amino acids, which must be present at the same time; small highly specific RNA molecules called **transfer (tRNA)** to which amino acids are covalently bound and which play a part in the decoding of the genetic code; ATP is required as a source of energy for this highly non-spontaneous (energy consuming) process; various proteins and enzymes; for example, the set of enzymes which specifically join the amino acids to the tRNA molecules.

Transfer RNA (tRNA) : The Adapter Molecules

Transfer RNA (tRNA) molecules are the information adapter molecules. It is the direct interface between amino-acid sequence of a protein and the information in DNA. Therefore, it decodes the information in DNA. There are 20 or more different types of tRNA molecules. Most tRNA molecules are about 76 nucleotides in length, but they range from 60 to 95. All tRNA's from all organisms have a similar structure, indeed a human tRNA can function in yeast cells.

There are 4 arms and 3 loops. The acceptor—it is so named because amino acid becomes attached to this arm. The TψC arm is named for the presence of this triplet sequence and ψ stands for Pseudouridine. The anticodon arm always contain anticodon triplet, in the centre of the loop. And the DHU arm contains the modified uridine, the dihydrouridine. Sometimes tRNA molecules have an extra or variable loop present between DHU loop and anticodon loop.

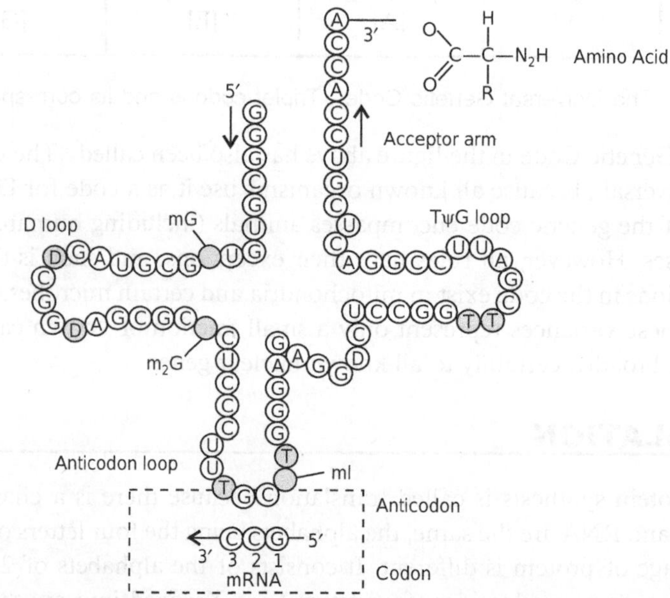

The primary structure of yeast alanine tRNA (tRNA^{Ala}) the first such sequence determined. This molecule is synthesized from the nucleotides A, C, G, and U, but some of the nucleotides, shown in red, are modified after synthesis : D = dihydrouridine, I = inosine, T = thymine, ψ = pseudouridine, and m = methyl group. The primary structure can be diagrammed as a cloverleaf consisting of four base-paired stems and three loops : the D loop (for dihydrouridine, a virtually constant constituent of this loop) ; the anticodon loop ; and the TψCG loop (for thymidylate, pseudouridylate, cytidylate, and guanylate, which are aloost always present in sequence in this loop). As discussed later, attachment of an amino acid to the acceptor arm yields an aminoacyl-tRNA.

Figure 5.6 Structure of a typical tRNA : Yeast alanine tRNA (tRNA^{Ala}).

Transfer RNA is synthesized in two parts. The body of the tRNA is transcribed from a tRNA gene. The acceptor stem is the same for all tRNA molecules and is added after the body is synthesized. It is replaced often during lifetime of a tRNA molecule. Nine hydrogen bonds hold the molecule into an 'L' shaped tertiary structure (shown on the right above). The anticodon is on a loop at the opposite end of the molecule to the amino acid binding region.

The acceptor stem is the site at which a specific amino acid is attached by an amino-acyl-tRNA synthase. Amino acids are covalently joined to the acceptor arm of the molecule that is not looped. tRNA molecules become 'charged' or **aminoacylated** in a two step reaction in which an enzyme first attaches most of a ATP molecule (AMP) to the amino acid to create a temporary intermediate. In the second step the tRNA molecule displaces the AMP from the amino acid and joins with it, forming an amino acid-tRNA complex. The enzymes recognize unique features on every different tRNA molecule and will only join the correct amino acid to the correct tRNA. The anticodon reads the information in an mRNA sequence by base pairing. The anticodon of the tRNA pairs with the codon of mRNA.

Charging the tRNA: Attachment of Amino Acid to tRNA

Activation of amino acids is carried out by a two-step process catalyzed by ***aminoacyl-tRNA synthetases***. Each tRNA, and the corresponding amino acid are recognized by individual aminoacyl-tRNA synthetases. This means there exists at least 20 different aminoacyl-tRNA synthetases; there are actually at least 21 since the initiator met-tRNA of both prokaryotes and eukaryotes is distinct from non-initiator met-tRNAs.

Activation of amino acids requires energy in the form of ATP and occurs in a two-step reaction catalyzed by the aminoacyl-tRNA synthetases. First the enzyme attaches the amino acid to the α-phosphate of ATP with the concomitant release of pyrophosphate. This is termed an **aminoacyl-adenylate intermediate**. In the second step the enzyme catalyzes transfer of the amino acid to either the 2′- or 3′-OH of the ribose portion of the 3′-terminal adenosine residue of the tRNA generating the **activated aminoacyl-tRNA**. Although these reaction are freely reversible, the forward reaction is favoured by the coupled hydrolysis of PP_i.

Accurate recognition of the correct amino acid as well as the correct tRNA is different for each aminoacyl-tRNA synthetase. Since the different amino acids have different R groups, the enzyme for each amino acid has a different binding pocket for its specific amino acid. It is not the anticodon that determines the tRNA utilized by the synthetases. Although the exact mechanism is not known for all synthetases, it is likely to be a combination of the presence of specific modified bases and the secondary structure of the tRNA that is correctly recognized by the synthetases.

It is absolutely necessary that the discrimination of correct amino acid and correct tRNA be made by a given synthetase prior to release of the aminoacyl-tRNA from the enzyme. Once the product is released there is no further way to **proof-read** whether a given tRNA is coupled to its corresponding tRNA. Erroneous coupling would lead to the wrong amino acid being incorporated into the polypeptide since the discrimination of amino acid during protein synthesis comes from the recognition of the anticodon of a tRNA by the codon of the mRNA and not by recognition of the amino acid. This was demonstrated by reductive desulphuration of cys-tRNA^{cys} with Raney nickel generating ala-tRNA^{cys}. Alanine was then incorporated into an elongating polypeptide where cysteine should have been.

We have seen the formation of charged aminoacyl-tRNAs and now the mRNAs to convert nucleotide sequences to amino acid sequences we need to bring the two together accurately and efficiently. This is the job of the ribosomes. Ribosomes are composed of proteins and rRNAs. All living organisms need to synthesis proteins and all cells of an organism need to synthesize proteins, therefore, it is not hard to imagine that ribosomes are a major constituent of all cells of all organisms. The make up of the ribosomes, both rRNA and associated proteins are slightly different between prokaryotes and eukaryotes.

Ribosome Structure and Function

Ribosomes are cytoplasmic organelles found in prokaryotes and eukaryotes. They are large complexes of proteins and three (prokaryotes) or four (eukaryotes) rRNA (ribosomal ribonucleic acid) molecules called subunits made in the nucleolus. The main function of ribosomes is to serve as the site of mRNA translation (protein synthesis, the assembly of amino acids into proteins); once the two (large and small) subunits are joined by the mRNA from the nucleus, the ribosome translates the mRNA into a specific sequence of amino acids, or a polypeptide chain. Ribosomes are found in cells in two ways: free and bound. In an electron micrograph as shown above, ribosomes appear as dark granules. Ribosomes exist at various locations within the cell; however, the location of ribosomes depends on the function of the cell :

Free Ribosomes

- Found in the cytosol.
- May occur as a single ribosome or in groups known as *polyribosomes* or *polysomes*.
- Occur in greater number than bound ribosomes in cells that retain most of their manufactured protein.
- Responsible for proteins that go into solution in the cytoplasm or form important cytoplasmic structural or motile elements.

Bound Ribosomes

- Found bound to the exterior of the endoplasmic reticulum (ER) constituting rough ER.
- Occur in greater number than free ribosomes in cells that secrete their manufactured proteins (*e.g.*, pancreatic cells, producers of digestive enzymes).
- Responsible for proteins that become a part of membranes or packaged into vesicles for storage in the cytoplasm or export to the cell exterior.

Ribosomes are also located in mitochondria and chloroplasts of eukaryotic cells; they are always smaller than cytoplasmic ribosomes and are comparable to prokaryotic ribosomes in both size and sensitivity to antibiotics; however, the sedimentation values *s* (*s* = Svedberg units: a measure of the rate of sedimentation of a component in a centrifuge, related both to the molecular weight and the 3-D shape of the component) vary somewhat in different phyla. Prokaryotic and eukaryotic ribosomes perform the same functions by the same set of chemical reactions; however, eukaryotic ribosomes are much larger than prokaryotic ones and most of their proteins are different. Mitochondrial and chloroplast ribosomes resemble bacterial ribosomes.

Prokaryotic and eukaryotic ribosomes are very similar in their form. Small prokaryotic and eukaryotic ribosomal subunits have a *head* and a *base* with an armlike *platform* extending from one side. Ribosomes are small, but complex structures, roughly 20 to 30 nM in diameter, consisting of two unequally sized subunits, referred to as *large* and *small sub units.* A subunit is composed of a complex between RNA molecules and proteins; each subunit contains at least one ribosomal RNA (rRNA) subunit and a large quantity of ribosomal proteins. The subunits together contain up to 82 specific proteins assembled in a precise sequence.

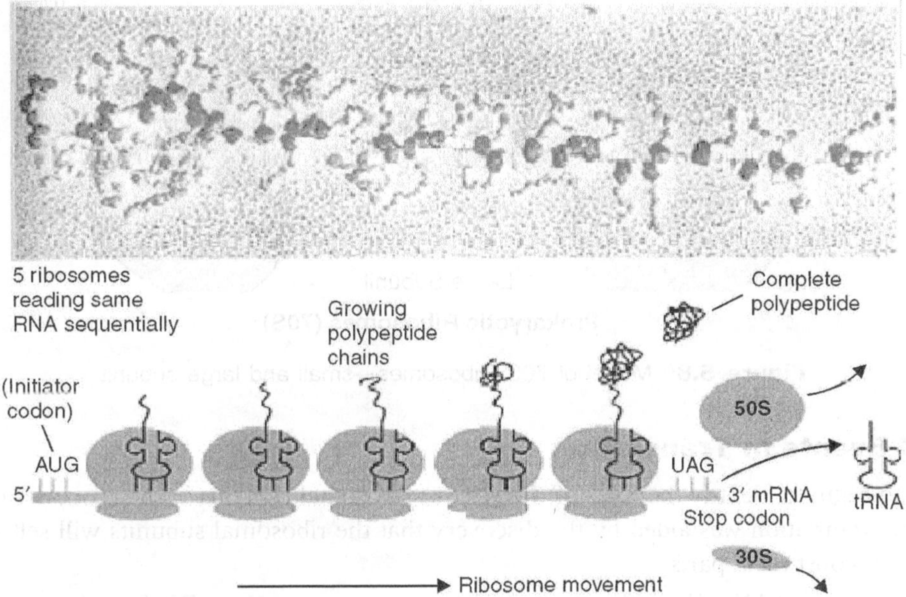

Figure 5.7 Electron micrograph showing translation : Polyribosome and its diagrammatic sketch.

The prokaryotic ribosome in *E. coli* has a size of 70S. The two subunits have distinct and recognizable 3-D shapes. Approximately two-thirds of the *E. coli* ribosome consists of rRNA with the rest consisting of ribosomal proteins. Overall, the 50S and 30S subunits combined are 70S due to the 3-D shape of the ribosomes. In general, eukaryotic ribosomes are larger and more complex than prokaryotic ribosomes. The size of the ribosome and the molecular weights of the rRNA molecules differ from organism to organism. Simple eukaryotes have the smallest ribosomes, although they are larger than *E. coli* ribosomes, while mammals have the largest ribosomes. All eukaryotic ribosomes have many common structural and chemical features. For example, the mammalian ribosome has a size of 80S consisting of a large 60S subunit and a small 40S subunit. The 80S mammalian ribosome consists of about equal weights of rRNA and ribosomal proteins.

There are three sites on a ribosome—A site or the acceptor site, P site or the peptide site and E site or exit site. A site is the entry site for an aminoacyl-tRNA, P site is the binding site for the peptidyl-tRNA carrying the nascent polypeptide under synthesis and E site is the exit for the leaving tRNA.

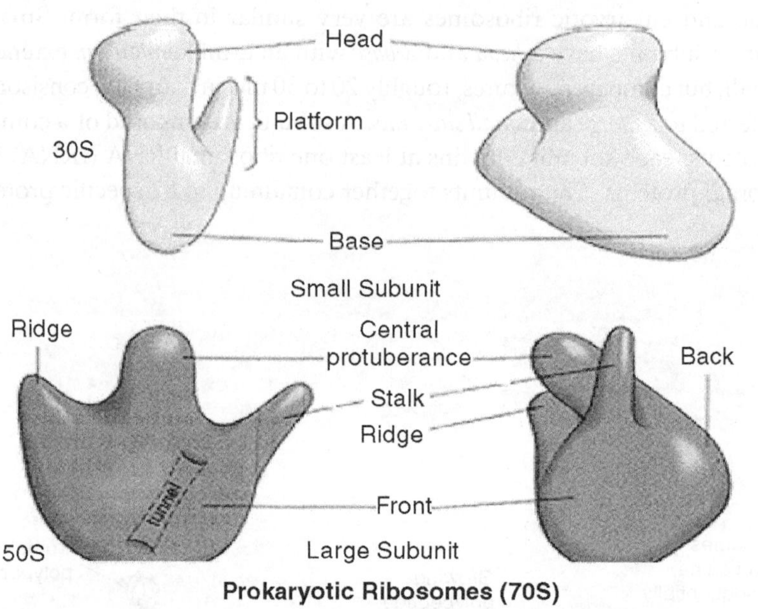

Figure 5.8 Model of 70S Ribosomes—small and large subunit.

Order of Events in Translation

The ability to begin to identify the roles of the various ribosomal proteins in the processes of ribosome assembly and translation was aided by the discovery that the ribosomal subunits will self assemble *in vitro* from their constituent parts.

Following assembly of both the small and large subunits onto the mRNA, and given the presence of charged tRNAs, protein synthesis can take place. To reiterate the process of protein synthesis :

1. Synthesis proceeds from the N-terminus to the C-terminus of the protein.
2. The ribosomes 'read' the mRNA in the 5′ to 3′ direction.
3. Active translation occurs on polyribosomes (also termed polysomes). This means that more than one ribosome can be bound to and translate a given mRNA at any one time.
4. Chain elongation occurs by sequential addition of amino acids to the C-terminal end of the ribosome bound polypeptide.

Translation proceeds in an ordered process. First accurate and efficient **initiation** occurs, then chain **elongation** and finally accurate and efficient **termination** must occur. All three of these processes require specific proteins, some of which are ribosome associated and some of which are separate from the ribosome, but may be temporarily associated with it.

Initiation

Initiation of translation in both prokaryotes and eukaryotes requires a specific initiator tRNA, $tRNA^{met}_i$, that is used to incorporate the initial methionine residue into all proteins. In *E. coli* a specific version of $tRNA^{met}_i$ is required to initiate translation, ($tRNA^{fmet}_i$). The methionine attached to this initiator tRNA is formylated. Formylation requires N^{10}-formy-THF and is carried out after the methionine is attached to the tRNA. The fmet-$tRNA^{fmet}_i$ still recognizes the same codon, AUG, as regular $tRNA^{met}$. Although $tRNA^{met}_i$ is specific for initiation, in eukaryotes it is not a formylated $tRNA^{met}$.

The initiation of translation requires recognition of an AUG codon. In the polycistronic prokaryotic RNAs this AUG codon is located adjacent to a Shine-Delgarno element in the mRNA. The Shine-Delgarno element is recognized by complimentary sequences in the small subunit rRNA (16*S* in *E. coli*). In eukaryotes initiator AUGs are generally, but not always, the first encountered by the ribosome. A specific sequence context surrounding the initiator AUG aids ribosomal discrimination. This context is $^{A}/_{G}CC^{A}/_{G}CCAUG^{A}/_{G}$ in most mRNAs.

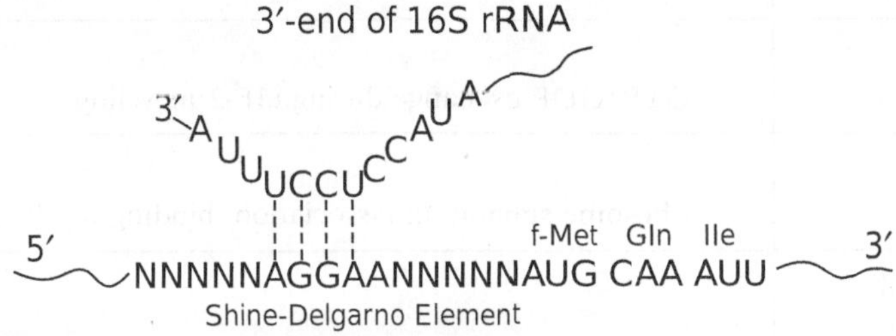

Figure 5.9 Shine Delgarno sequence and its complement in 16S rRNA.

Eukaryotic Initiation Factors and Their Functions

The specific non-ribosomally associated proteins required for accurate translational initiation are termed as initiation factors. In *E. coli* they are IFs in eukaryotes they are eIFs. Numerous eIFs have been identified.

Specific Steps in Translational Initiation

Initiation of translation requires four specific steps :

1. A ribosome must dissociate into its 40S and 60S subunits.
2. A ternary complex termed the preinitiation complex is formed consisting of the initiator, GTP, eIF-2 and the 40S subunit.
3. The mRNA is bound to the preinitiation complex.
4. The 60S subunit associates with the preinitiation complex to form the 80S initiation complex.

The initiation factors eIF-1 and eIF-3 bind to the 40S ribosomal subunit favouring antiassociation to the 60S subunit. The prevention of subunit reassociation allows the preinitiation complex to form.

The first step in the formation of the preinitiation complex is the binding of GTP to eIF-2 to form a binary complex. eIF-2 is composed of three subunits, a, b and g. The binary complex then binds to the activated initiator tRNA, met-$tRNA^{met}$ forming a ternary complex that then binds to the 40*S* subunit forming the 43S preinitiation complex. The preinitiation complex is stabilized by the earlier association of eIF-3 and eIF-1 to the 40S subunit.

The cap structure of eukaryotic mRNAs is bound by specific eIFs prior to association with the preinitiation complex. Cap binding is accomplished by the initiation factor eIF-4F. This factor is actually a complex of 3 proteins; eIF-4E, A and G. The protein eIF-4E is a 24 kDa protein which physically recognizes and binds to the cap structure. eIF-4A is a 46 kDa protein which binds and hydrolyzes ATP and exhibits RNA helicase activity. Unwinding of mRNA secondary structure is necessary to allow access of the ribosomal subunits. eIF-4G aids in binding of the mRNA to the 43S preinitiation complex.

TABLE 5.1 Initiation Factor

Initiation Factor	Activity
eIF-1	repositioning of met-tRNA to facilitate mRNA binding
eIF-2	ternary complex formation
eIF-2A	AUG-dependent met-$tRNA_{met}^{i}$ binding to 40S ribosome
eIF-2B (also called GEF)	GTP/GDP exchange during eIF-2 recycling
eIF-3 composed of ~10 subunits	ribosome subunit antiassociation, binding to 40S subunit
eIF-4F composed of 3 subunits: eIF-4E, eIF-4A, eIF-4G	mRNA binding to 40S subunit, ATPase-dependent RNA helicase activity
eIF-4A	ATPase-dependent RNA helicase
eIF-4E	5′ cap recognition
eIF-4G	acts as a scaffold for the assembly of eIF-4E and -4A in the eIF-4F complex
eIF-4B	stimulates helicase, binds simultaneously with eIF-4F
eIF-5	release of eIF-2 and eIF-3, ribosome-dependent GTPase
eIF-6	ribosome subunit antiassociation

Once the mRNA is properly aligned onto the preinitiation complex and the initiator met-$tRNA_{met}$ is bound to the initiator AUG codon (a process facilitated by eIF-1) the 60S subunit associates with the complex. The association of the 60S subunit requires the activity of eIF-5 which has first bound to the preinitiation complex. The energy needed to stimulate the formation of the 80S initiation complex comes from the hydrolysis of the GTP bound to eIF-2. The GDP bound form of eIF-2 then binds to eIF-2B, which stimulates the exchange of GTP for GDP on eIF-2. When GTP is exchanged eIF-2B dissociates from eIF-2. This is termed the eIF-2 cycle (see Figure 5.10). This cycle is absolutely required in order for eukaryotic translational initiation to occur. The GTP exchange reaction can be affected by phosphorylation of the α-subunit of eIF-2. At this stage the initiator met-$tRNA^{met}$ is bound to the mRNA within a site of the ribosome termed the P-site, for peptide site. The other site within the ribosome to which incoming charged tRNAs bind is termed the A-site, for amino acid site.

Elongation

The process of elongation, like that of initiation requires specific non-ribosomal proteins. In *E. coli* these are EFs and in eukaryotes **eEFs**. Elongation of polypeptides occurs in a cyclic manner such that at the end of one complete round of amino acid addition the A site will be empty and ready to accept the incoming aminoacyl-tRNA dictated by the next codon of the mRNA. This means that not only does the incoming amino acid need to be attached to the peptide chain but the ribosome must move down the mRNA to the next codon. Each incoming aminoacyl-tRNA is brought to the ribosome by an **eEF-1a-GTP complex**. When the correct tRNA is deposited into the A site the GTP is hydrolyzed and the eEF-1a-GDP complex dissociates. In order for additional translocation events the GDP must be exchanged for GTP. This is carried out by **eEF-1bg** similarly to the GTP exchange that occurs with eIF-2 catalyzed by eIF-2B.

The peptide attached to the tRNA in the P site is transferred to the amino group at the aminoacyl-tRNA in the A site. This reaction is catalyzed by ***peptidyltransferase***. This process is termed **transpeptidation**. The elongated peptide now resides on a tRNA in the A site. The A site needs to be freed in order to accept the next aminoacyl-tRNA. The process of moving the peptidyl-tRNA from the A site to the P site is termed, **translocation**. Translocation is catalyzed by **eEF-2** coupled to GTP hydrolysis. In the process of translocation the ribosome is moved along the mRNA such that the next codon of the mRNA resides under the A site. Following translocation eEF-2 is released from the ribosome. The cycle can now begin again.

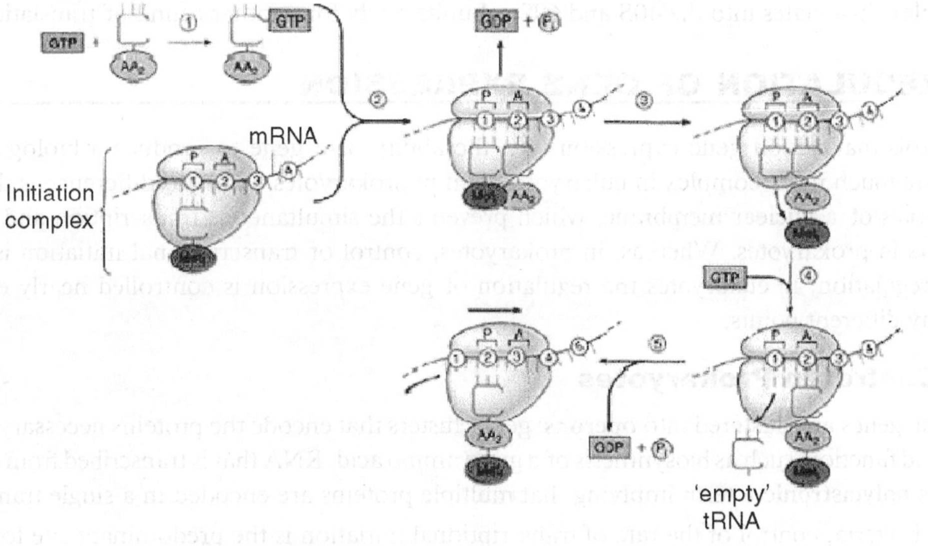

Figure 5.10 Translation—Elongation of polypeptide.

Termination

Like initiation and elongation, translational termination requires specific protein factors identified as releasing factors, RFs in *E. coli* and **eRFs** in eukaryotes. There are 2 RFs in *E. coli* and one in eukaryotes. The signals for termination are the same in both prokaryotes and eukaryotes. These signals are termination codons present in the mRNA. There are 3 termination codons, UAG, UAA and UGA.

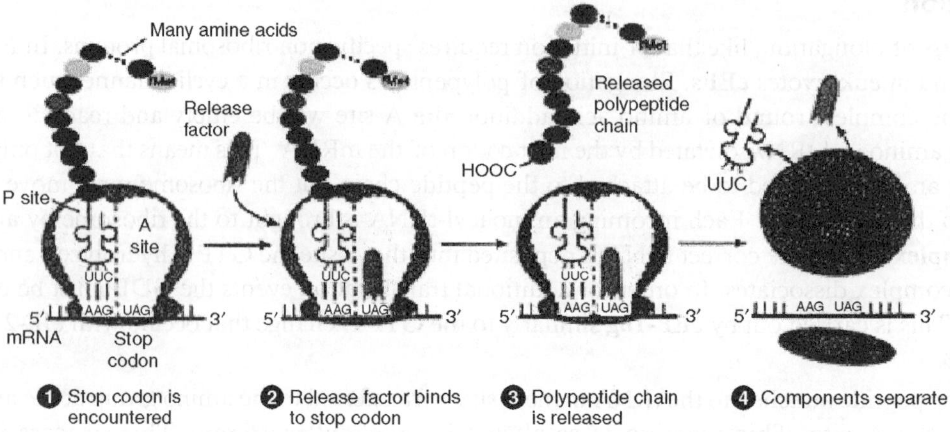

Figure 5.11 The process of termination that marks the end of translation.

In *E. coli* the termination codons UAA and UAG are recognized by RF-1, whereas RF-2 recognizes the termination codons UAA and UGA. The eRF binds to the A site of the ribosome in conjunction with GTP. The binding of eRF to the ribosome stimulates the peptidyl transferase activity to transfer the peptidyl group to water instead of an aminoacyl-tRNA. The resulting uncharged tRNA left in the P site is expelled with concomitant hydrolysis of GTP. The inactive ribosome then releases its mRNA and the 80S complex dissociates into the 40S and 60S subunits ready for another round of translation.

5.7 REGULATION OF GENE EXPRESSION

The controls that act on gene expression (*i.e.*, the ability of a gene to produce a biologically active protein) are much more complex in eukaryotes than in prokaryotes. A major difference is the presence in eukaryotes of a nuclear membrane, which prevents the simultaneous transcription and translation that occurs in prokaryotes. Whereas, in prokaryotes, control of transcriptional initiation is the major point of regulation, in eukaryotes the regulation of gene expression is controlled nearly equivalently from many different points.

Gene Control in Prokaryotes

In bacteria, genes are clustered into **operons**: gene clusters that encode the proteins necessary to perform coordinated function, such as biosynthesis of a given amino acid. RNA that is transcribed from prokaryotic operons is **polycistronic** a term implying that multiple proteins are encoded in a single transcript.

In bacteria, control of the rate of transcriptional initiation is the predominant site for control of gene expression. As with the majority of prokaryotic genes, initiation is controlled by two DNA sequence elements that are approximately 35 bases and 10 bases, respectively, upstream of the site of transcriptional initiation and as such are identified as the -35 and -10 positions. These 2 sequence elements are termed **promoter sequences**, because they *promote* recognition of transcriptional start sites by ***RNA polymerase***. The consensus sequence for the -35 position is TTGACA, and for the -10 position, TATAAT. (The -10 position is also known as the **Pribnow-box**.) These promoter sequences are recognized and contacted by **RNA polymerase**.

The activity of RNA polymerase at a given promoter is in turn regulated by interaction with accessory proteins, which affect its ability to recognize start sites. These regulatory proteins can act both positively (activators) and negatively (repressors). The accessibility of promoter regions of prokaryotic DNA is in many cases regulated by the interaction of proteins with sequences termed operators. The operator region is adjacent to the promoter elements in most operons and in most cases the sequences of the operator bind a repressor protein. However, there are several operons in *E. coli* that contain overlapping sequence elements, one that binds a repressor and one that binds an activator.

As indicated above, prokaryotic genes that encode the proteins necessary to perform coordinated function are clustered into operons. Two major modes of transcriptional regulation function in bacteria (*E. coli*) to control the expression of operons. Both mechanisms involve repressor proteins. One mode of regulation is exerted upon operons that produce gene products necessary for the utilization of energy; these are catabolite-regulated operons. The other mode regulates operons that produce gene products necessary for the synthesis of small biomolecules such as amino acids. Expression from the latter class of operons is attenuated by sequences within the transcribed RNA.

A classic example of a catabolite-regulated operon is the ***lac* operon**, responsible for obtaining energy from b-galactosides such as lactose. A classic example of an attenuated operon is the ***trp* operon**, responsible for the biosynthesis of tryptophan.

The *lac* Operon

The *lac* operon (see diagram below) consists of one regulatory gene (the *i* gene) and three structural genes (*z*, *y*, and *a*). The *i* gene codes for the repressor of the *lac* operon. The *z* gene codes for **β-galactosides** (β-gal), which is primarily responsible for the hydrolysis of the disaccharide, lactose into its monomeric units, galactose and glucose. The gene ***'y'*** codes for **permease**, which is responsible for the high the permeability of bacterial membranes for the β-galactosides. The '*a*' gene encodes a **transacetylase**.

During normal growth on a glucose-based medium, the *lac* repressor is bound to the operator region of the *lac* operon, preventing transcription. However, in the presence of an inducer of the *lac* operon, the repressor protein binds the inducer and is rendered incapable of interacting with the operator region of the operon. RNA polymerase is thus able to bind at the promoter region, and transcription of the operon ensues.

The *lac* operon is repressed, even in the presence of lactose, if glucose is also present. This repression is maintained until the glucose supply is exhausted. The repression of the *lac* operon under these conditions is termed **catabolite repression** and is a result of the low levels of cAMP that result from an adequate glucose supply. The repression of the *lac* operon is relieved in the presence of glucose if excess cAMP is added. As the level of glucose in the medium falls, the level of cAMP increases. Simultaneously there is an increase in inducer binding to the *lac* repressor. The net result is an increase in transcription from the operon.

The ability of cAMP to activate expression from the *lac* operon results from an interaction of cAMP with a protein termed **CRP** (for **cAMP receptor protein**). The protein is also called **CAP** (for **catabolite activator protein**). The cAMP-CRP complex binds to a region of the *lac* operon just upstream of the region bound by RNA polymerase and that somewhat overlaps that of the repressor-binding site of the operator region. The binding of the cAMP-CRP complex to the *lac* operon stimulates RNA polymerase activity 20-to-50-fold.

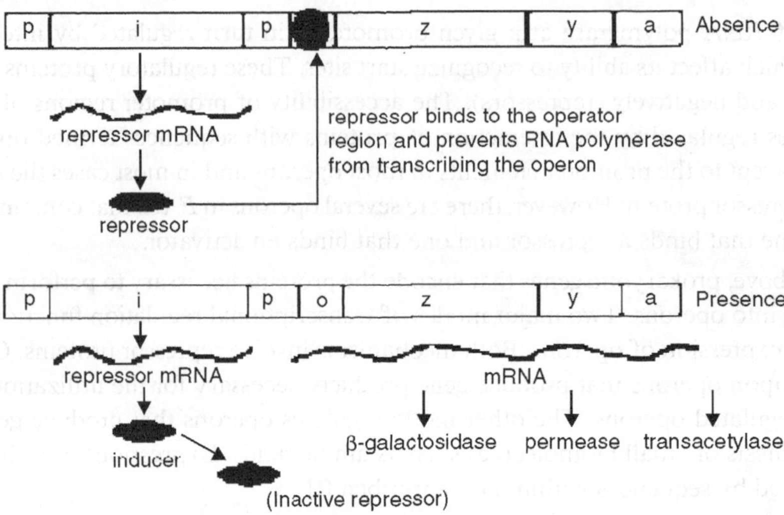

Figure 5.12 Regulation of the *lac* operon in *E. coli*—The repressor of the operon is synthesized from the *i* gene. The repressor protein binds to the operator region of the operon and prevents RNA polymerase from transcribing the operon. In the presence of an inducer (such as the natural inducer, allolactose) the repressor is inactivated by interaction with the inducer. This allows RNA polymerase access to the operon and transcription proceeds. The resultant mRNA encodes the β-galactosidase, permease and transacetylase activities necessary for utilization of β-galactosides (such as lactose) as an energy source. The *lac* operon is additionally regulated through binding of the cAMP-receptor protein, CRP (also termed the catabolite activator protein, CAP) to sequences near the promoter domain of the operon. The result is a 50 fold enhancement of polymerase activity.

The *trp* Operon

The *trp* operon (see Figure 5.13) encodes the genes for the synthesis of tryptophan. This cluster of genes, like the *lac* operon, is regulated by a repressor that binds to the operator sequences. The activity of the *trp* repressor for binding the operator region is enhanced when it binds tryptophan; in this capacity, tryptophan is known as a **co-repressor**. Since the activity of the *trp* repressor is enhanced in the presence of tryptophan, the rate of expression of the *trp* operon is graded in response to the level of tryptophan in the cell.

Expression of the *trp* operon is also regulated by **attenuation**. The attenuator region, which is composed of sequences found within the transcribed RNA, is involved in controlling transcription from the operon after RNA polymerase has initiated synthesis. The attenuator of sequences of the RNA is found near the 5′ end of the RNA termed the leader region of the RNA. The leader sequences are located prior to the start of the coding region for the first gene of the operon (the *trpE* gene). The attenuator region contains codons for a small leader polypeptide, which contains tandem tryptophan codons. This region of the RNA is also capable of forming several different stable stem-loop structures.

Depending on the level of tryptophan in the cell—and hence the level of charged trp-tRNAs—the position of ribosomes on the leader polypeptide and the rate at which they are translating allows different stem-loops to form. If tryptophan is abundant, the ribosome prevents stem-loop 1-2 from forming and thereby favours stem-loop 3-4. The latter is found near a region rich in uracil and acts as the transcriptional terminator loop as in the RNA synthesis. Consequently, RNA polymerase is dislodged from the template. The operons coding for genes necessary for the synthesis of a number of other amino acids are also regulated by this attenuation mechanism. It should be clear, however, that this type of transcriptional regulation is not feasible for eukaryotic cells.

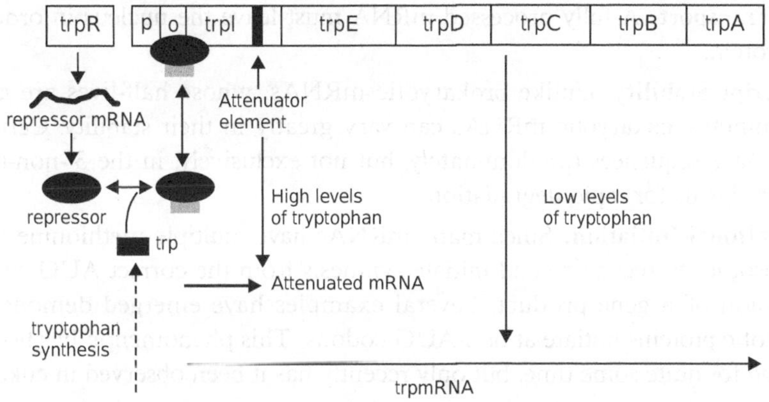

Figure 5.13 *trp* Operon—Regulation of the *trp* operon in *E. coli*. The *trp* operon is controlled by both a repressor protein binding to the operator region as well as by translation-induced transcriptional attenuation. The *trp* repressor binds the operator region of the *trp* operon only when bound to tryptophan. This makes tryptophan a co-repressor of the operon. The *trpL* gene encodes a non-functional leader peptide which contains several adjacent *trp* codons. The structural genes of the operon responsible for tryptophan biosynthesis are *trpE, D, C, B* and *A*. When tryptophan level are high some binds to the repressor which then binds to the operator region and inhibits transcription. The mechanism of attenuation of the *trp* operon is diagrammed below.

Gene Control in Eukaryotes

In eukaryotic cells, the ability to express biologically active proteins comes under regulation at several points :

1. **Chromatin Structure.** The physical structure of the DNA, as it exists compacted into chromatin, can affect the ability of transcriptional regulatory proteins (termed transcription factors) and RNA polymerases to find access to specific genes and to activate transcription from them. The presence of the histones and CpG methylation most affect accessibility of the chromatin to RNA polymerases and transcription factors.
2. **Transcriptional Initiation.** This is the most important mode for control of eukaryotic gene expression (see below for more details). Specific factors that exert control include the strength

of promoter elements within the DNA sequences of a given gene, the presence or absence of enhancer sequences (which enhance the activity of RNA polymerase at a given promoter by binding specific transcription factors), and the interaction between multiple activator proteins and inhibitor proteins.

3. **Transcript Processing and Modification.** Eukaryotic mRNAs must be capped and polyadenylated, and the introns must be accurately removed. Several genes have been identified that undergo tissue-specific patterns of alternative splicing, which generate biologically different proteins from the same gene.
4. **RNA Transport.** A fully processed mRNA must leave the nucleus in order to be translated into protein.
5. **Transcript Stability.** Unlike prokaryotic mRNAs, whose half-lives are all in the range of 1—5 minutes, eukaryotic mRNAs can vary greatly in their stability. Certain unstable transcripts have sequences (predominately, but not exclusively, in the 3′-non-translated regions) that are signals for rapid degradation.
6. **Translational Initiation.** Since many mRNAs have multiple methionine codons, the ability of ribosomes to recognize and initiate synthesis from the correct AUG codon can affect the expression of a gene product. Several examples have emerged demonstrating that some eukaryotic proteins initiate at non-AUG codons. This phenomenon has been known to occur in *E. coli* for quite some time, but only recently has it been observed in eukaryotic mRNAs.
7. **Post-translational Modification.** Common modifications include gycosilation, acetylation, fatty acylation, disulphide bond formations, etc.
8. **Protein Transport.** In order for proteins to be biologically active following translation and processing, they must be transported to their site of action.
9. **Control of Protein Stability.** Many proteins are rapidly degraded, whereas others are highly stable. Specific amino acid sequences in some proteins have been shown to bring about rapid degradation.

Control of Eukaryotic Transcription Initiation

Transcription of the different classes of RNAs in eukaryotes is carried out by three different polymerases. RNA pol I synthesizes the rRNAs, except for the 5S species. RNA pol II synthesizes the mRNAs and some small nuclear RNAs (snRNAs) involved in RNA splicing. RNA pol III synthesizes the 5S rRNA and the tRNAs. The vast majority of eukaryotic RNAs are subjected to post-transcriptional processing.

The most complex controls observed in eukaryotic genes are those that regulate the expression of RNA pol II-transcribed genes, the mRNA genes. Almost all eukaryotic mRNA genes contain a basic structure consisting of coding exons and non-coding introns and basal promoters of two types and any number of different transcriptional regulatory domains (see Figure 5.15). The basal promoter elements are termed **CCAAT-boxes** (pronounced *cat*) and **TATA-boxes** because of their sequence motifs. The TATA-box resides 20 to 30 bases upstream of the transcriptional start site and is similar in sequence to the prokaryotic Pribnow-box (consensus TATA$^{T}/_{A}$A$^{T}/_{A}$, where $^{T}/_{A}$ indicates that either base may be found at that position).

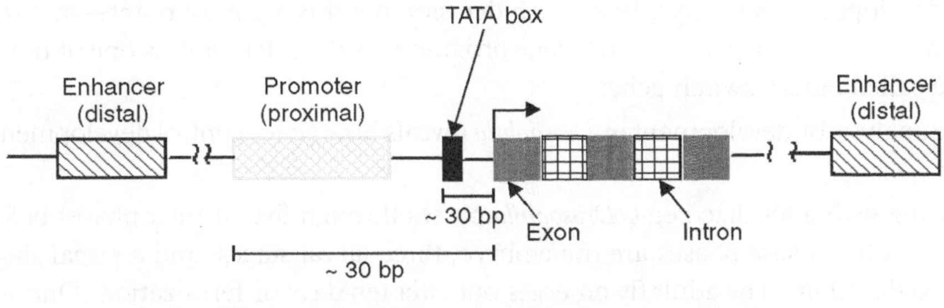

Figure 5.14 Structure of a typical eukaryotic mRNA gene.

Numerous proteins identified as **TFIIA, B, C**, etc. (for **t**ranscription **f**actors regulating RNA pol **II**), have been observed to interact with the TATA-box. The CCAAT-box (consensus $GG^{T}/_{C}CAATCT$) resides 50 to 130 bases upstream of the transcriptional start site. The protein identified as **C/EBP** (for **C**CAAT-box/**E**nhancer **B**inding **P**rotein) binds to the CCAAT-box element.

There are many other regulatory sequences in mRNA genes, as well, that bind various transcription factors. These regulatory sequences are predominantly located upstream (5′) of the transcription initiation site, although some elements occur downstream (3′) or even within the genes themselves. The number and type of regulatory elements to be found varies with each mRNA gene. Different combinations of transcription factors also can exert differential regulatory effects upon transcriptional initiation. The various cell types each express characteristic combinations of transcription factors ; this is the major mechanism for cell-type specificity in the regulation of mRNA gene expression.

5.8 GENETIC BASIS OF DIFFERENTIATION AND DEVELOPMENT

All multicellular organisms developed from a single cell, the fertilization product of two gametes—the zygote. Embryonic development involves a series of cell division cells and development events, morphogenesis, and cell differentiation, which ultimately give rise to a multicellular adult organism. A fully developed adult organism is a collection of millions of cells organised in to a cohesive and coordinated unit. The development of such an organism from a single cell, like any other life activities is governed by a large number of genes which express in a very organised and time bound manner.

Researchers study development in model organisms to identify general principles : *science as a process.* Development is the attainment of differentiated state of various cells in multicellular organisms. At differentiated state of an organism, the cells show the differential gene expression.

Differential Gene Expression

All the different types of cells in an organism have the same DNA. But different cell types make different proteins, usually as a result of transcriptional regulation. Maternal molecules in the cytoplasm (maternal influence or cytoplasmic influence) and signals from other cells direct transcriptional regulation the cell—cell interactions.

Genetic and Cellular Mechanisms of Pattern Formation

In embryo development, as the development progress there is a progressive restriction in transcriptional pattern of the genome. During the development certain genes act as switch points controlling the number of alternative development pathways. Each such decision point is a binary point—that is there are two alternatives and the performance of a switch gene programmes the cell to follow one of the two pathways. Such genes are called binary switch genes.

Genetic analysis of development in *Drosophila* reveals how genes control development: an overview is as follows :

1. Starting with a fertilized egg, *Drosophila* passes through five distinct phases before it becomes an adult fly. These phases are the embryo, three larval stages, and a pupal stage, which give rise to the adult. The adult fly emerges out after ten days of fertilization. During the divisions of the fertilized egg, the cytoplasm divided into a series of maternally divided molecular gradients. These gradients have an important role in deciding the developmental fate of the nuclei that migrates into specific regions of the embryo. Gradients of maternal molecules in the early embryo control axis formation. The first group of genes that are required for development of Drosophila are transcribed by maternal genome during oogenesis. The mRNA or those proteins are transported into the oocyte and are stored there. These proteins and RNA are used in the early phases of development and convert the egg into an embryo with anterior-posterior and dorsal-ventral axes. These are the developments influenced by the maternal genes.
2. The maternal genes through their products trigger a cascade of zygotic gene activations, which sets up the segmentation pattern in *Drosophila.*

- Homeotic genes direct the identity of body parts. As segmentation boundaries are being established by the action of segmentation genes, a group of selector genes is activated. The expression these genes determine the structures to be formed by each segment such as antennae, mouthparts, legs, thorax, wings etc. These genes are known as homeotic genes.
- Molecular biology studies of these developments have shown that each of the homeotic genes encodes a transcription factor having a DNA binding domain. This domain is encoded by a 180 bp sequence called home box. Home obox genes have been highly conserved in evolution.

The work in the development of Drosophila has inspired others to study the molecular basis of development and differentiation in other organisms also. In nematodes like *Coenorabiditis* was observed that neighbouring cells instruct other cells to form particular structures or to follow a particular pattern of development. This is an example for developmental control by cell signalling and induction. Similarly in plant development process depends on cell signalling and transcriptional regulation. A lot of research studies of *Arabidopsis* has contributed to the understanding of organ identity in plants.

5.9 HOUSE KEEPING GENES

It was observed that the level of gene expression qualitatively and quantitatively is highly variable from cell to cell and also from tissue to tissues. Some genes are tightly regulated at different levels of its expression depending upon the stages of its development, age and environmental factors. Expression of these genes often regulated at the level of transcription. Some of these genes are induced at certain specific conditions, by altered environmental factors temperature or certain chemicals or by certain biological factors like insect or pathogen attack. Such factors or chemical responsible for the induction or expression of certain genes or gene groups are known as inducers.

But there are some genes, which express irrespective of the factors or inducers. Such genes code for proteins those are essential for maintaining the normal metabolic activities like respiration and energy transactions, protein synthesis and metabolism, essential cellular structural components like cell membranes and cell wall. These genes are expressed constitutively at relatively low levels in all cells of the body system. These genes that code for the basic metabolic enzymes and cellular components that are essential for maintaining the life activities of a living system is known as the **housekeeping genes**.

Genes which code for the formation of enzymes of glucose metabolism, enzymes of DNA replication, enzymes of metabolic pathways that leads to the production of amino acids, nucleotides, ATPase, etc. are examples of housekeeping genes.

5.10 MUTAGENESIS

The genetic material, DNA has got high fidelity and so the genetic information is transferred from generation to generation without changing its integrity. But changes may occur in the quality of the genetic material which transferable to the next generation. Such heritable changes occurring in the DNA is known as mutation. Mutation may be due to a chemical change in the DNA or by recombination. Mutations form an important cause of diversity for evolution to act upon. In a population high genetic variability of the forces of Natural selection can select for advantageous genetic changes and result in speciation. The mutations that are happening in the in germ line cells responsible for the production of gametes is passed onto the next generation. Generally mutations are random and recessive in nature. It can be both harmful and beneficial to the organisms, particularly very useful for the genetic improvement of crops and microbial strains. Some mutations cause genetic disorders and even results in the development of cancers. It was Hugo de Vries who proposed the theory of mutation in 1901 based on his observations on evening primrose. In 1928 H.J. Muller used X-rays for inducing mutations in fruit fly, for the first time.

Types of Mutations

Mutations may be spontaneous or induced. Spontaneous mutations are natural, which occur naturally by changes in the DNA sequences during replication or by recombination of genetic materials or by some environmental factors—chemical or physical factors. The following are the major types of mutations observed in living system.

1. Spontaneous mutations and Induced mutations. They can be subdivided into mutations occurring during replication or post replication periods. They are caused by errors occurring during DNA replication by incorporating the incorrect nucleotide in the growing DNA chain; this however is a very infrequent event because of the very effective DNA repair systems, which evolved in all organisms. During the post replication periods the mutations are mainly caused by spontaneous physicochemical alterations of a base. There are two possibilities how this can be brought about:

- By isomerization of bases such as adenine (amino into imino form) or thymine (keto into enol form). In both cases a AT/GC transition will occur. This event is not recognized by the cellular repair functions and can therefore escape their activity.

- By spontaneous deamination of bases such as cytosine to uracil or 5-methylcytosine to thymine (note: 5% of cytosine in bacteria and bacteriophages are methylated). Although cells have evolved a repair mechanism to remove uracil from DNA it will of course not remove thymine because it is a normal component of native DNA. Therefore, a GC pair is converted into an AT pair.

  Induced mutations those happen in the genome as a whole in the DNA as described above but induced by some environmental factors called mutagens. These can be physical mutagens like UV light, X-rays, gamma rays etc. or chemical mutagens like alkylating agents like nitrosoguanidine and ethylmethane sulphonate or acridine compounds like acriflavine.

  Genetic recombination also can cause large-scale genetic changes, which includes exchange of polynucleotide fragments between chromosomes or insertion or transpositions of fragments of DNA from one position to another position in a chromosome.

2. Somatic and Germline mutations. Eukaryotic organisms have two primary cell types—germ and somatic. Mutations can occur in either cell type. If a gene is altered in a germ cell, the mutation is termed a germinal mutation. Because germ cells give rise to gametes, some gametes will carry the mutation and it will be passed onto the next generation when the individual successfully mates. Typically germinal mutations are not expressed in the individual containing the mutation. The only instance in which it would be expressed is if it negatively (or positively) affected gamete production.

Somatic cells give rise to all non-germline tissues. Mutations in somatic cells are called somatic mutations. Because they do not occur in cells that give rise to gametes, the mutation is not passed along to the next generation by sexual means. To maintain this mutation, the individual containing the mutation must be cloned. Two examples of somatic clones are navel oranges and red delicious apples. Horticulturists first observed the mutants. They then grafted mutant branches onto the stocks of 'normal' trees. After the graft was established, cuttings from that original graft were grafted onto tree stocks. In this way the mutation was maintained and proliferated. Most tissues are derived from a cell or a few progenitor cells. If a mutation occurs in one of the progenitor cells, all of its daughter cells will also express the mutation. For this reason, somatic mutations generally appear as a sector on the mutated individual.

Cancer tumors are a unique class of somatic mutations. The tumor arises when a gene involved in cell division, a protooncogene, is mutated. All of the daughter cells contain this mutation. The phenotype of all cells containing the mutation is uncontrolled cell division. This results in a tumor that is a collection of undifferentiated cells called tumor cells.

Mutations Based on Functions

1. Morphological mutants or Phenotypic mutations. This type of mutation generates a visible phenotypic or morphological alteration. Plant height mutations could change a tall plant to a short one, or from having smooth to round seeds.

The molecular explanations of these mutations are threefold: **Frameshift mutation** (addition or deletion of 1 or 2 nucleotides); ethidium bromide or other intercalating agents are effective in producing this type of mutation. **Chain-termination mutations** (nonsense mutations) by mutation of an amino acid encoding nucleotide triplet into a stop-codon leading to a premature dissociation of the ribosomal translation machinery from the mRNA.

Missence mutation leading to the substitution of one nucleotide by another one and resulting in the replacement of one often crucial amino acid in the polypeptide by another one with different physicochemical properties. An example of chemical agent causing missense mutation is nitrous acid.

2. Biochemical mutations. These types of mutations produce one or more a lesions in one specific step of an enzymatic pathway. This is because the specific enzyme, which carries out the reaction, has changed. For bacteria, biochemical mutants need to be grown on a media supplemented with a specific nutrient. Such mutants are called **auxotrophs**. Often though, morphological mutants are the direct result of a mutation in a biochemical pathway. In humans, albinism is the result of a mutation in the pathway from converts the amino acid tyrosine to the skin pigment melanin. Similarly, cretinism results when the tyrosine to thyroxin pathway is mutated. Therefore, in a strict genetic sense, if appropriate experiments are performed, a morphological mutation can be explained at the biochemical level.

3. Conditional mutations. The expression of these types of mutations is depended on the individual's specific environmental conditions. If these mutated genes expression is restricted to a particular condition and not in any other conditions, it is called the **restrictive condition**. But if the individual grows in any other environment **(permissive condition)**, the wild type phenotype is expressed. These are called conditional mutations. Mutations that only expressed at a specific temperature (temperature sensitive mutants), usually elevated, can be considered to be **conditional mutations**.

4. Lethal mutations. As the term implies, the mutations lead to the death of the individual. Death does not have to occur immediately, it may take several months or even years. But if the expected longevity of an individual is significantly reduced, the mutation is considered a lethal mutation.

5. Loss-of-function mutations. Wild type alleles typically encode a product necessary for a specific biological function. If a mutation occurs in that allele, the function for which it encodes is also lost. The general term for these mutations is **loss-of-function mutations**. The degree to which the function is lost can vary. If the function is entirely lost, the mutation is called a **null mutation**. It is also possible that some function may remain, but not at the level of the wild type allele. These are called **leaky mutations**. 'Loss of function mutations' are typically recessive. When a heterozygote consists of the wild-type allele and the loss-of-function allele, the level of expression of the wild type allele is often sufficient to produce the wild type phenotype. Genetically this would define the loss-of-function mutation as recessive. Alternatively, the wild type allele may not compensate for the loss-of-function allele. In those cases, the phenotype of the heterozygote will be equal to that of the loss-of-function mutant, and the mutant allele will act as a dominant.

6. Gain-of-function mutations. Although it would be expected that most mutations would lead to a loss of function, it is possible that a new and important function could result from the mutation. In these cases, the mutation creates a new allele that is associated with a new function. Any heterozygote containing the new allele along with the original wild type allele will express the new allele. Genetically this will define the mutation as a dominant. This class of mutations is called **'gain-of-function mutations'**.

7. Resistant mutations. These mutations are mainly seen in microbial populations. Certain microbes of the population become resistant to certain antibiotics like ampicillin and streptomycin resistance.

8. Regulatory mutations. These mutations occur in the regulatory regions of a gene and they loose its ability to control the expression of gene or genes.

Genome, Chromosome and Gene Mutations

Genome and chromosome mutation include the change in the number, rearrangements in the chromosomes or change in a single gene. The first two involves changes in the entire genome or in the whole chromosome. But the third one occurs in individual gene. Both chromosomal mutations and gene mutations can occur spontaneously or can be induced by physical or chemical mutagens.

Non-Disjunction and Down's Syndrome. One well-known example of mutation is non-disjunction. Non-disjunction is when the spindle fibers fail to separate during meiosis, resulting in gametes with one extra chromosome and other gametes lacking a chromosome.

If this non-disjunction occurs in chromosome 21 (Figure 5.15) of a human egg cell, a condition called ***Down's syndrome*** occurs. This is because the mutant possesses 47 chromosomes as opposed to the normal chromosome compliment in humans of 46.

The Genome Mutation or Polyploidy

Polyploidy is a change in the quantity of total genome or this type of mutation affects chromosome content of an organism. Humans are diploid creatures, meaning for every chromosome in our body, there is another one to match it. If an organism possesses multiples of haploid number of chromosomes, it is called euploid. Human and other eukaryotes are diploid (2n). The following are the various types of euploid.

- Haploid creatures have one of each chromosome.
- Diploid creatures have two of each chromosome.
- Triploid creatures have three of each chromosome.
- Polyploid creatures have three or more of each chromosome.

They can be represented by n where n equals haploid, 2n equals diploid and so on. It is possible for a species, particularly plant species, to produce offspring that contains more chromosomes than its parent. This can be a result of non-disjunction, where normally a diploid parent would produce diploid offspring, but in the case of non-disjunction in one of the parents, produces a polyploid. In the case of triploid, although the creation of particular triploids in species is possible, they cannot reproduce themselves because of the inability to pair homologous chromosomes at meiosis, therefore preventing the formation of gametes. Polyploids have important application in plant breeding and agriculture and they are responsible for the creation of thousands of species in today's planet, and will continue to do so. It is also responsible for increasing genetic diversity and producing species showing an increase in size, vigour and an increased resistance to disease. **Autopolyploids** arise due to doubling of 2n genome (diploids) to a 4n genome (autotetraploids). These are also very important in crop breeding as it increase the size of plant parts like seeds, fruits, leaves etc. The haploids can be converted into homozygous diploids by a process called **diploidization** by using chemical mutagenic agents like colchicine. These homozygous diploids can be used as parents in hybridization.

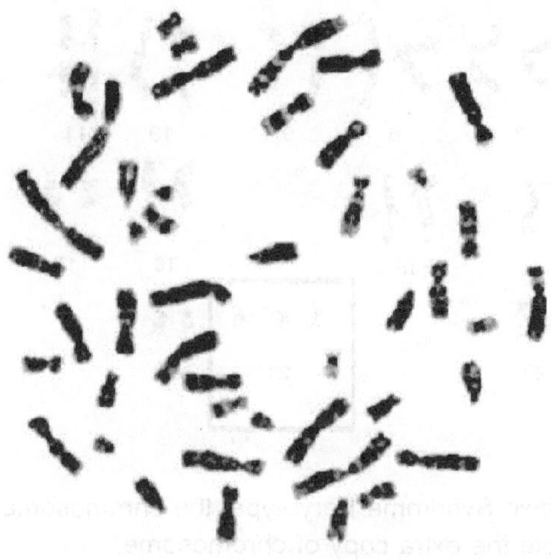

Figure 5.15 Normal Karyotype of human.

There are some types of chromosome multiplication, which does not involve the complete set. They are known as **aneuploids**. In the case of aneuploids the chromosome number is abnormal and it differs from the wild type by only addition or deletion of one or a small number of chromosomes. The following are some of the common types of aneuploids.

- Monosomic —One less than diploid number (2n – 1)
- Trisomic —One more than diploid number (2n + 1)
- Nullisomic —Two less than diploid number (2n – 2)

Unequal separation chromosomes due to non-disjunction during mitosis and meiosis or the chromosome duplication are the reasons for the formation of aneuploidy. Aneuploids are important for understanding the functions of individual chromosomes and also important in engineering specific crop genotypes. In human and animals it can cause genetic disorders. For example, Down Syndrome results when three, rather than the normal two, copies of chromosome 21 are present in each cell. An abnormal number of chromosomes result when a mistake called **nondisjunction** occurs in the production of eggs or sperm. (Figure 5.16).

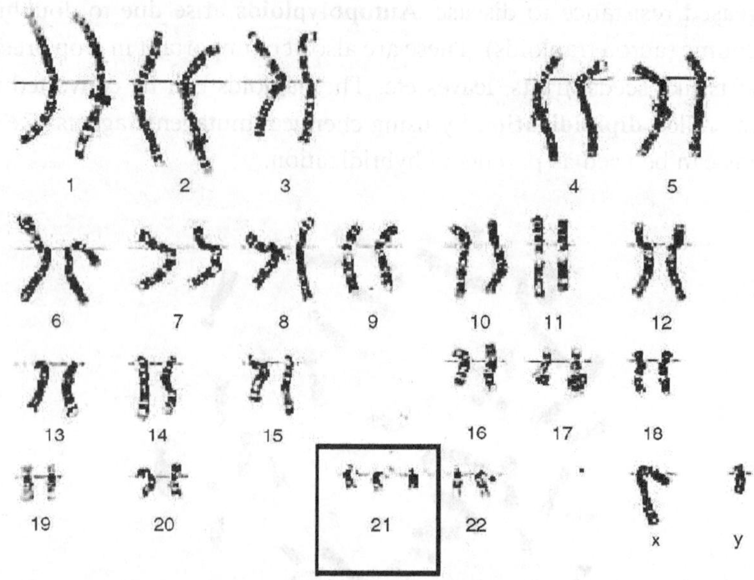

Figure 5.16 In the Down Syndrome Karyotype, the chromosomes have been arranged in their pairs. Note the extra copy of chromosome.

Chromosome Mutations

The fundamental structure of a chromosome is subject to mutation, which will most likely occur during crossing over at meiosis. There are a number of ways in which the chromosome structure can change, as indicated below, which will detrimentally change the genotype and phenotype of the organism. However, if the chromosome mutation effects an essential part of DNA, it is possible that the mutation will abort the offspring before it has the chance of being born. The following indicates types of chromosome mutation where whole genes are moved.

Deletion of a Gene. As the name implies, genes of a chromosome are permanently lost as they become unattached to the centromere and are lost forever.

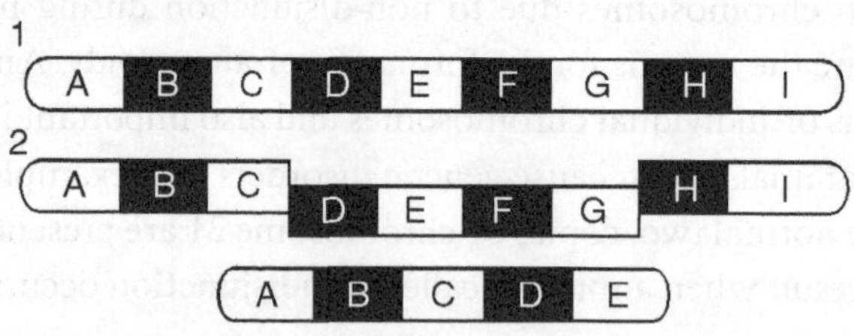

Figure 5.17 Deletions

1. Normal chromosome before mutation.
2. Genes not attached to centromere become loose and lost forever.
3. New chromosome lacks certain genes which may prove fatal depending on how important these genes are.

Duplication of Genes. In this mutation, the mutant genes are displayed twice on the same chromosome due to duplication of these genes. This can prove to be an advantageous mutation as no genetic information is lost or altered and new genes are gained.

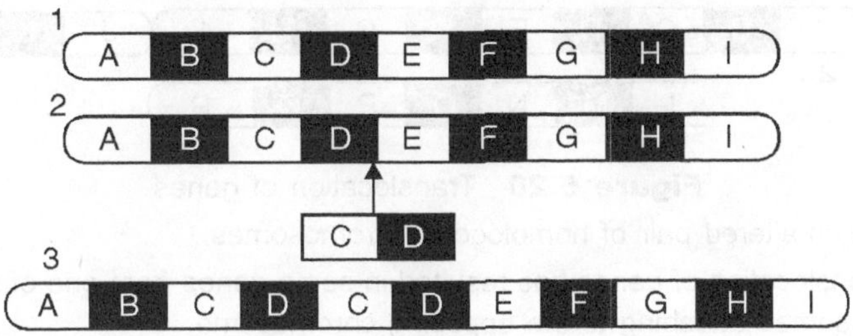

Figure 5.18 Duplication of genes

1. Normal chromosome before mutation.
2. Genes from the homologous chromosome are copied and inserted into the genetic sequence.
3. New chromosome possesses all its initial genes plus a duplicated one, which is usually harmless.

Inversion of Genes. This is where the order of a particular order of genes are reversed as seen below.

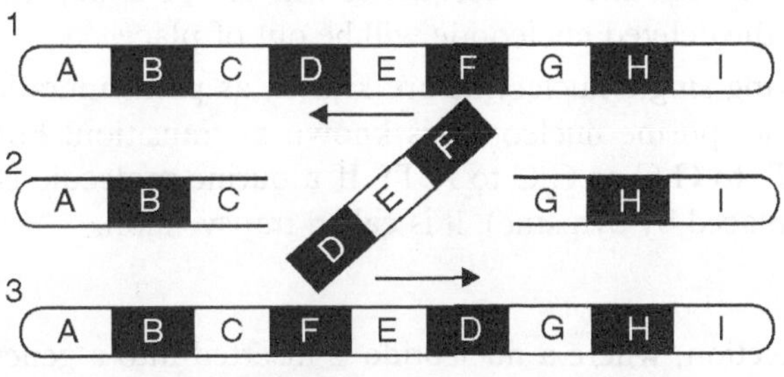

Figure 5.19 Inversion of genes

1. Normal chromosome un-altered.
2. The connection between genes break and the sequence of these genes are reversed.
3. The new sequence may not be viable to produce an organism, depending on which genes are reversed. Advantageous characteristics from this mutation are also possible.

Translocation of Genes. This is where information from one of two homologous chromosomes breaks and binds to the other. Usually this sort of mutation is lethal.

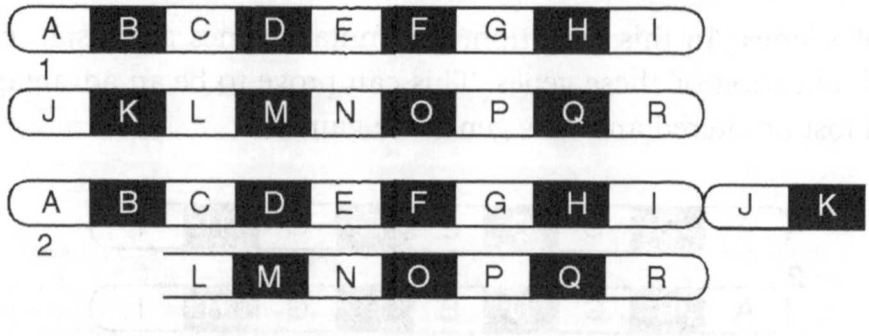

Figure 5.20 Translocation of genes

1. An un-altered pair of homologous chromosomes.
2. Translocation of genes has resulted in some genes from one of the chromosomes attaching to the opposing chromosome.

Gene Mutation (Alteration of a DNA Sequence)

The previous examples of mutation have investigated changes at the chromosome level. The sequences of nucleotides on a DNA strand are also susceptible to mutation.

Deletion

Here, certain nucleotides are deleted, which affects the coding of proteins that use this DNA sequence. If for example, a gene coded for alanine, with a genetic sequence of C-G-G, and the cytosine nucleotide was deleted, then the alanine amino acid would not be able to be created, and any other amino acids that are supposed to be coded from this DNA sequence will also be unable to be produced because each successive nucleotide after the deleted nucleotide will be out of place.

The changes involving single nucleotide are known as point mutation. A replacement of one purine nucleotide by another purine nucleotide is known as transition. For example, replacement of adenine by guanidine (A.T to G.C or G.C to A.T). If a purine molecule is replaced by a pyrimidine molecule (Guanidine is replaced by cytosine), it is called transversion.

Insertion

Similar to the effects of deletion, where a nucleotide is inserted into a genetic sequence and therefore alters the chain thereafter. This alteration of a nucleotide sequence is known as **frameshift** mutation. The effect of frameshift mutation is the change of reading frame of codons downstream from the mutation. This results in a completely changed gene product (protein) and finally the phenotype or physiology.

Inversion

Where a particular nucleotide sequence is reversed, and is not as serious as the above mutations. This is because the nucleotides that have been reversed in order only affect a small portion of the sequence at large.

Substitution

A certain nucleotide is replaced with another, which will affect any amino acid to be synthesised from this sequence due to this change. If the gene is essential, *i.e.* for the coding of haemoglobin then the effects are serious, and organisms in this instance suffer from a condition called sickle cell anemia.

All of the genetic mutations looked at through the last 2 pages more or less have a negative impact and are undesired, however, in some cases they can prove advantageous.

Genetic mutations increase genetic diversity and therefore have an important part to play. They are also the reason many people inherit diseases.

Silent mutations are when a point mutation occurs that changed the codon to another codon for the same amino acid so no change results in the gene product (protein).

Nonsense mutations change a codon to a stop leading to premature chain termination and a piece of random peptide or a truncated protein will be translated.

Molecular Mechanism of Mutations

Causes of Mutations

Spontaneous DNA mutations arise from a variety of sources including the errors during DNA replication. It can be caused by radiation (either UV or ionizing), alkylation, deamination, introduction of base analogues and tautomeric shifts. In **tautomeric shifts** the normal keto form of each base changes to enol or amino from which results in transitions by mispairing. Mutation results in the tautomeric shift between the keto and enol forms of thymine since one form pairs with A and one with G. If it happens to be in the form that pairs with G, at the moment of replication then G will be transcribed in A's place making the mutation. **Depurination**—the removal of purine base by breaking glycosidic bond between sugar and the base to form a gap known as apurinic site. The apurinic site is filled with another base resulting in transition, transversion or without any change. Another mechanism of point mutation is by **deamination**. In this mechanism, cytosine is deaminated to uracil and if it is not repaired, it pairs with adenine in the next cycle of DNA replication.

Point mutations and small deletions

Wild-type sequences						
Amino acid	N-Phe	Arg	Trp	Ile	Ala	Asn-C
mRNA	5′-UUU	CGA	UGG	AUA	GCC	AAU-3′
DNA	3′-AAA	GCT	ACC	TAT	CGG	TTA 5′
	5′-TTT	CGA	TGG	ATA	GCC	AAT 3′
Missense						
	3′-AAT	GCT	ACC	TAT	CGG	TTA-5′
	5′-TTA	CGA	TGG	ATA	GCC	AAT-3′
	N-Leu	Arg	Trp	Ile	Ala	Asn-C
Nonsense						
	3′-AAA	GCT	ATC	TAT	CGG	TTA-5′
	5′-TTT	CGA	TAG	ATA	GCC	AAT-3′
	N-Phe	Arg	Stop			
Frameshift by addition						
	3′-AAA	GCT	ACC	ATA	TCG	GTT A-5′
	5′-TTT	CGA	TGG	TAT	AGC	CAA T-3′
	N-Phe	Arg	Trp	Tyr	Ser	Gln
Frameshift by deletion						
		GCTA				
		CGAT				
	3′-AAA	CCT	ATC	GGT	TA-5′	
	5′-TTT	GGA	TAG	CCA	AT-3′	
	N-Phe	Gly	Stop			

Figure 5.21 Various types of gene mutations.

Radiations are two types—ionizing radiations and non ionizing radiations. Ionizing radiations like X-rays, gamma rays and thermal neutrons cause mutations by ionization of DNA molecule. Non-ionizing radiations like UV light causes dimerization of adjacent bases in the DNA strand, especially the thymine dimers known as photodimers.

Chemical mutagens include alkylating agents, base analogues, intercalating agents etc. can cause chemical modification of the nitrogen bases leading to the mutations. Alkylating agents add alkyl group—methyl or ethyl group to the appropriate nitrogen bases (mainly guanidine). The alkylated base undergoes mispairing results in mutation during DNA replication, if not corrected.

A base analog is a compound sufficiently similar to one of the four DNA bases but have different pairing properties. For example, 5-bromouracil is the analog of thymine but sometimes pairs with guanine and 2-aminopurine is the analog of adenine but sometimes pairs with cytosine. The incorporation of a base analog will result in mispairing and ultimately result in complete change of base pair during DNA replication.

Certain DNA binding chemicals like ethidium bromide can intercalate and slip into the space between the sequence of nitrogen bases and increase the distance between them. This can lead to the additions and deletions in the DNA strand. Most of the chemical mutagens have mutational specificity, as each mutation is due to a specific chemical reaction with DNA.

Nitrous acid is another chemical that affects DNA complementation. The acid randomly modifies the base adenine so that it will pair with cytosine instead of thymine. This change is made evident during DNA replication when a new base pair appears in daughter cells in a later generation.

Even though mutations are randomly happening in large numbers, most of them can be reversed before it get stabilized and transferred to the next generation. Mutations can also be reversed spontaneously or can be induced by mutagens. For example, if an alkylating agent like EMS is creating a transition mutation like GC to AT, it can be reversed by 5-bromouracil as it causes both type of transitions—GC to AT and AT to GC. Because of this same reason, mutations caused by base analogues 5-bromouracil can be reversed by the same mutagens.

There are a number of methods developed for the screening of suspected mutagenic compounds. It mainly includes cytological and cytochemical techniques, which are time consuming. In 1970, Bruce Ames developed a very simple but effective assay technique to check the genotoxicity. This test is known as Ames test. The test consists of two auxotrophic strains of *Salmonella typhimurium*. Ames devised a method to simulate the human metabolism in the bacterial cultures by adding the rat liver enzymes to the bacterial suspension containing the suspected compound. The liver enzymes are involved in the detoxification of harmful compounds. The mutageneity of the unknown compound can be determined by reversion of histidine mutations. The reversion of histidine mutation can be detected by the growth and formation of colonies on the nutrient medium lacking histidine. Bacterial growth cannot happen unless histidine mutation is reversed.

5.11 DNA REPAIR

As a major defense against environmental damage to cells DNA repair is present in all organisms examined including bacteria, yeast, Drosophila, fish, amphibians, rodents and humans. DNA repair is involved in processes that minimize cell killing, mutations, replication errors, persistence of DNA damage and genomic instability. Abnormalities in these processes have been implicated in cancer and aging (Figure 5.22). The genetic material has to be guarded against the various types of damage to keep the high fidelity. There are various types of enzymatic machineries to check and repair the possible errors in the total DNA molecules at different stages of cell cycle. The error checking and repair occurs just before the starting the DNA replication, it operates during the replication and checking continues even after the replication. The DNA repair enzyme can recognize the damaged sequences and it repair the damage very promptly. There are some mechanisms, which even neutralize the potentially damaging compounds before they even react with the DNA. Example for one such system is the enzymes that detoxify the superoxide radicals, which can do oxidative damage to DNA.

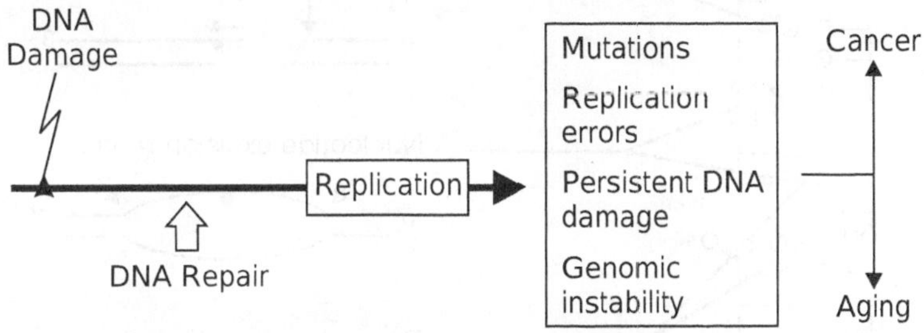

Figure 5.22 The functions of DNA repair.

The repair mechanisms for DNA damage can be broadly classified into two: Light repair system and Dark repair system. The repair mechanism that operates in the presence of light is the light repair mechanism and light is not a factor for the dark repair mechanism to operate. It can function at all times.

The **light repair** system is called the photoreactivation mechanism. It can remove the photodimers or the thymine dimmers induced by the UV radiations of the sunlight. The photoreactivating enzyme known as photolyase can bind to the thymine dimmers and split it in the presence of light and folic acid, which is the coenzyme of the photolyase. Folic acid takes the light energy and helps the enzyme to split the photodimers. Photodimers can also be removed by dark repair systems.

The dark repair system consists of enzymes that are able to recognize damaged DNA strands and correct it. These repair mechanisms exist to correct most of the chemical changes that come to the bases. We will discuss five types of dark repair :

1. *Base Excision Repair* (BER) is the removal of the specific damaged bases. The enzymes, DNA glycosylases scan the DNA for to recognize the wrong bases and repair them. For example, uracil if it is present in DNA, it can be efficiently removed by uracil—DNA glycosylase. Now the AP endonucleases can recognize the apurinated or apyrimidinated (AP) sites and repair them.
2. *Nucleotide Excision Repair* (NER) is the removal of an entire nucleotide or a fragment of the single DNA strand along with the modified base. This is used over base excision repair when the mutation results in change of more than just the base but a chemical change to the sugar or phosphate as well. This is a more general form of DNA repair. It can recognize several kinds of DNA damages including the photodimers. In the first step, an enzyme recognize the damaged

nucleotides and makes single strand nicks few base pairs away on either side of the damaged part. In the second step, the segment of DNA containing the damaged nucleotide is removed by leaving a gap. A DNA polymerase, using the opposite strand as the template fills this gap and finally the DNA ligase joints the nick. DNA polymerase III is the main enzyme involved in the cutting removal and synthesis of the damaged DNA strand.

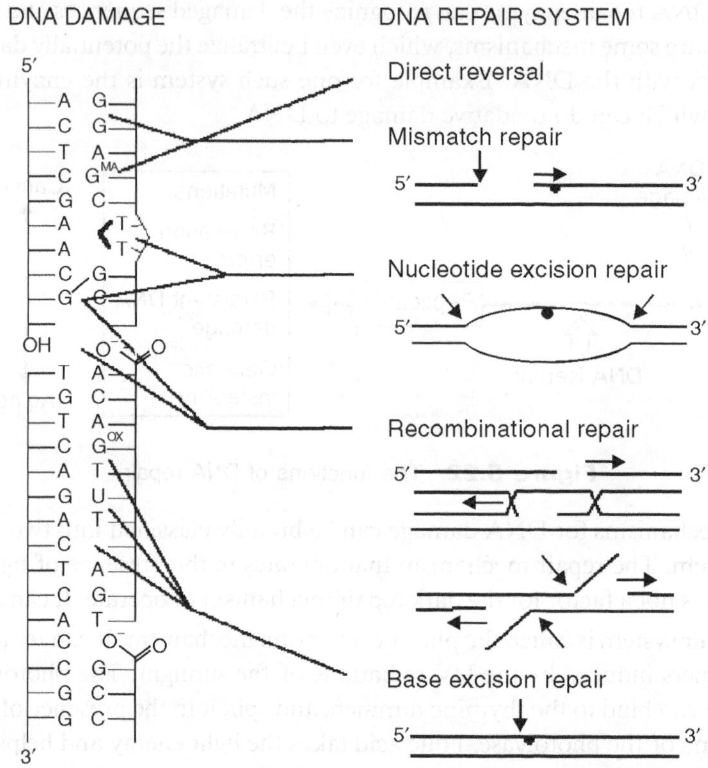

Figure 5.23 Various types of DNA damage and the DNA repair mechanisms.

1. *Mismatch Repair* (MMR). Post replication mismatch repair is the normal checking and removal of mismatched nitrogen bases or nucleotides by DNA polymerase during DNA replication or just after that.
2. *Recombination Repair* (RR). When the DNA damage is very extensive, which include a large part of gene or DNA, and then it has to be corrected by recombination method. If a major part of the gene is damaged or lost its correct sequence in both the strands, it cannot be repaired by normal mechanisms. Because damage has occurred in both strands of the DNA and template is not available for correcting the error. Then the damage correcting system will check the entire genome to see the similar sequences so that, it can be copied and exchanged with the damaged part of the DNA. This mechanism damage correction is called the retrieval system.
3. **SOS Repair.** When there is very extensive DNA damage which inactivates the DNA and the repair mechanisms are not able to repair it, then the DNA cannot replicate and finally it leads

to the death of the cell. The DNA cannot replicate because the modified bases formed by mutation like that of thymine dimmers will block the functioning of the DNA polymerases. Under such situation SOS induction occurs. This is the last resort to minimize the degree of mutations and allow for its survival. Therefore, this repair pathway is called error prone repair. SOS modifies the DNA replication complex also, so that the DNA replication proceeds even though the DNA contains some error. Since it is the last resort for the survival of the cell, it is called the SOS (Save Our Soul) mechanism.

Xeroderma Pigmentosum. It happens when there is a failure of the DNA polymerases to carry out their repair functions to remove the photodimers formed due to excessive exposure of skin in sunlight. The patient will have all forms of skin pigmentation problems. Because of the lack of ability to repair mutations they will usually develop skin cancer in either the first or second decades even if they are very careful to stay out of the sun.

5.12 GENETIC DISORDERS

In all organisms including humans, genes play an important role in the development of a disease. What ever may be the type of disease ultimately it ends in the expression of or lack of a gene. But the term genetic disease or genetic disorder indicates the mutation of one or more genes are primarily responsible for that particular disorder or diseased state. Genetic diseases like hemophilia, sickle cell anemia and cystic fibrosis etc. are caused by mutations. These changes are inherited and passed on from affected parents to their children. All these disease-causing mutations are recessive in nature and happen for single gene. But translocation also creates some types of genetic disorder including cancer. Some are due to aneuploidy (loss or gains of one or more chromosomes). Klinefelters syndrome is a genetic disorder in males due to the Trisomic condition, XXY and shows feminine characters. Turner's syndrome affects the females and is due to the presence of a Monosomic condition, XO and shows male features. Down syndrome affects both males and females and is due to a trisomy of the 21st chromosome and shows mental retardation and short stature.

In addition to these any damage or mutation on the genes responsible for DNA repair will cause a variety of genetic disorders and cancer susceptibilities. Xeroderma pigmentosum is one such disorder. Xeroderma pigmentosum develops when there is a failure of the DNA polymerases to carry out their repair functions. The patient will have all forms of skin pigmentation problems. Because of the lack of ability to repair mutations they will usually develop skin cancer in either the first or second decades even if they are very careful to stay out of the sun. A recessive mutation results in the absence or reduced activity of excision repair mechanism and therefore they are susceptible to UV light.

Even though the genetic disorders are due to gene mutations, all them may not be hereditary disease. If a genetic disorder is transmitted to the next generation then it is a hereditary disease. Mutations related to the development of cancers occur in somatic cells and are often associated with genes that control the cell cycle. Cancerous cells show uncontrolled cell growth resulting in tumor formation.

Sickle Cell Anemia (SCA) is characterized by episodes of pain, chronic hemolytic anemia and severe infections, usually beginning in early childhood. SCA is an autosomal recessive disease caused by a point mutation in the *hemoglobin beta* gene (*HBB*) found on chromosome 11. Carrier frequency of

HBB varies significantly around the world, with high rates associated with zones of high malaria incidence, since carriers are somewhat protected against malaria. A mutation in *HBB* results in the production of structurally abnormal hemoglobin (Hb), called HbS. Hb is an oxygen carrying protein that gives red blood cells (RBC) their characteristic colour. Under certain conditions, like low oxygen levels or high hemoglobin concentrations, in individuals who are homozygous for HbS, the abnormal HbS crystallize and clusters together, distorting the RBCs into sickled shapes. These deformed and rigid RBCs become trapped within small blood vessels and block them, producing pain and eventually damaging organs. Though, as yet, there is no cure for SCA, a combination of fluids, painkillers, antibiotics and transfusions are used to treat symptoms and complications. Hydroxyurea, an antitumor drug, has been shown to be effective in preventing painful crises. Hydroxyurea induces the formation of fetal Hb (HbF)—a Hb normally found in the foetus or newborn—which, when present in individuals with SCA, prevents sickling. A mouse model of SCA has been developed and is being used to evaluate the effectiveness of potential new therapies for SCA.

Figure 5.24 Normal and sickled RBC.

Thalassemia is another genetical disorder related to hamoglobin. This diseased condition occurs when one or more copies of one of the Hb genes have been recombined to give an unrecognizable sequence. The lack of the gene will lead to insufficient production of Hb and obvious problems transporting O_2 in the blood stream.

5.13 TRANSPOSONS

Genes are supposed to be static having a fixed position in the chromosomes. They will not change its position unless some chromosomes mutation occurs. But in 1940 Barbara McClintock demonstrated the presence of moving genes or jumping genes in maize. These genes move from one location to another position within the genome and are named as transposable elements or jumping gene. The transposons they isolated from maize were named as Ac and Ds elements. Later such mobile genes or transposable elements (transposons) discovered from a large number of organisms including microbes. These moving fragments of DNA are responsible for the variety of genetic instabilities including chromosome breakages and other mutations. Transposons can be used as marker genes in genetic mapping and genetic transformation experiments.

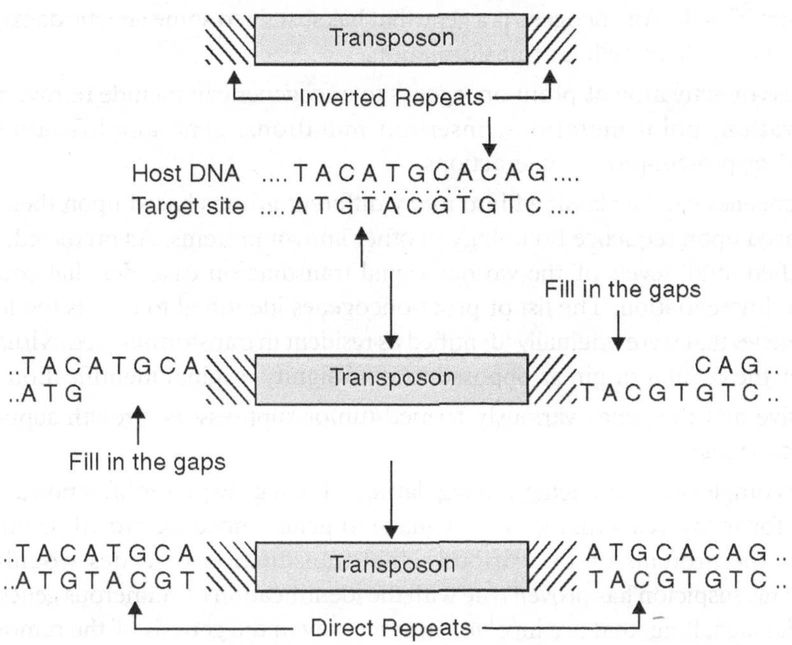

Figure 5.25 Cutting and pasting of transposable elements.

A transposable DNA fragment has unique nucleotide sequences at each of its ends, which enable its transfer to other locations. The transposable elements are always a part of chromosomes, plasmids or a viral DNA. Their ends are characterized by short sequences of inverse repeats or palindromic sequences. An enzyme known as transposase encoded by the same element or another element of the same cell carries out the process of transpositions. This enzyme makes use of the repeat DNA sequences at the ends to transfer all the DNA of the element to another location within the chromosome or to another chromosome (Figure 5.25). Many transposable elements while its transfer leave a copy behind by replication. The simplest of the transposable elements have an insertional sequence with a single transposase gene. Certain other types of transposable elements known as T_n elements have antibiotic resistance genes, in addition to transposase gene and repeat sequences. The transposable elements present in plants are found to be more complex than that present in microbes.

5.14 GENETICS OF CANCER

Most, if not all, cancer cells contain genetic damage that appears to be the responsible event leading to tumorigenesis. The genetic damage present in a parental tumorigenic cell is maintained (*i.e.*, not correctable) such that it is a heritable trait of all cells of subsequent generations. Genetic damage found in cancer cells is of two types :

1. Dominant and the genes have been termed **proto-oncogenes**. The distinction between the terms proto-oncogene and oncogene relates to the activity of the protein product of the gene. A proto-oncogene is a gene whose protein product has the capacity to induce cellular transformation given it sustains some genetic insult. An oncogene is a gene that has sustained some genetic damage and, therefore, produces a protein capable of cellular transformation.

The process of activation of proto-oncogenes to oncogenes can include retroviral transduction or retroviral integration, point mutations, insertion mutations, gene amplification, chromosomal translocation and/or protein-protein interactions.

Proto-oncogenes can be classified into many different groups based upon their normal function within cells or based upon sequence homology to other known proteins. As predicted, proto-oncogenes have been identified at all levels of the various signal transduction cascades that control cell growth, proliferation and differentiation. The list of proto-oncogenes identified to date is too lengthy to include here. Proto-oncogenes that were originally identified as resident in transforming retroviruses are designated as **c-indicative** of the cellular origin as opposed to **v-** to signify original identification in retroviruses.

2. Recessive and the genes variously termed tumor suppressors, growth suppressors, recessive oncogenes or anti-oncogenes.

Given the complexity of inducing and regulating cellular growth, proliferation and differentiation, it was suspected for many years that genetic damage to genes encoding growth factors, growth factor receptors and/or the proteins of the various signal transduction cascades would lead to cellular transformation. This suspicion has proven true with the identification of numerous genes, whose products function in cellular signalling, that are involved in some way in the genesis of the tumorigenic state. The majority of these proto-oncogenes were identified by either of two means: as the transforming genes (oncogenes) of transforming retroviruses or through transfection of DNA from tumor cell lines into non-transformed cell lines and screening for resultant tumorigenesis.

Viruses and Cancer

Tumor cells also can arise by non-genetic means through the actions of specific tumor viruses. Tumor viruses are of two distinct types. There are viruses with DNA genomes (*e.g.* papilloma and adenoviruses) and those with RNA genomes (termed **retroviruses**).

RNA tumor viruses are common in chickens, mice and cats but rare in humans. The only currently known human retroviruses are the **human T-cell leukemia viruses (HTLVs)** and the related retrovirus, **human immunodeficiency virus (HIV)**. Retroviruses can induce the transformed state within the cells they infect by two mechanisms. Both of these mechanisms are related to the life cycle of these viruses. When a retrovirus infects a cell its RNA genome is converted into DNA by the viral encoded RNA-dependent DNA polymerase (**reverse transcriptase**). The DNA then integrates into the genome of the host cell where it can remain being copied as the host genome is duplicated during the process of cellular division. Contained within the sequences at the ends of the retroviral genome are powerful transcriptional promoter sequences termed **long terminal repeats (LTRs)**. The LTRs promote the transcription of the viral DNA leading to the production of new virus particles.

At some frequency the integration process leads to rearrangement of the viral genome and the consequent incorporation of a portion of the host genome into the viral genome. This process is termed transduction. Occasionally this transduction process leads to the virus acquiring a gene from the host that is normally involved in cellular growth control. Because of the alteration of the host gene during the transduction process as well as the gene being transcribed at a higher rate due to its association with the retroviral LTRs the transduced gene confers a growth advantage to the infected cell. The end result of this process is unrestricted cellular proliferation leading to tumorigenesis. The transduced genes are termed oncogenes. The normal cellular gene in its unmodified, non-transduced form is termed a proto-oncogene since it has the capacity to transform cells if altered in some way or expressed in an uncontrolled manner. Numerous oncogenes have been discovered in the genomes of transforming retroviruses.

The second mechanism by which retroviruses can transform cells relates to the powerful transcription promoting effect of the LTRs. When a retrovirus genome integrates into a host genome it does so randomly. At some frequency this integration process leads to the placement of the LTRs close to a gene that encodes a growth regulating protein. If the protein is expressed at an abnormally elevated level it can result in cellular transformation. This is termed retroviral integration induced transformation. It has recently been shown that HIV induces certain forms of cancers in infected individuals by this integration induced transformation process.

Cellular transformation by DNA tumor viruses, in most cases, has been shown to be the result of protein-protein interaction. Proteins encoded by the DNA tumor viruses, termed **tumor antigens** or T antigens, can interact with cellular proteins. This interaction effectively sequesters the cellular proteins away from their normal functional locations within the cell. The predominant types of proteins that are sequestered by viral T antigens have been shown to be of the tumor suppressor type. It is the loss of their normal suppressor functions that results in cellular transformation.

Tumor Suppressor Genes

Tumour suppressor genes were first identified by making cell hybrids between tumor and normal cells. On some occasions a chromosome from the normal cell reverted the transformed phenotype. Several familial cancers have been shown to be associated with the loss of function of a tumor suppressor gene. Some of the tumor suppressor gene identified from different organisms includes the **retinoblastoma susceptibility gene (RB)**, **Wilms' tumors (WT1)**, **neurofibromatosis type-1 (NF1)**, **familial adenomatosis polyposis coli (APC or FAP)**, and those identified through loss of heterozygosity such as in colorectal carcinomas (called **DCC** for **deleted in colon carcinoma**) and **p53** which was originally thought to be a proto-oncogene. However, the wild-type p53 protein suppresses the activity of mutant alleles of p53, which are the oncogenic forms of p53.

Translocation and Chromosomes

Chromosomal mutations are also seen to be associated with various forms of cancers. One of the best-studied examples is translocation between chromosomes 9 and chromosome 22. This aberration is associated with chronic myelogenous leukemia (CML). Originally this translocation was described as abnormal chromosome 21 and named as Philadelphia chromosome. Now it is clearly known that Philadelphia chromosome results from the exchange of genetic material between chromosome 9 and 22 and is never seen in normal chromosomes, but only in white blood cells having this leukemia.

Role of Environmental Agents in Cancer

There is an important role for the environmental factors to play in the incidence of cancers developments in addition to the personal behaviour. Environmental factors such as UV light and other types ionizing and non-ionizing radiations, mutagenic chemicals like asbestos, and personal behaviours like use of tobacco, certain diets, drugs etc. are also trigger the development of cancers. Certain types of viruses also play important role in cancer development.

5.15 IMMUNOGENETICS

The vertebrates particularly the mammals have a well-developed immune system. It consists of lymphoid organs, various types of cells such as B and T lymphocytes, macrophages, and NK cells. These immune cells interact with each other and also with other cells through the receptors present on their membrane surface. The development of antibodies or the humoral immune response is the outcome of these interactions. This is also known as cell mediated immune response or CMI. The main purpose of immunogenetics is to study the genetic basis of receptors present on the cells of the immune system in addition to genetic basis of autoimmune disorder. The autoimmune disorder or disease is characterized by the damage of the immune system by its own organs and cells by generating an immune response. This is a highly disabling state and no cure is available for most of them.

The important molecules involved in the process of immune response are immunoglobulins, T-cell receptors and major histocompatibility antigens. *Tge* genes for histocompatibility antigens are grouped together and are present in a chromosome and forms the Major Histocompatibility Complex or MHC. It has been identified in different organisms. In mouse it is located in the 6th chromosome and is named as H2 complex. In human it is located in the 17th chromosome and is known as Human Leukocyte Antigen or HLA complex. The genes of these complexes are very closely linked, both in the case of mouse and human. For every locus there are alleles in the homozygous chromosomes. The number of alleles may be many in most of the cases. The complete set of alleles for all the loci situated in H2 or HLA are haplotype and is inherited as a single unit. Thus, a child inherits a haplotype from father and another from mother.

MHC has an important role for the successful transplantation of organs like kidneys, heart, bone marrow etc. between two individuals. The MHC or the HLA of the donor and recipient should closely match each other for the successful tissue or organ transplantation. In the cases of mismatches, the transplanted organ is liable to be rejected by the immune system of the recipient. Therefore, organ transplantation is usually preferred between blood relatives or close relatives after ensuring adequate MHC matching.

Certain autoimmune disorders are seen to be associated with certain genes of MHC. The presence of a particular allele in the MHC is found to be more prone to certain autoimmune disorders. For example, the development of ankylosing spondylosis is linked to the presence of the allele HLA-B27. In this state of the vertebrate tend to fuse and this results the inflammation and discomfort. Some alleles of this complex (HLA) make the person more susceptible to diabetes. All these information clearly indicates the relevance of immunogenetics in the medical application.

5.16 EVOLUTIONARY GENETICS

In the previous chapters we discussed about the involvement of genetics and mutation in the process of development of new traits and its combinations. These form an important driving force in the process of evolution. According to the law of speciation proposed by Charles Darwin, new species of organisms are developed from pre-existing ones by the natural selection. For the force of natural selection to play there should be enough number of variations created by mutations on a particular loci or gene. Thus, mutations and the gene frequency of different alleles in the population is an important factor for the natural selection to drive the speciation. If a new mutation has a selective advantage over its allele in that particular environment its frequency will increase rapidly. And the other may be eliminated from the population. This may cause a complete change in the appearance and behaviour of the organism.

Sexual barriers also can create new species. Mutations can directly cause sexual isolation or create sexual barrier because of the reasons like incompatibility or sterility with certain genotypes. These type sexual barriers check the free mixing of new forms genes created by mutations in the population. Geographical isolation also can create a sexual barrier and prevent the mixing of the variation among the population and gradually evolve into a different species. The process of natural selection changes the gene and gene frequencies in the population. New variations are produced by rare genetic recombinations or various types of spontaneous mutations and these variations are a key factor in the process of speciation. If the variation is not genetic, that is not resulted out of gene change the selection will not have any consequences. Such variations may be epigenetic induced by the environmental factors, such as the acquired characters like muscle strength features obtained by regular exercises, and such other characters are not transmitted to the next generation and, therefore, does not have any impact on *evolution*.

• REVIEW QUESTIONS •

1. Describe the molecular basis sickle cell anemia.
2. What is meant by DNA repair mechanism ? What is the need of repair mechanism ?
3. What is meant photodimers? Describe two mechanisms by which it can be repaired.
4. If a point mutation occurs in the coding region of a gene what will be the nature of its protein.
5. What is meant by a frame shift mutation ?
6. Explain what is Xeroderma pigmentosum.
7. Give definitions for the following :

   (*a*) Mutation (*b*) CentiMorgan

   (*c*) Missense mutation (*d*) Aneuploids

   (*e*) Photolyase (*f*) Conditional mutations

   (*g*) Hardy—Weinberg law(*h*) Dominant and recessive allele

   (*i*) Incomplete dominance(*j*) Transposase.
8. What are the cytoplasmic genes ? How it affects the heredity of organisms ?
9. Why the inheritance of cytoplasmic genes is named as maternal inheritance ?
10. What is meant by a linkage group ?
11. Describe the various mechanisms of DNA repair.
12. Describe about the quantitative inheritance.
13. How can Mendel's law of inheritance be explained by the chromosomal movements during meiosis ?
14. What is meant by phenotypes does and genotypes ?
15. What is SOS response ? When it start functioning ?

16. Define the following terms :
    - (*a*) Autotetraploids
    - (*b*) Monohybrid cross and monohybrid ratio
    - (*c*) Lethal mutation
    - (*d*) Reverse mutation
    - (*e*) Auxotrophs
    - (*f*) Trisomics
    - (g) Haemophilia
    - (*h*) Genetic mapping
    - (*i*) Alkylating agents
    - (*j*) Petite yeast
    - (*k*) AP site
    - (*l*) Tautomeric shifts
    - (*m*) Base excision repair
    - (*n*) Down's syndrome
    - (o) Sex determination
    - (*p*) Sex linked inheritance.
17. Describe the inheritance of haemophilia.
18. What is the contribution of linkage and crossing over towards evolution ?
19. What is meant by spontaneous mutation and induced mutations ?
20. What are the various types of gene mutations ?
21. Explain the biochemical mutations.
22. Differentiate between euploids and aneuploids.
23. What are the various types of chromosome mutations ?
24. Explain the cytoplasmic inheritance in *Mirabilis jalapa.*
25. 5-bromouracil can reverse the mutation effect created by EMS. Explain.
26. Explain the photoreactivation mechanism of DNA repair.

• SUGGESTED READING •

1. Genes VII, Benjamin Lewin. Oxford University Press, New York. 2000.
2. The Metabolic and Molecular Basis of Inherited Disease, Scriver, C.R., Beaudet, A.L., Sly, W.S., and Valle, D. (Eds.) Seventh Edition., McGraw Hill, New York.
3. DNA Repair and Mutagenesis, Friedberg, E.C., Walker, G.C., and Siede, W., ASM Press, Washington, D.C. 1995, pp. 1-698.
4. Principles of Genetics, E. W. Sinnot, L. C. Dunn, T. Dobshansky. Tata McGraw-Hill Publishing Company, New Delhi.
5. Molecular Biology of the Gene, J. D. Watson, N. H. Hopkins, J. W. Roberts, J. A. Steitz and A. M. Weiner, The Benjamin/Cummings Publishing Company, Inc. Menlo Park, California.
6. Theory and Problems of Genetics, W. D. Stansfield, McGraw-Hill Book Company, New York.
7. Principles of Genetics, E. J. Gardener, M. J. Simmons and D. P. Snusted, John Wiley & Sons, Inc. New York.
8. Molecular Cell Biology, H. Lodish, A. Berk, S. L. Zipursky, P. Matsudaira, D. Baltimore and J. Darnell, W. H. Freeman and Company, New York.
9. Genetics: Analysis and Principles, R. J. Brooker, Addison Wesley Longman. Inc., New York.
10. Genetics: A Molecular Approach, T. A. Brown, Nelson Thrones Ltd., Cheltenham, London.

6 GENETIC ENGINEERING

6.1 INTRODUCTION

The field of modern biotechnology started when recombinant human insulin produced by bacteria was first marketed in the United States in 1982. The effort leading up to this landmark event began in the early 1970's when research scientists developed protocols to construct new types of bacterial plasmids or vectors, by cutting out and pasting pieces of DNA together to create a new piece of DNA (recombinant DNA) that could be inserted into the host bacterium like *Escherichia coli*. We have also observed that yeasts can be made to produce vaccines like Hepatitis B, plants having special properties like resistance to certain diseases, pests and herbicides and plants with superior nutritive qualities can be generated very efficiently. These excellent goals of genetic engineering were achieved because of the advent of recombinant DNA technology. Recombinant DNA technology is one of the few techniques that made the conventional biotechnology into the "Modern Biotechnology". Paul Berg, Herbert Boyer, Annie Change and Stanley Cohen are the team of scientist, who made the first recombinant DNA molecule in 1973.

Simply defined, it is the art of cutting and pasting genes. There are, however, many new applications of this technology invented each year, and it is impossible for any textbook to be completely up-to-date. This technique encompasses a number of methodologies or tools that enables to construct new combinations of DNA(recombinant DNA or rDNA) in the laboratory for different purposes. The rDNA molecule thus constructed can be introduced into an appropriate host cell, where they can be multiplied and generate many copies. This forms the basic concept of the process known as gene cloning or DNA cloning. This chapter we will examine the basic tools, methodologies and applications of recombinant DNA techniques in various fields of biological research.

The technique of gene cloning can generate unlimited copies of a DNA molecule (*e.g.*, recombinant DNA) by replication in a host cell and it was first developed in 1970. The following are some of the major applications of rDNA technique.

- Genetic mapping
- DNA sequencing
- Mutation studies
- Transformation
- Genetic engineering
- Recombinant DNA Libraries
- Restriction Enzyme Site Analysis
- Analysis of Gene Transcripts
- Polymerase Chain Reaction (PCR)

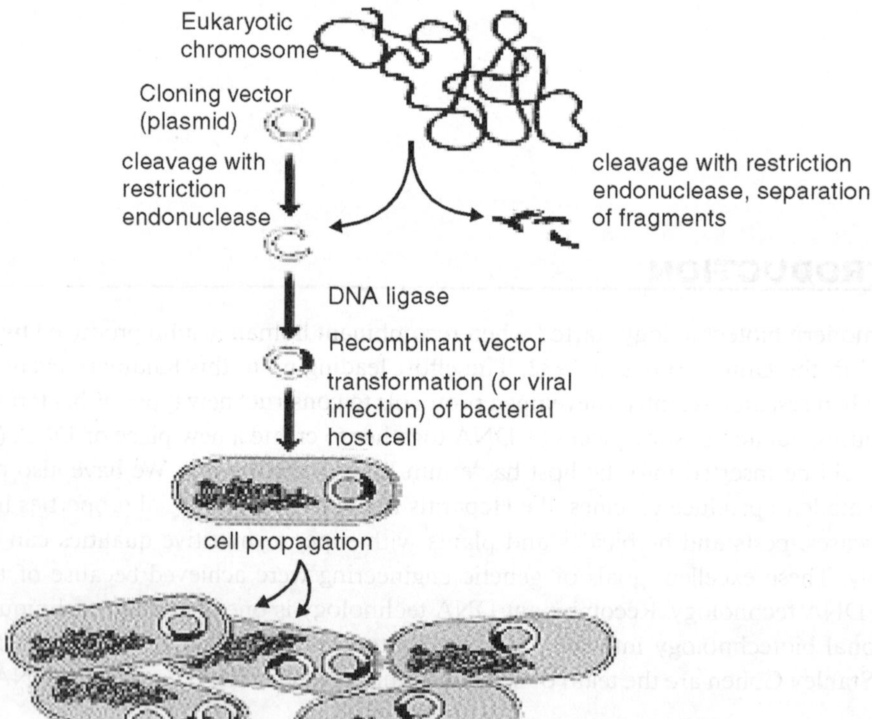

Figure 6.1 Different steps in a typical rDNA experiment.

The goal of rDNA technique is gene cloning to generate large amounts of pure DNA that can be manipulated and studied. The following are the basic steps involved in the process of rDNA technique for gene cloning.

1. Isolate DNA from organism (*e.g.*, extraction).
2. Cut DNA with **restriction enzymes** to generate DNA fragments of desired size, which may have a specific type of DNA sequence or gene that has to be cloned.

3. Ligate or splice each piece of DNA into a cloning vector to create a recombinant DNA molecule. A cloning vector is an artificial DNA molecule, capable of replicating in a host organism (*e.g.*, bacteria).
4. Transform recombinant DNA (cloning vector + DNA fragment) into a host that will replicate and transfer copies to progeny.

6.2 TOOLS OF RECOMBINANT DNA TECHNOLOGY

The generation of recombinant DNA molecules, we require the DNA fragment that has to be cloned known as the insert DNA and a vehicle DNA known as the vector to carry the DNA into a host cell and multiply. Since the process of making a recombinant DNA require the precise cutting and stitching of DNA molecules, it involves a number of molecular tools- the enzymes to cut and modify the DNA, finally resulting in a recombinant DNA molecule. The major enzymes required for the making of a rDNA molecule are the following :

1. Type II Restriction Endonucleases
2. DNA ligase
3. Alkaline phosphatase

The role of each of these enzymes in the making of an rDNA molecule is explained in the following parts.

1. Restriction Endonucleases

In 1968 a group of three scientists—H.O. Smith, K.W. Wilcox and T.J. Kelley isolated and characterized the first sequence specific restriction endonuclease. The discovery of these enzymes marked the beginning of the recombinant DNA research and sequence specific modification of DNA molecules. Since these restriction endonucleases are responsible for converting the DNA molecules into a specific number of fragments by cutting the molecule at certain specific sequences known as restriction sites, these enzymes are known as molecular scissors. These enzymes along with another equally important DNA modifying enzyme—the DNA ligase made it possible to create entirely new kinds of DNA molecules and, equally important, to manipulate the functioning of the genes located on these new molecules.

Growth Restriction of Bacteriophages

It had been known since the 1950's that phage particles that grow well and efficiently infect one strain of bacteria are often unable to grow well and infect other strains of the same bacterial species. A series of studies showed that phage particles that efficiently grow and infect host cells have DNA molecules that have been chemically modified by the addition of methyl groups to some of their adenine and/or cytosine bases, while the DNA of poorly infecting phage particles does not show this pattern of chemical modification or 'methylation'. Phage particles with unmethylated DNA do not grow and infect efficiently because their DNA molecules are cleaved and degraded by enzymes of the host cell, while methylated DNA is protected from this degradation. This phenomenon of degrading unmethylated DNA destroys the growth ability of the phage, and is responsible for the pattern of growth restriction described above.

In the late 1960's, two scientists—Stewart Linn and Werner Arber isolated two types of enzymes responsible for phage growth restriction in *Escherichia coli* bacteria. One of these enzymes methylate

DNA, while the other cleaved unmethylated DNA at a wide variety of specific locations along the length of the molecule. The first type of enzyme was called a 'methylase' while the other was called "restriction endonuclease". These restriction nucleases were important to scientists who were gathering the tools needed to cut DNA molecule at specific sites. Because of this scientists could cut DNA molecules in a predictable and reproducible way and thus laid the foundation of recombinant DNA technique.

These two enzyme systems—restriction endonucleases and DNA methylases forms the basic components of the defense mechanism in bacterial systems, known as Restriction-Modification system. The first part of this system is the restriction endonucleases that selectively recognizes specific DNA sequences and fragment the DNA having this sequences leading to its complete degradation. Since these enzymes are responsible for the restricted growth and multiplication of bacterial viruses in bacteria they are named as restriction endonucleases. The second part is the modification system, which recognize the restriction sites and modifies those sites by adding methyl groups to one or two bases within those restriction sites. Thus, it modifies the restriction sites by methylation, so that the restriction enzymes were not able to cut those sites. This usually happens for the DNA of the host cells. Thus, these modification systems protect the DNA of the host cell from the action of its own restriction systems.

The first sequence specific endonuclease was isolated and characterized by a group of three scientists, H.O. Smith, K.W. Wilcox and T.J. Kelley of Johns Hopkins University, in the year 1968. Working with *Haemophilus influenzae* bacteria, this group isolated an enzyme, called *Hind*II, which always cut DNA molecules at a particular point within a specific sequence of six base pairs. This sequence is :

5′ G T (T or C) (A or G) A C 3′

3′ C A (A or G) (T or C) T G 5′

They found that the *Hind*II enzyme always cuts directly in the centre of this sequence. Wherever this particular sequence of six base pairs occurs unmodified in a DNA molecule, *Hind*II will cleave both DNA backbones between the 3rd and 4th base pairs of the sequence. Moreover, *Hind*II will only cleave a DNA molecule at this particular site. For this reason, this specific base sequence is known as the "recognition sequence" for *Hind*II. *Hind*II is only one example of the class of enzymes known as restriction nucleases. In fact, more than 900 restriction enzymes, some sequence specific and some not, have been isolated from over 230 strains of bacteria since the initial discovery of *Hind*II.

These restriction enzymes generally have names that reflect their origin—The first letter of the name comes from the genus and the second two letters come from the species of the prokaryotic cell from which they were isolated. For example, *Eco*RI comes from *Escherichia coli* RY13 bacteria, while *Hind*II comes from *Haemophilus influenzae* strain Rd. Numbers following the nuclease names indicate the order in which the enzymes were isolated from single strains of bacteria. Nucleases are further described by addition of the prefix 'endo' or 'exo' to the name: The term 'endonuclease' applies to sequence specific nucleases that break nucleic acid chains somewhere in the interior, rather than at the ends, of the molecule. Nucleases that function by removing nucleotides from the ends of the molecule are called 'exonucleases.' There are three classes of restriction endonucleases. They are type I, type II and type III, each is characterized by a slightly different mode of action on DNA. Type II restriction endonucleases are used in the recombinant DNA techniques as they can recognize specific DNA sequences and cleave at a site that comes within that sequences. These recognition sequences are called the restriction sites.

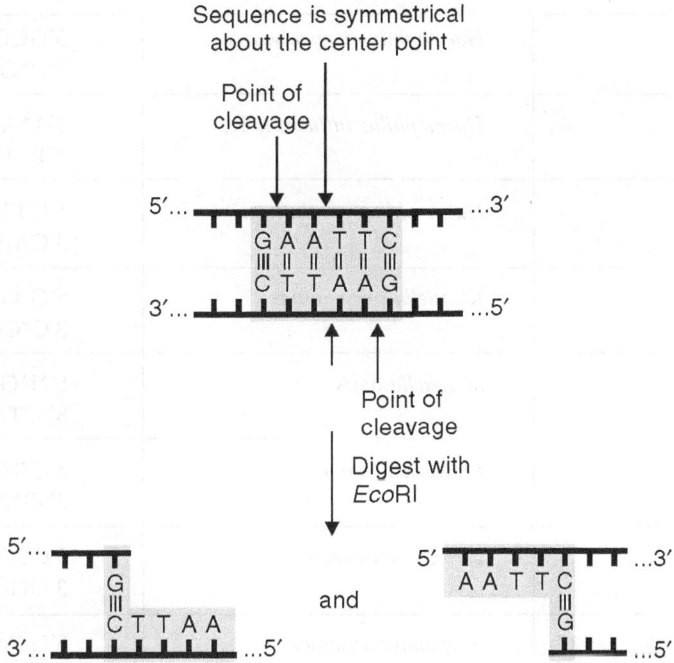

Figure 6.2 The restriction site sequence of EcoRI is a palindrome of a hexamer and it cuts between the bases G and A on both the strands resulting in sticky ends or cohesive ends.

TABLE 6.1 Examples of Type II Restriction endonucleases used in recombinant DNA experiments

Enzyme	Organism from which derived	Target sequence (cut at*) 5′ → 3′
Ava I	*Anabaena variabilis*	5′C*C/TCGA/GG 3′ 3′GG/AGCT/C*C5′
Bam HI	*Bacillus amyloliquefaciens*	5′G*GATCC3′ 3′CCTAG*G5′
Bgl II	*Bacillus globigii*	5′A*GATCT3′ 3′TCTAG*A5′
Eco RI	*Escherichia coli* RY 13	5′G*AATTC3′ 3′CTTAA*G
Eco RII	*Escherichia coli* R245	5′*CCA/TGG3′ 3′GGT/ACC*5′
Hae III	*Haemophilus aegyptius*	5′GG*CC3′ 3′CC*GG5′

Hha I	*Haemophilus haemolyticus*	5′GCG*C3′ 3′C*GCG5′
Hind III	*Haemophilus influenzae* Rd	5′A*AGCT T3′ 3′T TCGA*A5′
Hpa I	*Haemophilus parainfluenzae*	5′GTT*AAC3′ 3′CAA*TTG5′
Kpn I	*Klebsiella pneumoniae*	5′G GTAC*C3′ 3′C*CATG G5′
Mbo I	*Moraxella bovis*	5′N*GATC 3′ 3′ CTAG*N5′
Pst I	*Providencia stuartii*	5′C TGCA*G 3′ 3′ G*ACGT C 5′
Sma I	*Serratia marcescens*	5′CCC*GGG 3′ 3′GGG*CCC5′
Sst I	*Streptomyces stanford*	5′G AGCT*C3′ 3′ C*TCGA G 5′
Sal I	*Streptomyces albus* G	5′G*TCGA C 3′ 3′C AGCT*G 5′
Taq I	*Thermophilus aquaticus*	5′ T*CG A 3′ 3′ A GC*T 5′
Xma I	*Xanthamonas malvacearum*	5′ C*CCGG G 3′ 3′ G GGCC*C 5′

The restriction sites or the recognition sequences are usually 4 to 8 nucleotides in length and are palindromic. The palindromic sequences read the same on both the strands of the DNA in a 5′—3′ direction. For example, the restriction site of *Eco*RI is 5′GAATTC3′ and the cleave site is between the G and A on the complementary strands, which is demonstrated in figure 6.2.

The restriction site recognition sequence for different restriction enzymes are unique and, therefore, different endonucleases yield different sets of cuts or DNA fragments. But one endonuclease will always cut a particular base sequence the same way, no matter what DNA molecule it is acting on. Once the cuts have been made, the DNA molecule will break into fragments.

Type I and III Restriction Endonucleases. They have both endonuclease and methylase activity on a single protein. Type I REs cleave the DNA at random site located at 1000 base pairs from the recognition site. Type III RE does the same at 24-24 bp away from recognition sites.

Type II Restriction Endonuclease. They cleave DNA at specific site within the recognition sequence. This property has made type II RE as most important tool in molecular biology.

The type II restriction enzymes cut DNA and produce two types of fragments. Some restriction enzymes produce fragments with blunt ends, whereas others produce fragments with sticky (overhanging or staggered) ends. Some enzymes like *Hind*II and *Sma*I cuts the DNA symmetrically on both the strands at exactly the same nucleotide position leaving blunt ends as shown in figure 6.3.

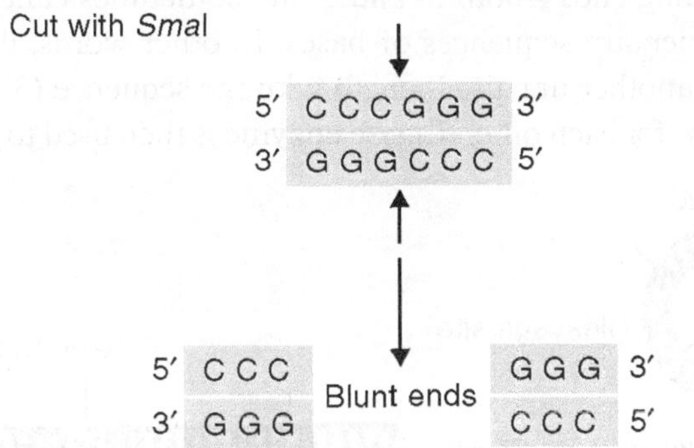

Figure 6.3 Restriction Cleavage with *Sma*I resulting in blunt ends.

Many endonucleases cleave the DNA backbones in positions that are not directly opposite each other as in the case of *Eco*RI as illustrated in figure 6.2.

Restriction Digestion with *Eco*RI

When the restriction endonuclease encounters its respective restriction site sequence (5′GAATTC 3′), it cleaves each backbone between the G and the closest A base residues. Once the cuts have been made, the

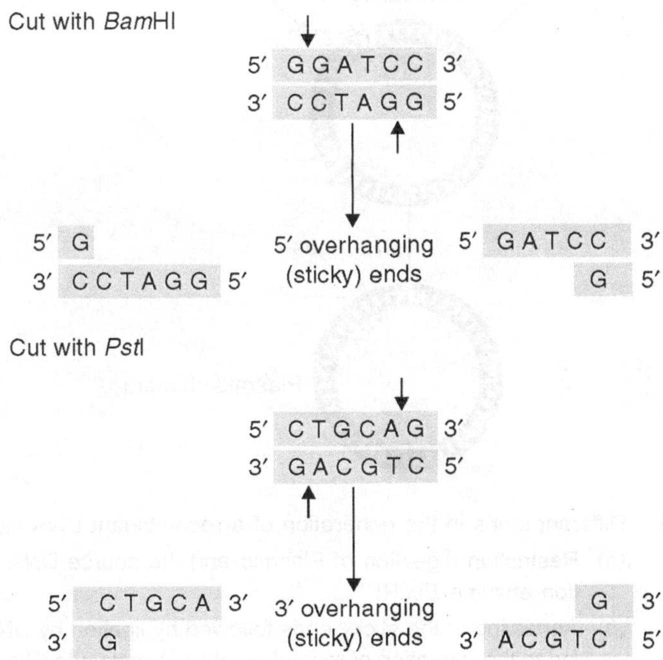

Figure 6.4 Restriction digestion with BamHI resulting in 5′ overhanging ends and with Pst I resulting in 3′ overhanging ends.

resulting fragments are held together only by the relatively weak hydrogen bonds that hold the complementary bases to each other. The weakness of these bonds allows the DNA fragments to separate from one each other. Each resulting fragment has a protruding 5′ end composed of unpaired bases (Figure 6.2). Other enzymes also create cuts in the DNA backbone in the same manner, which result in protruding 3′ ends. Protruding ends—both 3′ and 5′ are sometimes called "sticky ends" because they tend to bond with complementary sequences of bases. In other words, if an unpaired length of bases (5′ A A T T 3′) encounters another unpaired length with the sequence (3′ T T A A5′) they will bond to each other— they are 'sticky' for each other. Ligase enzyme is then used to join the phosphate backbones

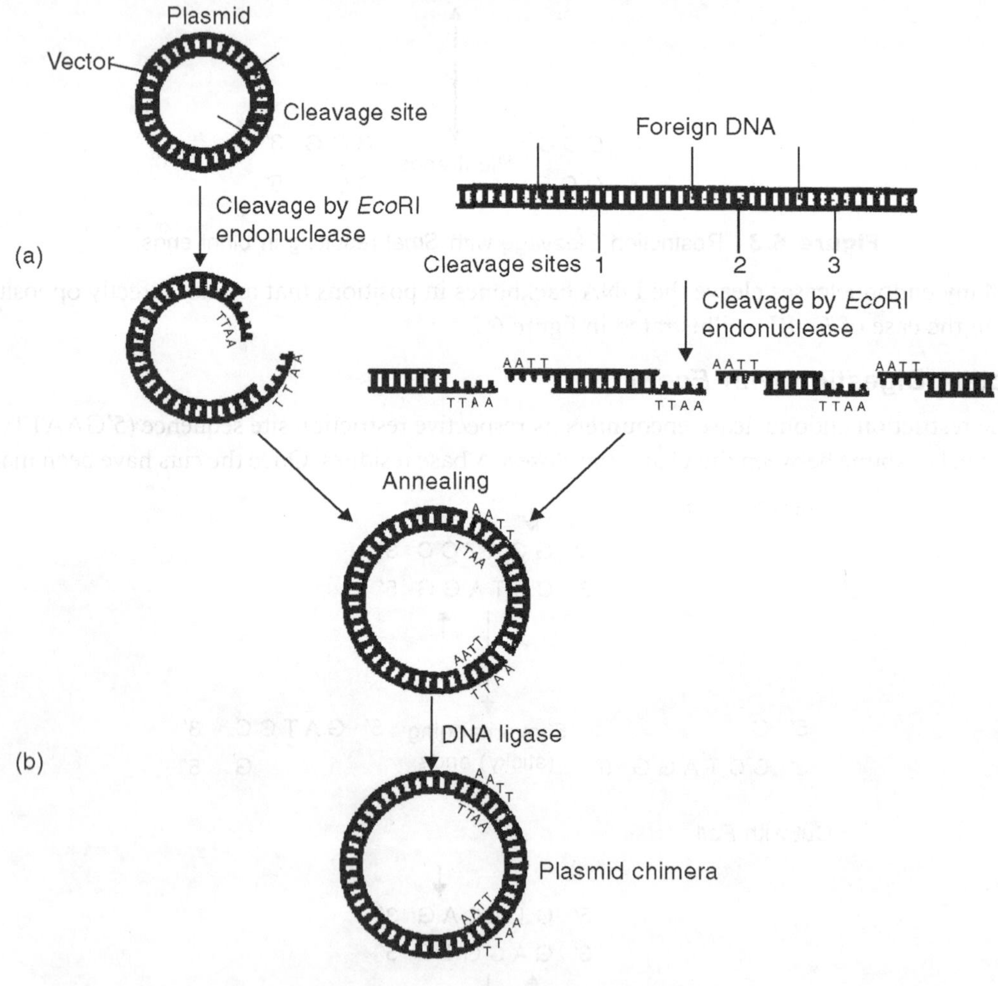

Figure 6.5 Different steps in the generation of a recombinant DNA molecule

(*a*) Restriction digestion of Plasmid and the source DNA by the restriction enzyme *Eco*RI

(*b*) Annealing of the sticky ends followed by ligation by DNA ligase leading to the formation of recombinant DNA molecule (Plasmid chimera).

of the two molecules. The cellular origin, or even the species origin, of the sticky ends does not affect their stickiness. Any pair of complementary sequences will tend to bond, even if one of the sequences comes from a length of human DNA, and the other comes from a length of bacterial DNA. In fact, it is this quality of stickiness that allows production of recombinant DNA molecules, molecules which are composed of DNA from different sources, and which has given birth to a powerful technology and industry—the genetic engineering or the recombinant DNA. Examples of some other restriction enzymes, its mode of cutting and generation of 5′ overhangs and 3′ overhangs are illustrated in figure 6.4.

Sticky ends are useful for DNA cloning because complementary sequences anneal and can be legated directly by DNA ligase. If two different DNA samples cleaved with the same types of restriction enzymes are mixed together in the presence of DNA ligase, a recombinant DNA molecule can be generated. This is possible because of the presence of the same type of sticky ends. The complementary sequences of the sticky ends from the unrelated DNA samples will anneal together and is finally joined by the DNA ligase enzyme to form the recombinant DNA molecule. Figure 6.5 illustrates the generation of a recombinant DNA molecule from two different sets of DNA samples.

6.3 OTHER ENZYMES USED IN THE RECOMBINANT DNA TECHNIQUE

Type II Restriction endonuclease, DNA ligase and alkaline phosphatase are the main types of enzymes used in the process of gene cloning or recombinant DNA technique. The role of Restriction endonuclease in rDNA technique has been discussed above. The role of other two major enzymes—DNA ligase and alkaline phosphatase (AP) are the following :

DNA Ligase. This enzyme is responsible for the formation of phosphodiester linkage between two adjacent nucleotides and thus joints two double stranded DNA fragments. The phosphodiester linkage is formed between the 3′OH group (hydroxyl group at the 3′ carbon) and 5′ PO_4 group of the pentose sugar of adjacent nucleotides of two different DNA fragments or the nicks created in a double stranded DNA molecule. In rDNA experiments DNA ligase is used to joints two different DNA fragments (plasmid/vector and the foreign DNA) that are annealed by the sticky ends. There are different types of ligase enzymes from different sources. But the most frequently used one is the T_4 DNA ligase produced by the T_4 bacteriophage.

Alkaline phosphatase. In rDNA experiment there is always the problem of self-ligation and the reformation of the original plasmid or the vector. This reduces the efficiency of the recombinant DNA formation. Self-ligation is the process of annealing the sticky ends of the linearised vector without inserting the foreign DNA fragment. On ligation it will result in the original vector without any insert DNA. This can be prevented by using alkaline phosphatase. We know that the presence of 3′OH group and 5′ PO_4 group is a prerequisite for the ligase enzyme to act. If any one group of this is absent DNA ligase cannot function. Alkaline phosphatase can remove the phosphate group from the 5′ end of the DNA molecule resulting in a 5′OH group. Now the only chance of ligation is between the vector and the foreign DNA resulting a recombinant DNA molecule as the insert DNA has the 5′PO_4 group. Thus, the self-ligation can be prevented and the efficiency of rDNA formation can be increased. This enzyme is isolated from some bacteria (Bacterial alkaline phosphatase—BAP) or calf intestine (Calf intestine alkaline phosphatase—CAP).

In addition to these there are several other enzymes that are used for certain special purposes. Table 6.2 gives a summary of all enzymes that have some role in the recombinant DNA experiments.

TABLE 6.2 Various Enzymes used in Recombinant DNA Technique

Enzyme(s)	Function
Type II restriction endonucleases	Cleave DNAs at specific base sequences
DNA ligase	Joints two DNA molecules or fragments
DNA polymerase I (*E. coli*)	Fills gaps in duplexes by stepwise addition of nucleotides to 3′ ends
Reverse transcriptase	Makes a DNA copy of an RNA molecule
Polynucleotide kinase	Adds a phosphate to the 5′-OH end of a polynuclectide to label it or permit ligation
Terminal transferase	Adds homopolymer tails to the 3′-OH ends of a linear duplex
Exonuclease III	Removes nucleotide residues from the 3′ ends of a DNA strand
Eacteriophage λ exonuclease	Removes nucleotides from the 5′ ends of a duplex to expose single stranded 3′ ends
Alkaline phosphatase	Removes terminal phosphates from either the 5′ or 3′ end (or both).

6.4 VECTORS : THE VEHICLE FOR CLONING

One of the major applications of rDNA experiments is the gene cloning or DNA cloning. Vector is another major component to make an rDNA molecule for gene cloning. Vectors act as a vehicle for carrying the foreign DNA into a host cell for multiplication. Usually small circular DNA molecules of bacterial origin are used as a cloning vector. A DNA molecule should possesses the following essential characteristics to act as a cloning vector :

1. Origin of Replication. Required for autonomous replication of the plasmid using the host's replication machinery. Almost all commonly used plasmids are based on the ColE1 origin of replication (ori). Naturally occurring origins of replication are negatively regulated to keep copy number down (typically 5 to 10 copies per cell) to reduce load on the host's replication machinery.

While high copy number is disadvantageous in a natural system, it is a desirable feature in a cloning vector—since the whole idea is to be able to easily isolate substantial quantities of particular DNA sequences. Therefore, considerable work has gone into engineering the ColE1 ori such that the negative regulatory mechanisms that limit copy number of plasmid are disabled. Modern plasmid vectors are therefore often called 'runaway replicons' and are present at 100 to 1000 copies per cell.

2. Selectable Markers. Selectable markers are essential for the identification of bacteria containing recombinant plasmids. Selection can be divided into two types—**Positive selection and Negative selection.**

Positive selection is used to identify bacteria that contain a plasmid. The most common markers used for positive selection are the antibiotic resistance genes carried by the original R factors. While many antibiotics and resistance genes are available, the commonly used ones fall into two general classes: **Antibiotics affecting cell wall synthesis** and **Antibiotics affecting translation.** Ampicillin is a beta-lactam based antibiotic that acts by inhibiting the synthesis of the bacterial peptidoglycan cell wall. Sensitive bacteria are not actively 'killed', but on cell division are unable to synthesize the cell wall and suffer from osmotic lysis. The enzyme beta-lactamase is secreted into the periplasmic space where it breaks down the antibiotic, allowing cell wall synthesis to proceed. The antibiotics tetracycline, kanamycin and chloramphenicol all act by inhibiting translation. The covalent modification (phosphorylation, acetylation) of these antibiotics blocks their interaction with the translation apparatus. Positive selection is particularly important when introducing plasmids into bacteria by transformation. At best, only 1 in about 10,000 bacteria picks up a plasmid that carries the antibiotic resistance. A strong positive selection system is essential to eliminate the 9,999 bacteria that did not pick up a plasmid from the one that did. By plating the transformation products directly on antibiotic plates, all untransformed bacteria die and only those containing the plasmid (and antibiotic resistance marker) grow to form colonies.

A second selection system is necessary to distinguish between plasmids that merely recircularized from those that carry a foreign DNA insert. Usually it is a negative selection method. In order to identify those plasmids carrying a foreign DNA fragment, the site of insertion is chosen such that insertion disrupts a selectable marker—a phenomenon we call insertional inactivation. It can be the insertional inactivation of antibiotic resistance gene or enzymes such as β-galactosidase, a product of *Lac* Z gene of *lac* operon. The insertion of the foreign DNA into the vector will disrupt the expression of these genes and will produce a colour difference for the colony from that of cells with intact vectors.

3. Multiple Cloning Sites (MCS) or Polylinker. A vector should have a site specific for cloning the foreign DNA fragment provided with one restriction site for most of the commonly used unique restriction endonucleases. All these unique restriction sites are grouped together in a small region of the vector and that location of the DNA molecule is known as the Multiple Cloning Site (MCS) or the polylinker. The presence of unique restriction sites at the MCS gives flexibility in the choice of restriction enzymes.

4. Small Size. Relatively small size of the vector is more desirable because it increases the transformation efficiency and is easy to manipulate. Small size also helps purification procedures to obtain intact plasmids.

A large number of vectors with the above characteristics have been developed and there are six different types of cloning vectors commonly used in the recombinant DNA experiments. They are the following :

1. Plasmid cloning vectors
2. Phage cloning vectors
3. Cosmid cloning vectors
4. Yeast artificial chromosomes (YACs)
5. Bacterial artificial chromosomes (BACs)
6. Animal and plant vectors (Shuttle vectors).

1. Plasmid Cloning Vectors

Plasmid cloning vectors are derived from the bacterial plasmids. Plasmids are the circular extra chromosomal self-replicating double stranded DNA molecules. They are usually found in bacteria but also present in yeasts. Usually plasmids are not essential for the normal growth and multiplication of the organism, but they are responsible for certain special characteristics of the organisms. For example, the genes for antibiotic resistance or resistance to certain toxins are present in the plasmids of the organisms. The copy number of the plasmids can vary from 1 or two to multiple copies of the range 100—1000 per cells. The naturally occurring plasmids have been modified or genetically engineered to serve as vectors of various types. Plasmid derived vectors are the most widely used, versatile and easily manipulated ones. ColE1 of *E. coli* is an example of naturally occurring plasmids. The *ori* in almost all plasmid vectors are that of ColE1.The following are some of the examples of plasmid cloning vectors :

pBR322

This was the first widely used, purpose built, plasmid vector. It has a number of useful features (Figure 6.6).

Origin of replication. It carries a fragment of the plasmid pMB1 that acts as an origin for DNA replication and thus ensures multiplication of the vector.

Size. It is relatively small, 4363 bp. This is important because transformation efficiency is inversely proportional to size and above 10 Kbp is very low. Thus there is 'room' in pBR322 for an insert of at least 6 Kbp.

Copy number. Reasonably high (~ 15 copies per cell) which can be increased 200 fold by treatment with a protein synthesis inhibitor—chloramphenicol amplification.

Selectable marker. pBR322 carries two antibiotic resistance genes.

Cloning sites. It carries a number of unique restriction sites. Some of these are located in one or other of the antibiotic resistance genes *e.g.* sites for *Pst* I, *Pvu* I and *Sac* I are found in *amp*R and *Bam*HI and *Hind* III in *tet* R. Cloning into one of these sites inactivates the gene allowing recombinants to be differentiated from non-recombinants—insertional inactivation.

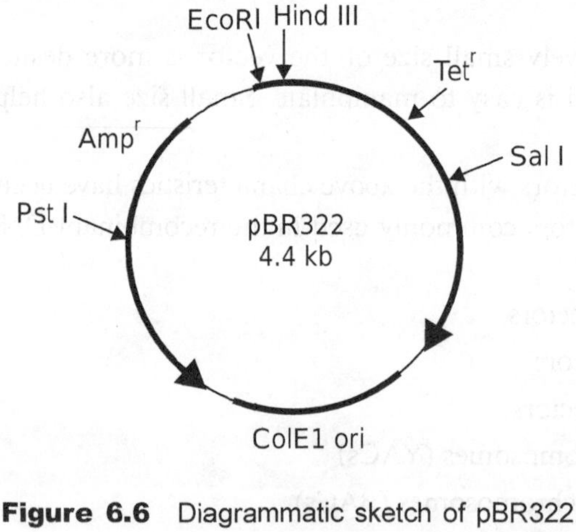

Figure 6.6 Diagrammatic sketch of pBR322.

The pUC Series

Another popular plasmid vector is the pUC series (Figure 6.7), which are extensively used as cloning as well as expression vectors. These vectors have three important additional features compared to pBR322.

High copy number. A mutation within the origin of replication produces 500-600 copies of the plasmid per cell without amplification.

Blue → white screening. This is a special form of insertional activation that can be used during the primary selection of transformants, rather than requiring a second round of screening. It utilises the N-terminal portion of the *E. coli* beta-galactosidase—the product of *lac* Z gene (alpha-peptide) encoded by the vector in a form of intermolecular complementation that restores beta-galactosidase activity to a defective enzyme (omega-peptide) encoded by the host. If a chromogenic substrate (X-gal) and a beta-galactosidase inducer (IPTG) are included in the plates on which the primary transformants are selected, non-recombinant molecules will catabolise the clourless substrate to give blue colonies.

A synthetic polylinker. This is a piece of man-made DNA that contains several unique restriction sites. It has been inserted within the portion of the vector encoding the beta-galactosidase alpha-peptide in such a way that it does not affect its expression. However, inserting a foreign DNA fragment into any one of the polylinker restriction sites almost invariably disrupts the enzyme activity. Thus recombinant colonies remain white but non-recombinants turn blue. The polylinker sequences differ between different members of the pUC family allowing different choices of cloning enzymes to be used. Moreover, the close proximity of many sites within each plasmid allows the cloning of fragments cut with different enzymes at each end. As well as being more flexible this allows the orientation of the insert within the plasmid to be predetermined—forced cloning.

These vectors can replicate only in *E. coli* cells, since they carry the ColE1 *ori* (origin of replication). There are some other vectors, which are constructed in such a way that they can exist and multiply both in eukaryotic cells and *E. coli* and such vectors are known as **shuttle vectors**. They are attached with two types of *ori* and selectable markers. One functions in the eukaryotic cells and the other works in the bacterial cell (*E. coli*). **YEp**, the yeast episomal plasmid is an example for shuttle vectors. There is a naturally occurring plasmid vector in plants; the plasmid of *Agrobacterium tumifaciens* called Ti plasmids. This is suitably modified to function as a shuttle vector, which now carries the genes that has to be introduced into the plant genome instead of Ti (tumor inducing) genes.

Expression vectors. In most of the cases the main purpose of the rDNA experiment and cloning is to sufficiently multiply or amplify the inserted DNA fragment. But sometimes the aim of the process will be to produce large quantity of protein encoded by the inserted gene. This can be accomplished by incorporating the necessary regulatory elements along with gene in the vector. Such vectors should be provided with signals necessary for initiation and termination of transcription such as suitable promoter and terminator sequences; and signals for translation initiation such as start codon ribosome binding site into the vector upstream to the multiple cloning site (MCS). These vectors, which have the regulatory elements and other machinery for the expression of the cloned gene, can be used for the production of recombinant proteins and are known as expression vectors. Vectors of pUC series are example of expression vectors.

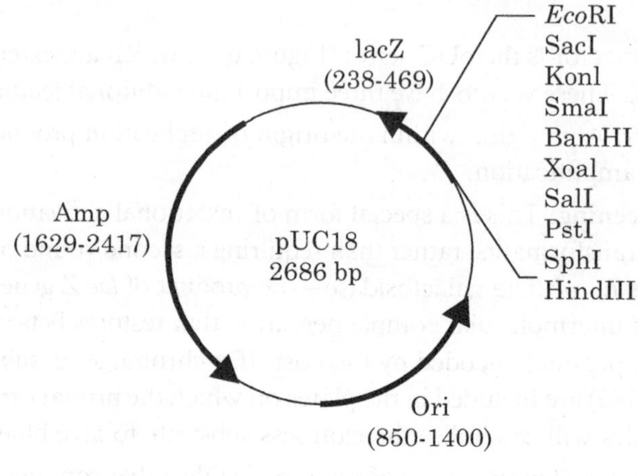

Figure 6.7 Diagrammatic sketch of pUC18. It differs from the pUC19 only in the region of polylinker site in the arrangement of restriction sites.

2. Bacteriophage Cloning Vectors

These are derived from viruses that specifically infect bacteria. Such viruses usually contain a comparatively small DNA genome surrounded by a protein coat. These viruses that infect bacteria are

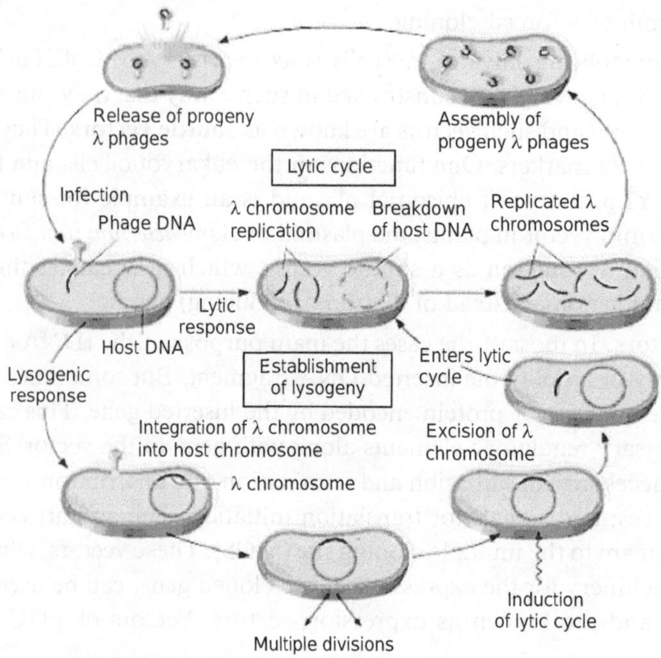

Figure 6.8 The bacteriophage infection and life cycle—Lysogeny and lytic cycle.

called bacteriophages. Those phages, which most frequently infect *E.coli*, are used for cloning purposes. Bacteriophages infect bacterial cells by injecting its DNA into the host cytoplasm. This DNA in the host cell is selectively replicated and expressed the protein required for the assembly of new phage particles. These processes result in the production of large number of phages, which breaks the cells by a process known as lytic cycle. These new viral particles will re-infect the neighbouring cells and the cycle continues. In some cases the phage DNA after entering into the host cell, is integrated along with the host genome and is separated at a later stage and starts the lytic cycle as explained above. The process of lytic cycle and phage multiplication is shown in Figure 6.8 below. The natural way of transferring the viral DNA into a specific bacterial host has attracted scientists and they modified these viral genome DNA to use as vectors for gene cloning. Bacteriophage lambda (λ phage) and M13 are extensively modified for the development of a phage based cloning vectors.

Phage λ Cloning Vectors

These are the engineered version of λ bacteriophage that infects *E. coli*. It has a double stranded linear DNA genome having a size of 48,514 bp (48.5 kb). About 12 bases at the ends of this DNA molecule is

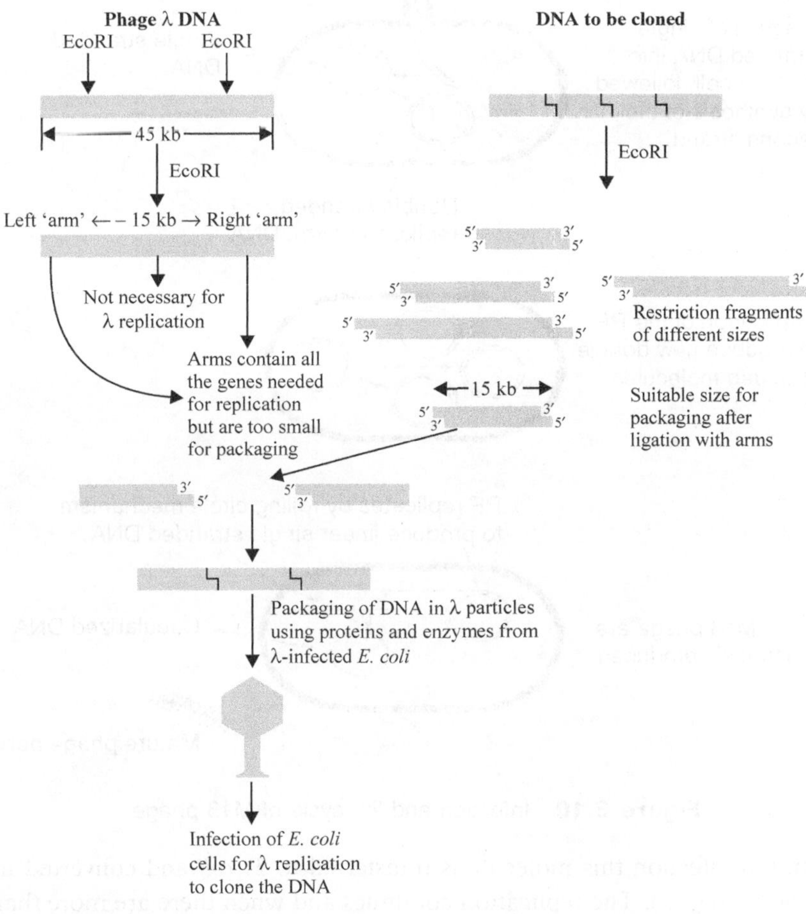

Figure 6.9 Phage λ based cloning vectors for cloning inserts of large sizes.

unpaired and complementary and therefore are sticky or cohesive and is known as cos sites (cohesive end sites). These cohesive sites are very essential for packaging the DNA into the viral particle during the lytic cycle. But a major portion of the central region of the viral genome DNA is not required for the infection and lytic cycle in *E. coli* cells. The λ phage vector is designed in such a way that the central region of the λ chromosome (linear) is cut with a restriction enzyme and is replaced with the foreign DNA digested with the same restriction enzyme. This recombinant DNA is packaged in phage heads to form virus particles (Figure 6.9). Phages with both ends of the λ chromosome and a 37-52 kb insert replicate by infecting *E. coli*. Phages replicate using *E. coli* and the lytic cycle (Fig. 6.8) and produce large quantities of 37-52 kb cloned DNA. A large number of unique restriction sites are available here also.

M13 based Cloning Vectors

The M13 family of vectors is derived from bacteriophage M13. This is a male-specific (infects *E. coli* having F pili), lysogenic filamentous phage with a circular single-stranded DNA genome about 6407 bp

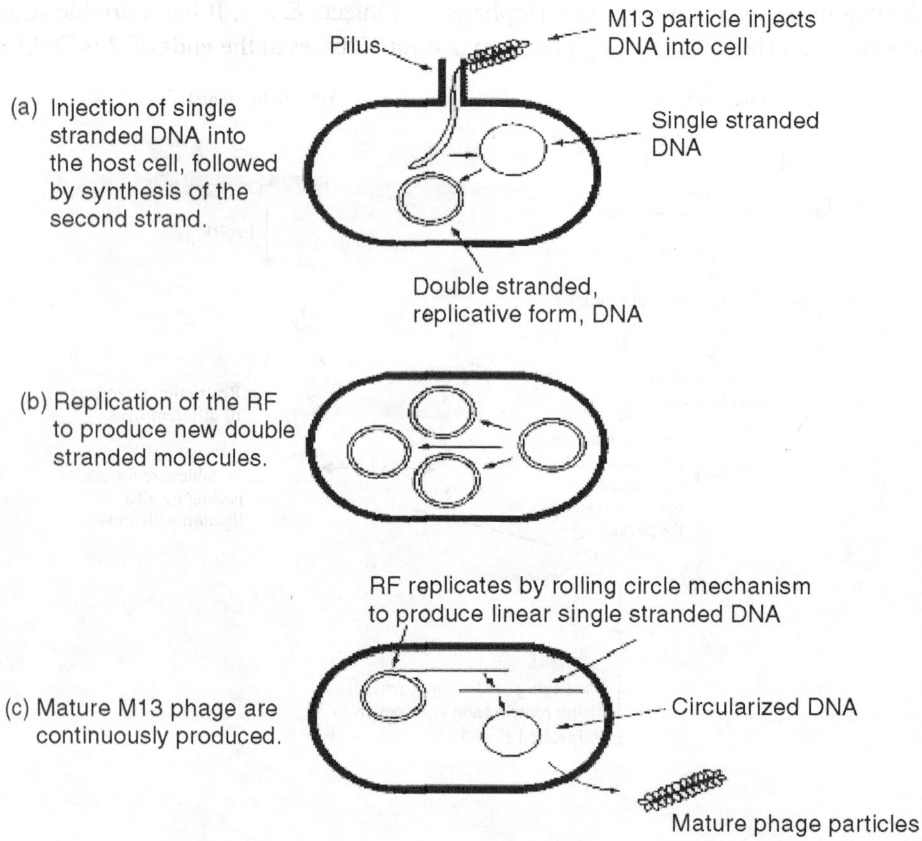

Figure 6.10 Infection and life cycle of M13 phage.

(6.4 kb) in length. On infection this molecule is transferred to *E. coli* and converted into the double-stranded replicative form (RF). The replication continues and when there are more than 100 copies of

DNA in the cell, the DNA replications become asymmetric and produce copies of the original single-stranded molecule, which are packaged into infective particles and extruded from the cell (Figure 6.10). The host is never lysed but continues to grow throughout infection, although at a significantly reduced rate. This difference in growth rate between infected and uninfected cells produces characteristic 'plaques' when M13 phage are plated on a suitable host.

The advantages of M13 based vectors are that contain the same polylinker and alpha-peptide fragments as the pUC series and recombinants can be selected by the blue → white colour test. The RF form of M13 vectors can be isolated by standard plasmid DNA preparation procedures and foreign DNA can be inserted into them as if they were conventional plasmids. The size of the genome is small in size, below 10 kb and so is easy to handle. The specific use of M13 is as an aid to DNA sequencing. Once cloned into M13, large amounts of the single stranded form of any given fragment can be easily isolated from the mature phage that are extruded from infected cells. This is an ideal template for a dideoxy sequencing reaction. If an oligonucleotide complementary to the region just downstream of the polylinker is used as the primer for such experiments, the same reaction components can be used to determine the sequence of any fragment that has been cloned into M13.

3. Cosmid Vectors

Cosmid vectors were among the first large insert cloning vehicles developed. It has been constructed by certain features of plasmid and *cos* site of λ phage. A cosmid can be defined as a plasmid that contains a *cos* site from the lambda phage genome. The vector replicates as a plasmid (it contains a ColE1 origin of replication), and uses Amp^r for positive selection and employs lambda phage packaging to select for recombinant plasmids carrying foreign DNA inserts up to 45 kb in size. The simplest cosmid vector has a ColE1 origin of replication, selectable markers including antibiotic resistance gene and β-galactosidase gene (a part of *lac Z* gene), suitable polylinker sites and lambda cos site (Figure 6.11).

Ligation of cosmid vector and foreign DNA fragments of sizes up to 45 kb is similar to ligation into a lambda substitution vector. The desired ligation product is a concatemer of 45 kb foreign DNA fragment and 5 kb cosmid vector sequences. This concatemer is then packaged into viral particles (remember packaging is cos site to cos site) and these are used to infect *E. coli* where the cosmid vector replicates using the ColE1 origin of replication (*ori*). Phage packaging serves only to select for recombinant molecules and to transfer these long DNA molecules (50 kb total) into the bacterial host (50 kb fragments transform very inefficiently while phage infection is very efficient).

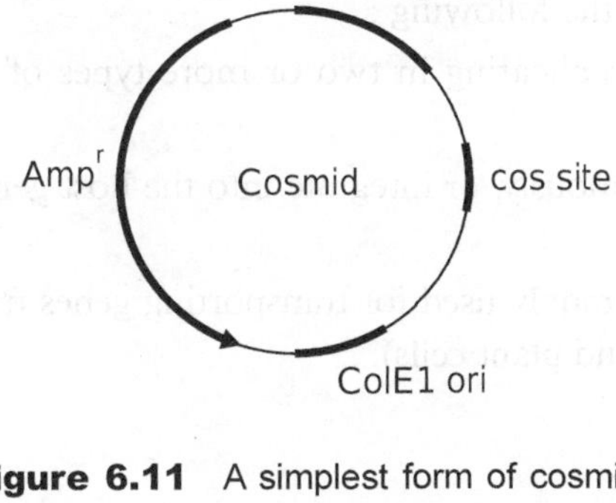

Figure 6.11 A simplest form of cosmid.

4. Yeast Artificial Chromosomes (YACs)

Yeast Artificial Chromosomes (YACs) are the vectors that enable formation of artificial chromosomes with the foreign DNA fragments and cloned into yeast. These are used for the cloning of very large DNA fragments in the range of 500—1000 kb in size. These vectors are constructed with yeast telomeres sequences at each ends, yeast centromere sequences and the yeast ARS (Autonomously replicating sequences) for replication. In addition to this there is the restriction sites for DNA insertion and selectable markers such as amino acid dependence and antibiotic resistance on each arm. The recombinant DNA is constructed by Ligating high-molecular weight to the YAC arms using appropriate restriction enzymes. These recombinant vectors or the artificial chromosomes are used for transformation into yeast cells by electroporation and the transformed cells are selected on selection growth medium.

5. Bacterial Artificial Chromosomes (BACs)

Bacterial Artificial Chromosomes (BACs) are the cloning vectors based on the extra-chromosomal plasmids of *E. coli* called F factor or fertility factor. These vectors enable the construction of artificial chromosomes, which can be cloned in *E. coli*. This vector is useful for cloning DNA fragments of size up to 350 kb, but can be handled like regular bacterial plasmid vectors, and is very useful for sequencing large stretches of chromosomal DNA. Like any other vectors, BACs contain ori sequences derived from *E. coli* plasmid F factor, Multiple Cloning Sites (MCS) having unique restriction sites and suitable selectable markers.

Both YACs and BACs were very useful in the genome sequencing projects like Human Genome Project.

6. Animal and Plant Vectors (Shuttle Vectors)

The genetic engineering of bacteria is possible because of the presence of special types of extra-chromosomal circular DNA that can replicate independently. Viruses that infect bacteria (M13 and Lambda Phage) have been used for constructing different types of vectors for the genetic transformation of bacteria. Similarly, there are some vectors developed for transforming plant and animal cells. Numerous vector systems have been developed for use in eukaryotic systems. The most common of these are often called '**shuttle vectors**' as they replicate in both prokaryotic and eukaryotic hosts. In general, DNA manipulations and characterizations are done in prokaryotic systems, and then the manipulated DNA is reintroduced to eukaryotic systems for functional analysis. The common features of such shuttle vectors or eukaryotic vectors are the following :

1. They are capable of replicating in two or more types of hosts including prokaryotic and eukaryotic cells.
2. They replicate autonomously, or integrate into the host genome and replicate when the host replicates.
3. These vectors are commonly used for transporting genes from one organism to another (*i.e.*, transforming animal and plant cells).

Mammalian Vectors

There are shuttle vectors developed for use in mammalian tissue culture. Like any other vector systems, mammalian shuttle vectors require sequences enabling them to replicate in mammalian tissue culture. These eukaryotic origins of replication are typically derived from well-characterized mammalian viruses such as Simian Virus 40 (SV-40) with SV-40 ori and Large T antigen system; or Epstein-Barr virus (mononucleosis). SV-40 was the vector used in the first cloning experiment in mammalian cell cultures in 1979. Since then a number of mammalian vectors based on viruses like Adenovirus and Pappiloma virus have been developed to transfer genes in mammals for various purposes including gene therapy. At present retrovirus based vectors are the most commonly used vectors for cloning genes in mammalian cells. Both systems eukaryotic and prokaryotic cells allow the vector to replicate within the host cell as a plasmid without integrating into the genome. In addition to the origin of replication, these shuttle vectors also carry antibiotic resistance genes, which function in eukaryotic cells (*i.e.* neomycin (G418) resistance, hygromycin resistance, methotrexate resistance etc.).

Finally, artificial chromosome vectors have been developed for use in mammalian cells known as mammalian artificial chromosomes (MACs) like the YACs and BACs. These eukaryotic vectors contain telomeric sequences (eukaryotic chromosomes are linear not circular molecules) and centromeres to ensure appropriate segregation of the artificial chromosomes as well as selectable markers. Such vector systems may gain in importance as we move towards manipulating the eukaryotic genome.

Plant Vectors

There are a number of shuttle vectors developed for the genetic transformation of plant systems. These shuttle vectors are based on certain viruses and bacteria, which are pathogenic to the plants. Plant viruses like Cauliflower Mosaic Virus (CMV), Tobacco Mosaic Virus (TMV) and Gemini Viruses were used for developing vectors for plant cell transformations, but with limited success. The most successful shuttle vectors developed for the plant system are those, which are based on the Ti plasmid of *Agrobacterium tumifaciens*, a bacterium that causes tumor formation in plants by a natural way of genetic engineering. This has been successfully exploited for the development of *A. tumifaciens* based vectors for the genetic engineering of plants. The success in plant genetic engineering and the development of a number of transgenic plants is mainly because of the development of vectors based on Ti plasmids and *A. tumifaciens.*

TABLE 6.3 Different types of cloning vectors and the size of the insert DNA that can be accommodated

Type of vectors	Insert DNA size (in kb)
Plasmid	0.5 —8
Bacteriophage Lambda	9 —25
Cosmid	30 —45
YAC	250 —1000
BAC	50 —300

6.5 HOST CELLS

The ultimate aim of the development of recombinant DNA molecule or vector is the multiplication of that rDNA by cloning in a suitable host cell. There are many types of host cells available for the purpose of cloning. The kind of host cell to be used mainly depends on the aim of the cloning experiment. The host cell can be bacteria yeast, plant cells or animal cells.

Escherichia coli, a gram-negative enteric bacterium of human is the most extensively studied and the most widely used organism in rDNA experiments. It is very simple, easy to handle and grow rapidly and is able to accept and maintain a range of vectors. The doubling time of *E. coli* under ideal growth conditions is 20 minutes. As the cell undergoes multiplication the rDNA within the cell also undergoes multiplication independent of its genome. Within a short time the number of cell and the quantity of the rDNA will be very high. If the recombinant plasmids are with appropriate promoter signals for expression, a large quantity of protein also can be purified from these systems. There are a large number of genetically defined strains, which can be used freely for the genetic engineering experiments and expression of recombinant proteins because they are disarmed and so are non infective or harmless.

If the recombinant protein that we want to produce is of eukaryotic origin, it is better to opt for a eukaryotic host system like yeast. The following are the reasons for using a eukaryotic host system for the expression of a protein of eukaryotic origin. The polypeptides after its synthesis have to undergo folding in the correct manner to attain its functionally active 3-dimensional structure. This may require the removal of some signal peptides or amino acids to initiate the process of folding to the correct 3-D structure. The correct 3-D structure is very essential for the production of functionally active protein. In addition to this there may be removal or addition or chemical modification of amino acids or the addition of other macromolecules like carbohydrates or lipids to make the protein functionally active. All these processes together are known as the posttranslational modification of proteins. Prokaryotic systems like *E. coli* lack all these enzymes and other machinery essential for the posttranslational modifications of proteins. The eukaryotic protein produced in bacteria may fold incorrectly resulting in the production of a functionally inactive molecule. Prokaryotic system does not have the molecular machinery necessary for the splicing of introns of eukaryotic genes resulting in the production of correct form of mRNA. This also can result in a completely altered protein, bacterial systems are used for the expression of a eukaryotic gene.

The most convenient eukaryotic system that can be used for the expression of eukaryotic genes is the different species of yeast like *Pichea pastoris*. There are several advantages in yeast as the expression system. Yeasts are the simplest of the eukaryotic systems, single celled, genetically and physiologically well characterized, easy to grow and manipulate. Like bacteria it can be cultivated in different volumes, either in flasks, laboratory fermentors or in the industrial scale. Plant and animal cells can also be used as host cells for the expression of cloned genes either in tissue culture or as cultured cells in fermentors or as a genetically engineered whole organism, transgenic animals and plants.

6.6 MAKING RECOMBINANT DNA

The first step in the construction of a recombinant DNA is the isolation and purification of vector and the DNA fragment containing the gene to be cloned. There are a number of standardized procedures for the isolation of plasmids and DNA fragment containing the gene. Digest the vector DNA with a suitable restriction endonuclease and make it linear with or without sticky ends, which is determined by the type of RE used. Isolate the DNA fragment carrying the gene by digesting the genome or the source DNA strand with the same RE used for cutting the vector. The linearised (digested) vector and the target DNA cut with the same restriction enzyme are incubated together in the presence of DNA ligase. During incubation the sticky ends of the two DNA strands come closer by the base complementation and become hydrogen bonded to each other. At this point a phosphodiester linkage established between them mediated by the DNA ligase enzyme. This results in the formation of recombinant DNA molecule of vector and the target DNA or the insert DNA. This is the basic strategy for the construction of a recombinant DNA molecule from a vector and a target DNA fragment.

The disadvantage of this strategy is that there is the possibility of re-annealing of the sticky ends of the linearised vector leading to the formation of the vector itself without any foreign DNA fragment, the non-recombinant vector (the original vector). This problem can be reduced considerably by adopting any one of the following strategies.

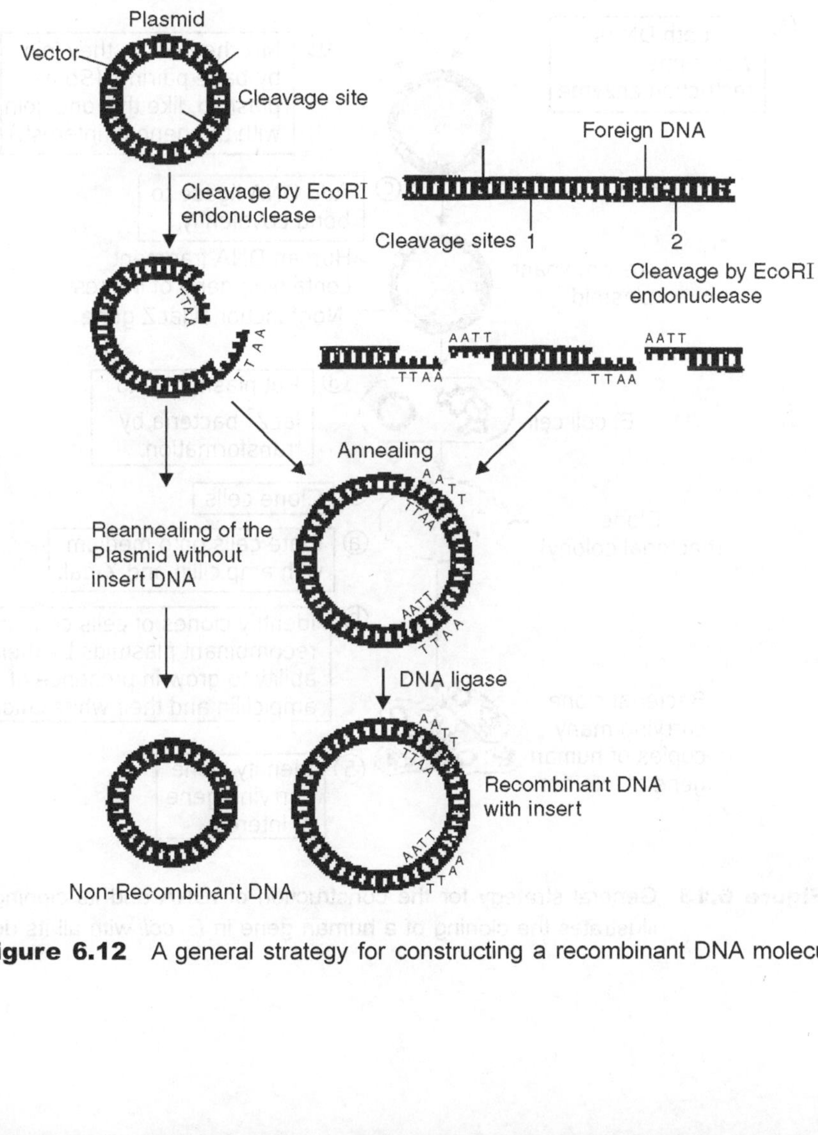

Figure 6.12 A general strategy for constructing a recombinant DNA molecule.

1. Treating the digested vector with alkaline phosphatase to remove the 5′ phosphate group as described earlier.
2. Linearisation of the vector (restriction digestion of the vector) and the isolation of the foreign DNA fragment with two restriction enzymes instead of one (double digestion). So that the two sticky ends of the vector will not be identical, which will prevent the re-annealing to some extend.

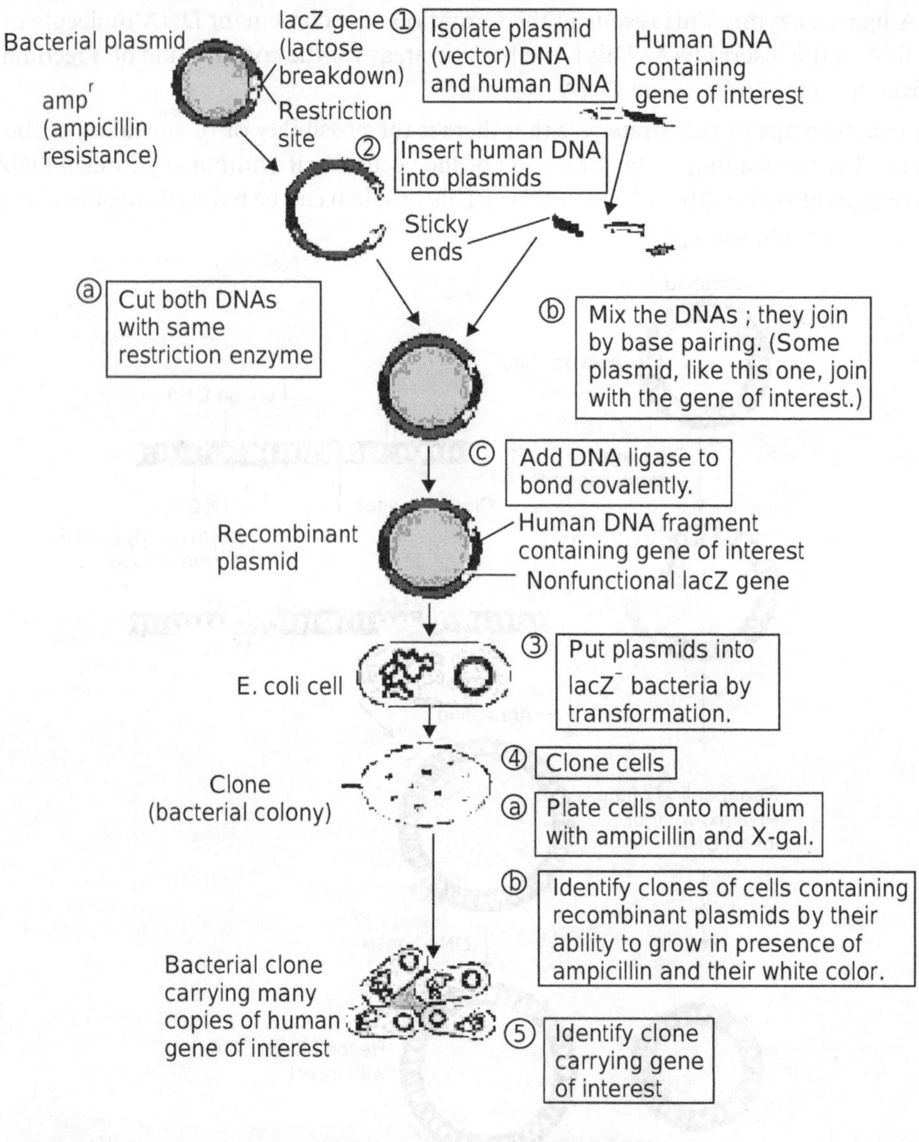

Figure 6.13 General strategy for the construction of rDNA and its cloning. The figure illustrates the cloning of a human gene in *E. coli* with all its details.

6.7 DNA LIBRARY

The isolation of a gene or the DNA strand that carries the gene is the first step in a gene cloning experiment. The construction of a DNA library is a prerequisite for this purpose. There are mainly three types DNA libraries—**genomic library, chromosomal libraries and cDNA library.** These DNA libraries can be used as the source for screening the target gene. These DNA libraries are generally known as the recombinant DNA libraries.

Genomic Library

A genomic library is a collection of cloned, restriction enzyme digested DNA fragments containing at least one copy of every DNA sequence in a genome. The entire genome of an organism is represented as a set of DNA fragments inserted into a vector molecule. These can be propagated or cloned in a suitable organism.

There are three ways to make a genomic library. The genomic DNA of the organism is extracted and is cut into fragments of suitable sizes by any of the following three methods. The genomic DNA is digested completely by a restriction enzyme and converts it into fragments of suitable sizes. The restriction enzyme cuts at all relevant restriction sites and produces a large number of short fragments with sticky ends. The disadvantage of this is that genes containing restriction sites within the reading frame may be cut into two or more fragments and may be cloned separately.

The genomic DNA can be fragmented un-enzymatically by means of mechanical shearing like sonication and it produces longer DNA fragments. The disadvantage in this case is that the ends of the fragments produced are not uniform and it needs enzymatic modification for inserting the fragments into a cloning vector.

The third method for fragmenting the genomic DNA is by partial enzymatic digestion with a restriction enzyme. By limiting amount of restriction enzyme and the time of exposure of DNA to the active enzyme, all the restriction sites present in the genome will not be acted upon by the restriction enzyme. So that the restriction digestion will be partial, resulting in long DNA fragments with overlapping sequences. After digestion the DNA fragments can be size selected by agarose electrophoresis. Like fragments of complete digestion, fragments have sticky ends and can be inserted directly.

The vector is also isolated from a host bacterium and cut with the same restriction enzyme used for the digestion of the genomic DNA. The linearised vector molecule is incubated with the fragments of genomic DNA along with DNA ligase. So that each of the genomic DNA fragments are inserted into each vector molecule resulting in recombinant DNA molecules. Each of the recombinant vector molecules will have different fragment of genomic DNA and all these recombinant DNA molecules are introduced into a suitable host organism such as bacterial cells or phage particle.

Chromosome Library

Chromosome library is the collection of cloned restriction enzyme digested DNA fragments from individual chromosomes. This is very important in the case genomic sequencing of eukaryotes and physical mapping of chromosomes, which indicates the locus of specific genes in specific chromosomes.

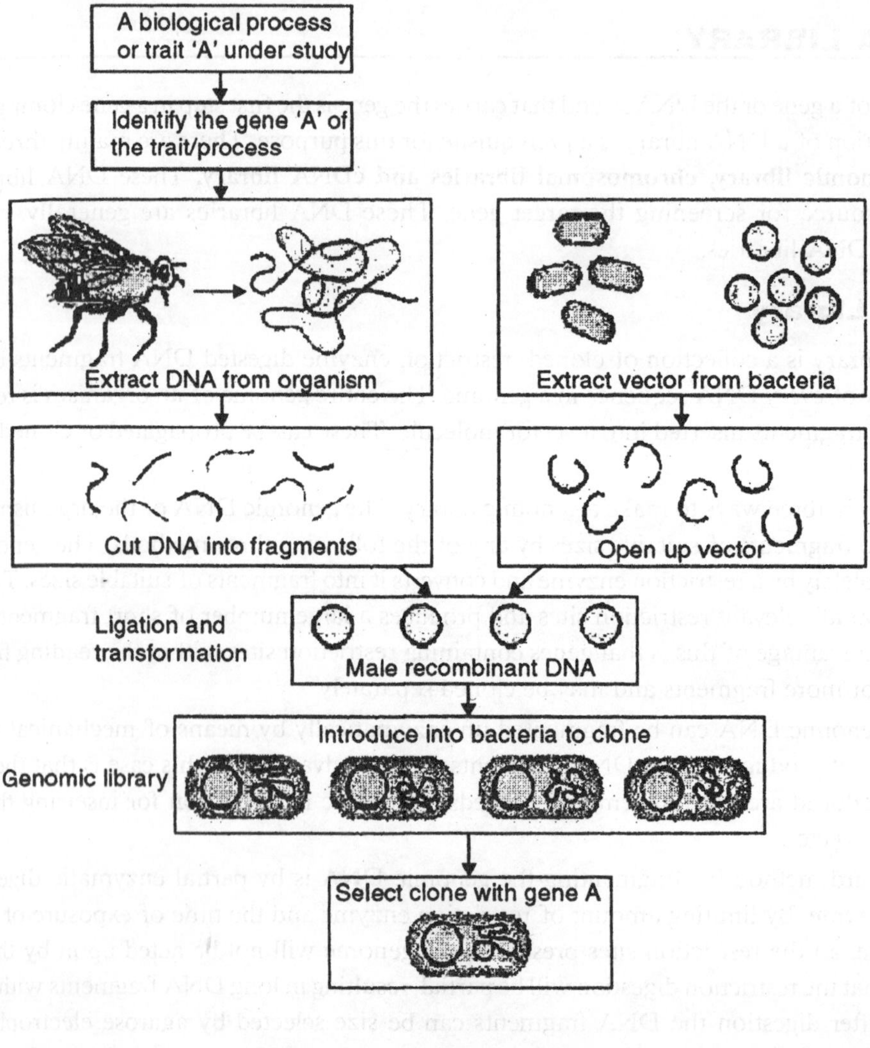

Figure 6.14 Schematic representation of the construction of a genomic library and identification of a gene in the library.

Complementary DNA Library (cDNA Library)

In the case of eukaryotic system, the genomic library may not be of much use if your aim is to isolate an intact gene and express the protein in bacterial host cell active form because of the following reasons:

1. Even though the genomic library represent the complete genome sequences, it is possible that certain genes may be fragmented and may be present in two different bacterial clones.
2. Eukaryotic genes usually contain intervening non-coding regions known as introns. The genes with introns can be expressed in prokaryotic systems as they lack the splicing machinery. So the gene will be expressed along with the introns and the result will be an entirely different protein, which is not active or functional.

3. The regulatory elements of the eukaryotic genes may not be properly understood by the prokaryotic systems and that may create problems in the expression proper protein in active form.

The complementary DNA library or the cDNA library can solve most of these problems, if our aim of gene cloning is to express the gene in bacterium. The cDNA library is the collection of clones of DNA, which are the complementary copies of messenger RNA (mRNA) isolated from the respective cells. Messenger RNA is the starting material for the construction of cDNA libraries. This DNA library represents only those genes, which are being expressed by a group of cells or tissues. Since the mRNAs are produced after splicing they are devoid of introns. Therefore, the complementary copy of these mRNA (cDNA) represents only the exons or the coding regions of the actual eukaryotic genes. This cDNA can be directly inserted into an expression vector like pUC18 or pUC19 and the protein can be expressed in the bacteria systems like *E. coli*.

The construction of a cDNA library starts with the isolation of mRNA from the specialised cells or tissue in which that particular mRNAs are abundant. The mRNA are purified from the total RNA by affinity chromatography using oligo dT columns, which selectively binds poly A tail of mRNA. The purified mRNAs are transcribed back into cDNA by an enzyme known as reverse transcriptase. In the first step single strand of the cDNA is produced and again by enzymatic method the single stranded DNA is converted into a double stranded cDNA molecule. These cDNA molecules are inserted into suitable vectors with the help of appropriate linkers or adapters, which provide the required restriction sites for annealing and ligation. Adapters or linkers are the specialised artificial oligonucleotides carrying restriction sites for a particular restriction enzyme. These can be attached to the ends of the cDNA molecule to provide the restriction site for a particular restriction enzyme. Once the recombinant DNA molecules with the insert cDNA are formed they are introduced into suitable bacterial host cells for cloning and establishing the cDNA library. These cDNA libraries can be used as the source for the isolation of a specific gene and its expression in a prokaryotic system. The expression vector should be provided with the regulatory sequences like the correct type of promoters.

6.8 TRANSGENICS—INTRODUCTION OF RECOMBINANT DNA INTO HOST CELLS

After the construction of the recombinant DNA molecule, it has to be introduced into a suitable host cell so that the DNA will multiply. There are several methods available for this, which depends on several factors including the type of vector used and the host cell that will receive the recombinant vector with the insert gene and the ultimate purpose of the process of gene cloning. The following are some of the common methods used for introducing the rDNA molecule into the host cell.

1. **Transformation** is the most commonly adopted method for introducing a recombinant DNA into a host cell. During the process of transformation, cells take up foreign DNA from its environment. But many cells including *E. coli*, yeast, and mammalian cells are not able to take up the DNA from its environment naturally. Certain chemical treatments can enhance the ability of cells to take up the foreign DNA or make the cell competent for transformation. This enhanced competency for transformation was found out by Mandel and Higa in 1970. They observed that *E. coli* cells become more competent to take up foreign DNA when these cells

were incubated in cold calcium chloride solution. This mechanism is still followed for the transformation of *E. coli* cells as a part of gene cloning.

2. **Transfection** is another method used for the transformation of cultured cells. In this method DNA (recombinant vector) is mixed with charged substances like calcium phosphate, cationic liposomes or DEAE dextran and overlayered on the host cells. This ultimately results in the uptake of the external DNA by these host cells.
3. **Electroporation** is another efficient method for introducing rDNA into a host cell. During the process of electroporation an electric current is used to create transient microscopic pores in the cell membrane of the host cell. Through these temporary openings foreign DNA enters the cell. This method is especially suitable for cells like yeast, mammalian cell and plant protoplasts.
4. **Microinjection** is specialised technique by which the DNA fragment or the gene can be directly injected into the nucleus of the plant and animal cells. This method carried out with out the use of any specialised vectors. The procedure involves the direct injection of the DNA into the nucleus of the host cell with the help of a glass microinjection tube or syringe. The whole mechanism is highly automated and sophisticated instruments attached with a microscope of higher magnification.
5. **Biolistic method** is developed for introducing foreign DNA into plant cells with the help of a gene or particle gun. During the biolistic method microscopic particles of gold or tungsten coated with the DNA of interest is bombarded into the cells at a high velocity. The bombardment of the particle is carried out with the help of a mechanical device called particle gun. Normally the method of introducing foreign gene into the plant cell is by using *Agrobacterium tumifaciens*. But this bacterium is host specific and may not infect and transfer the gene in certain cases, particularly in the case of monocots. Biolistic methods can be adopted in such cases.

In the case of **Bacteriophage vectors** no special methods or treatments are required for the inserting the DNA into the genome because the phage infection and the transfer of DNA into the genome is natural.

6.9 IDENTIFICATION OF RECOMBINANTS

The introduction of the recombinant DNA into a suitable host cell is followed by the selection of those cells, which contain the recombinant vectors. During the process of genetic transformation only very low percentage of the total cell population takes or receives the recombinant DNA. Therefore, it is very important to have an efficient screening method to select the transformed cells or those cells that have taken the foreign DNA. There are various selection methods or screening methods that are based on the expression or non-expression of some of the traits present in the vector or along with the cloned gene. Some of these traits are the resistance to certain antibiotics. If the antibiotic resistance gene is present along with the cloned gene, it is very easy to select the recombinant transformants directly on a medium supplemented with the respective antibiotic. In most of the cases there are two stages of selection. First one is the selection of transformed cells, *i.e.*, the cells that have taken a plasmid. The second one is to identify the transformed cells that have the recombinant plasmid.

The transformed cells that contain a plasmid can be identified by the positive selection method The most common markers used for positive selection are the antibiotic resistance genes present in the cloning vectors. For example, the Ampicillin resistance gene is present in pUC19 and other similar types of vectors. A positive selection method is very important when the DNA is introduced through a plasmid by transformation. Only a very small percentage of the cells pick up the plasmids and become antibiotic resistance. So a strong positive selection system can eliminate all the untransformed cells and allow the transformed cells to grow. By plating the transformation products directly on antibiotic plates, all untransformed bacteria die and only those containing the plasmid with antibiotic resistance marker gene can grow to form colonies.

By this selection method we will be able to get the transformed cells with the plasmids. But it is not possible to identify the transformed cells or colonies having recombinant plasmids from the transformed cells or colonies with parent plasmid without any insert DNA fragment (non-recombinant vector). A negative selection method can be adopted for identifying the presence of recombinant vector among the transformed cells. In order to identify those plasmids carrying a foreign DNA fragment, the multiple cloning site in the vector is engineered in such a way that insertion of a foreign DNA strand disrupts the expression of a selectable marker gene—a phenomenon known as **insertional inactivation**. Two types of selectable markers are used for negative selection. One is the insertional inactivation of antibiotic resistance and the second is insertional inactivation of an enzymatic activity.

In order to use antibiotic resistance as a negative as well as a positive selection system, the plasmid vector must carry two different antibiotic resistance genes. An example of such a vector is pBR322. pBR322 carries both an ampicillin and a tetracycline resistance gene. The phenotype of bacteria containing the intact plasmid is **Amp^r Tet^r** (Ampicillin resistant and Tetracycline resistant). Insertion of foreign DNA into the Pst I site located in the Amp^r gene results in an **Amp^s Tet^r** (ampicillin susceptible and tetracycline resistant) phenotype. Conversely, insertion of foreign DNA into the EcoRI, Hind III or Sal I sites located in the Tet^r gene results in an **Amp^r Tet^s** phenotype. In order to reveal these phenotypes, transformants are first plated on positive selective media (Tet or Amp respectively). Positively selected colonies are then restreaked on positive selective media (master plate from which we recover our desired vector + insert) and on the negative selection media. (Amp or Tet media respectively). The transformed cells with recombinant vector will be either **Amp^s - Tet^r** or **Amp^r - Tet^s** cells or colonies.

While the insertional inactivation of antibiotic resistance works, it requires a lot of manipulation—picking the positively selected bacteria and replating on negative selection media etc. In addition to the tedium of picking colonies, vectors like pBR322 also suffer from a paucity of convenient restriction sites at which to insert the foreign DNA fragments. These limitations promoted the development of a set of host-vector systems in which it is possible to positively select for bacteria carrying a plasmid and simultaneously select for insertional inactivation of enzymatic activity. This system is based on the beta-galactosidase gene (*lac Z* gene) of the *E. coli lac* operon. This method is based on the insertional inactivation of *lac Z* gene. The lac Z gene is a component of the lac operon and it produces an enzyme β-galactosidase. This enzyme can cleave a colourless, synthetic substrate, X-gal into a blue coloured product. If the β-galactosidase gene (*lac Z* gene) is disrupted and inactivated by inserting the foreign DNA fragment into it, X-gal will not be cleaved and the development of blue colour will be prevented. Thus the selection medium is provided with the X-gal, the host cells that carries normal vectors (without insert DNA) will produce blue coloured colonies and those cells which carries the recombinant plasmids will develop white coloured colonies. Here the positive selection (expression of antibiotic resistance) and the negative selection (inactivation of an enzyme activity—β-galactosidase) can be combined together and can be carried out in a single experiment. In addition to X-gal and antibiotic as the means of selection additives, the media also require a compound called IPTG as an inducer for the expression of *lac Z* gene.

The selection method just described is primarily used for the screening of colonies with recombinant plasmids in *E. coli*. But there are several other methods also to identify the clones having the recombinant plasmids. But it depends on the type of vector—host combinations used for the experiment. The presence of the desired insert DNA can be confirmed either by isolating the recombinant plasmids and digesting it with the same restriction enzyme used for making the recombinant vectors, by PCR, by direct DNA sequencing, or by molecular hybridization—Southern hybridization with DNA probes, Northern hybridization with RNA probes and Western hybridization with antibody proteins.

6.10 RESTRICTION FRAGMENT LENGTH POLYMORPHISM

In addition to the use of restriction enzymes in the generation of recombinant DNA molecules, these molecular scissors also have another important application—Restriction Fragment Length Polymorphism (RFLP) or the DNA fingerprinting. The DNA sample isolated from an organism has got a specific sequence and so the distribution of restriction site sequences for various types of restriction endonucleases. A specific type of restriction endonuclease can cut the DNA molecule into a set of fragments, which is reproducible for any DNA sample of that organism. The set of DNA fragments of various sizes produced by the restriction digestion of a DNA sample is known as the restriction pattern, which is reproducible for that individual and thus forms a molecular marker. The RFLP analysis is based on the variations in the DNA sequences even between individuals of a species and will result in the variations in the distribution of restriction sites for one or more types of restriction endonucleases. If two samples of DNA from two different organisms are subjected restriction digestion with the same type of restriction endonuclease, it can create two entirely different restriction patterns, which depends on the variation in the DNA sequence and number and distribution of the restriction sites of the enzyme used. An example is illustrated in figure 6.15.

There are two DNA samples 'A' and 'B' isolated from two different individuals. The sample 'A' has four restriction sites of a restriction endonuclease *Hind* III. The sample 'B' has three restriction sites for the same enzyme. On restriction digestion with *Hind*III, both the DNA samples will give a unique restriction fragment pattern. The number of restriction fragments corresponds to the number of restriction sites and the size of the fragments depends on the relative positions of the restriction sites. Thus, the individuals can be identified by the number and length of the fragments generated by the restriction digestion of the respective DNA samples by a specific restriction enzyme. Therefore, the Restriction Length Polymorphism is a powerful tool for the identification of organisms to the level of species, strains, varieties or individuals.

Different steps involved in the RFLP analysis or DNA finger printing are illustrated in figure 6.15. It is extensively used for the phylogenetic and forensic analysis. The technique of RFLP analysis can be summarized in the following steps. The DNA samples are digested by an appropriate restriction enzyme (single digestion) or by a mixture of two or more enzymes (double digestion, triple digestion etc.), producing a number of DNA fragments of varying sizes with single stranded overhangs (sticky ends) or blunt ends. DNA fragments can be separated according to their molecular weight on an agarose gel. In this process of gel electrophoresis DNA (which has a negative charge) migrates through the gel toward a positive electrode. Large pieces of DNA move through the tangle of agarose carbohydrates slower than do small fragments. DNA in the gel can be visualized under UV light after staining with the fluorescent dye, ethidium bromide. DNA bands in the gel can be transferred to a more durable nylon filter by capillary action (Southern blotting). DNA can be denatured (made single stranded), by heat, basic pH or enzymes. DNA probes are pieces of specific single stranded DNA fragments that have been tagged (labelled) with radioactivity or photoactive markers. Probes will bind their compliment and can be visualized either by autoradiography or by fluorescence.

In addition to RFLP analysis, restriction enzymes are used for the restriction mapping of new samples of DNA molecules like plasmids, viral genomes etc. the graphical representation of the relative positions of the restriction site sequences of various restriction endonucleases is known as the restriction map of a DNA sample.

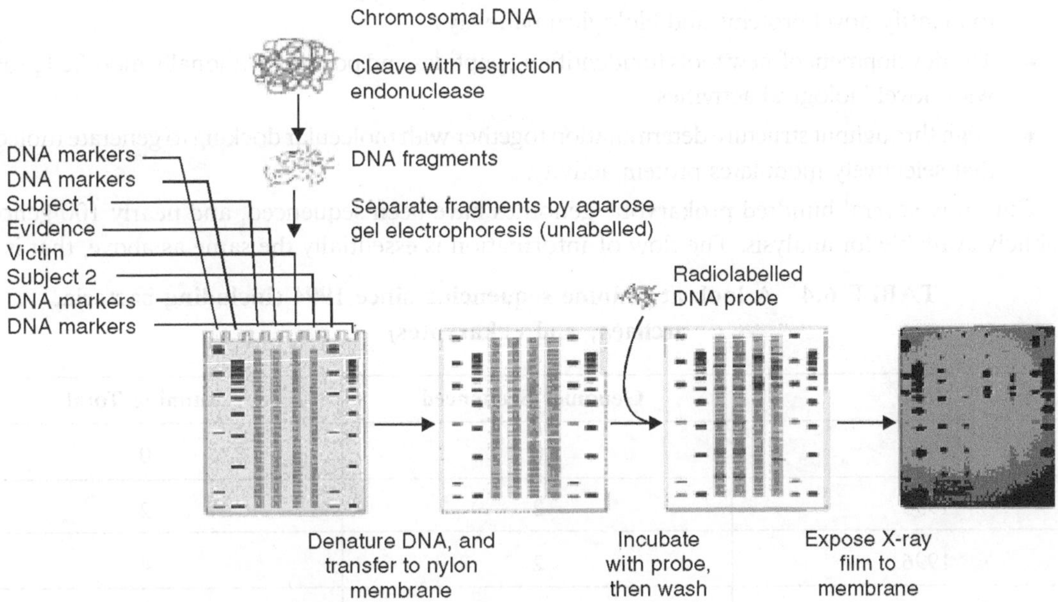

Figure 6.15 Different steps in RFLP analysis in forensic science. Two suspects are screened against a victim along with molecular markers.

6.11 GENOMICS

Genomics is simply the study of complete genomes. More recently, genomics has meant the analysis of the complete DNA sequence of an organism's genome. This field was not really possible until the publication of the genome of *Haemophilus influenzae* in 1995. Fred Sanger sequenced the first bacteriophage (phiX174, 5386bp long), for which he later won the Nobel prize. Although this was a dramatic improvement over the conventional methods, this was still very slow, compared to the amount of information in a single human cell. About a decade later, the human genome project was launched; this was an international effort, and the U.S. would pay about $200,000,000 per year for 20 years! Most of this investment was in technology to speed sequencing, which in fact has been realised. Within a few years, it is likely that it will be possible to read the entire DNA sequence of a human cell, in a few hours.

The completion of the human genome sequence opens an enormous opportunity for chemistry to understand the complex biological processes of living organisms at a molecular level. This opportunity coincides with a tremendous increase in our ability to design, synthesize, and analyze experiments due to combinatorial methods in biochemistry and genomics, enhanced computational capabilities, and highly sensitive analytical tools. We can apply these new methods to a number of problems in modern genomics, including :

- The synthesis of combinatorial chemical libraries and high throughput cell-based screening of these libraries for molecules with novel biological activities.
- The use of mRNA expression analysis together with yeast genetics and retroviral cDNA libraries to identify novel proteins and biological pathways.
- The development of new tools for identifying peptides and post-translationally modified proteins with novel biological activities.
- High throughput structure determination together with molecular docking to generate molecules that selectively modulates protein activity.

Currently several hundred prokaryotic genomes have been sequenced, and nearly 100 genomes are publicly available for analysis. The flow of information is essentially the same as above, that is :

TABLE 6.4 A look at genome sequencing since 1994 (including bacteria, archaea, and eukaryotes)

Year	Genomes Sequenced	Running Total
1994	0	0
1995	2	2
1996	2	4
1997	5	9
1998	8	17
1999	13	30
2000	23	53
2001	42	95
2002	>100	>200

6.12 GENOME MAPPING

The complete biological information required to build and maintain a living organism including human is designated as its genome. This biological information of the genome is encoded as the nitrogen bases or nucleotides of its **deoxyribonucleic acid (DNA)** and are divided into functional units called genes. These genes code for proteins and cause a number of biochemical reactions including the expression of a number of other genes. The structure of DNA and organization of genes were discussed in earlier chapters.

A genome map is the relative positions of genes in a genome. The graphical representation genes of genome are known as the Genome Map. The genome map help a scientist to locate a specific gene in a genome just like a road map. There are two types of genome map—the genetic map and the physical map.

The genetic map and the physical map provide the scientists the order and position the gene in question and its distance from other genes or markers in a genome, just like the address and map of the road that leads to a particular house that we are searching. The genetic map is just like the road map between cities and the physical map is similar to the map of a city specifying the exact distances between places giving almost complete information to find that place. Both genetic and physical maps provide the likely order of items along a chromosome. One could say that genetic maps serve to guide a scientist toward a gene, just like an interstate map guides a driver from city to city. On the other hand, physical maps mark an estimate of the true distance, in measurements called **base pairs (bp)**, between genes or locations of interest. Depending upon the quality of the genetic map and physical map of a genome one could easily locate a specific gene in the crowded city of genome.

Genetic Map

Genetic map is the graphical representation of various genes in its order and locations with the help of certain markers known as genetic markers. Just like certain places and buildings form the markers known as landmarks in the town map to locate a specific place of interest genetic markers form the land marks in the genome map directing certain destinations of interest. The genetic marker is a broad term, which includes any observable variation that results from an alteration, or mutation, at a single genetic locus. A genetic marker can be used as landmark in a genetic map if it is inherited from the parent to the offspring as per the standard genetic laws.

Markers can be a noticeable physical characteristic such as eye colour, drug resistance or a not so noticeable trait such as a disease. **DNA-based reagents** can also serve as markers. Markers for DNA based reagents are found mostly within the non-coding regions of genes and are used to detect unique regions on a chromosome. DNA markers are especially useful for generating genetic maps when there are occasional, predictable mutations that occur during **meiosis**, which cause a high degree of variations within the DNA content of the marker over many generations from individual to individual.

The following are the most commonly used genetic markers to design a genetic map :

1. **RFLPs**, or **Restriction Fragment Length Polymorphisms**, were among the first developed DNA markers. RFLPs are defined by the presence or absence of a specific site, called a restriction

site, for a bacterial restriction enzyme. This enzyme breaks apart strands of DNA wherever they contain a certain nucleotide sequence. RFLP also finds a number of other uses particularly in criminology.

2. **VNTRPs**, or **Variable Number of Tandem Repeat Polymorphisms**, occur in non-coding regions of DNA. This type of marker is defined by the presence of a nucleotide sequence that is repeated several times. In each case, the number of times a sequence is repeated may vary.
3. **Microsatellite polymorphisms** are defined by a variable number of repetitions of a very small number of base pairs within a sequence. Oftentimes, these repeats consist of the nucleotides, or bases, cytosine and adenosine. The number of repeats for a given microsatellite may differ between individuals, hence the term **polymorphism**—the existence of different forms within a population.
4. **SNPs**, or **Single Nucleotide Polymorphisms**, are individual point mutations, or substitutions of a single nucleotide, that do not change the overall length of the DNA sequence in that region. SNPs occur throughout an individual's genome.

   Genetic maps are also used to generate the essential framework needed for the construction of more detailed genome maps. These detailed maps, called **physical maps**, further define the DNA sequence between genetic markers and are essential to the rapid identification of genes.

Physical Map of Genome

Physical maps are the detailed genetic maps, which describe the position of genes in terms of base pairs or gene sequences between genetic markers or genes. Physical maps can be divided into three general categories. They are **chromosomal** or **cytogenetic maps**, **radiation hybrid (RH) maps**, and **sequence maps**. The different types of maps vary in their degree of **resolution or clarity.**

The lowest-resolution physical map is the **chromosomal** or **cytogenetic map**, which is based on the distinctive banding patterns observed by light microscopy of stained chromosomes. As with genetic linkage mapping, chromosomal mapping can be used to locate genetic markers defined by traits observable only in whole organisms. Because chromosomal maps are based on estimates of physical distance, they are considered to be physical maps. Yet, the number of base pairs within a band can only be estimated.

RH maps and **sequence maps**, on the other hand, are more detailed. RH maps are similar to linkage maps in that they show estimates of distance between genetic and physical markers, but that is where the similarity ends. **RH maps** are able to provide more precise information regarding the distance between markers than can a linkage map.

The physical map that provides the most detail is the sequence map. **Sequence maps** show genetic markers, as well as the sequence between the markers measured in base pairs.

RH Mapping (Radiation Hybrid Maps)

In this technique, scientists, instead of depending on natural recombination to separate two markers, use breaks induced by radiation to determine the distance between two markers. In RH mapping, DNA is exposed to a measured dose of radiation, and in doing so, controls the average distance between breaks in a chromosome. By varying the degree of radiation exposure to the DNA, one can control the induction of breaks between two markers that are very close together. RH mapping, like linkage mapping, shows an estimated distance between genetic markers. The ability to separate closely linked markers allows scientists to produce more detailed maps. RH mapping provides a way to localize almost any genetic marker, as well as other genomic fragments, to a defined map position. Actually RH maps are intermediate between linkage maps and sequence maps in clarity or resolution. Therefore, RH maps can be used as a bridge between linkage maps and sequence maps in locating and identifying the position of gene in the genome. This can help in the easy identification of the location(s) of genes involved in certain diseases such as spinal muscular atrophy and hyperekplexia (startle disease). RH maps are extremely useful for ordering markers in regions where highly polymorphic genetic markers are scarce or absent.

Sequence Mapping

Sequence tagged site, or STS mapping, is one of the latest physical mapping technique, which can give a high degree of resolution. An STS is a short DNA sequence that has been shown to be unique. To qualify as an STS, the exact location and order of the bases of the sequence must be known, and this sequence may occur only once in the chromosome being studied or in the genome as a whole if the DNA fragment set covers the entire genome. The following are some of the most commonly used sequences for the identification of gene locations in a genome.

Expressed sequence tags (ESTs) are short sequences obtained by analysis of complementary DNA (cDNA) clones. Complementary DNA is prepared by converting mRNA into double-stranded DNA and is thought to represent the sequences of the genes being expressed. An EST can be used as a STS if it comes from a unique gene and not from a member of a gene family in which all of the genes have the same, or similar, sequences.

Simple sequence length polymorphisms (SSLPs) are arrays of repeat sequences that display length variations. SSLPs that are polymorphic and have already been mapped by linkage analysis are particularly valuable because they provide a connection between genetic and physical maps.

Random genomic sequences are obtained by sequencing random pieces of cloned genomic DNA or by examining sequences already deposited in a database.

6.13 GENOME SEQUENCING PROJECTS

The genomes of a large number of organisms were sequenced and a number of them are in the final stages by different genome project groups. The genome projects are having a profound impact on health care discovery. The main purpose of genome projects are to access the entire genome sequences and that can be used to find out the probable genes and their functions of the various organisms. Some important genomes that are partially or completely sequenced include that of *E. coli*, *Mycobacterium tuberculosis*, yeast, *Arabidopsis*, *Drosophila*, mouse, humans, rice, wheat etc. Complete sequences are available for *Arabidopsis*, yeast, Human, *Mycobacterium* etc. By 2005 most of the important genomes would be sequenced and the sequence will be available for non-commercial purposes from the databases. The availability of these databases for researchers will help them to better understanding of the gene and its expression and thereby the better understanding of the fundamental biological processes. For example, the human genome project would greatly help the development of gene therapy. The final and the complete sequence of the human genome would unravel the mystery of heredity and reveal the sources of many diseases like diabetes, heart-diseases, cancers, hemophilia, Alzheimer's diseases and cystic fibrosis. It can also help unravelling the intricacies of development and differentiation. The genome sequencing of a number of important pathogenic organisms like malarial parasite will help to design methodologies to control the disease.

6.14 SNPs AND COMPARATIVE GENOMICS

Human genome project has proved beyond doubt that the genetic sequence of all human are identical to about 99.8 %. That means the DNA sequence in all individuals are identical. But this 0.2% difference is more than enough to make unique differences between individuals. But it became evident from the recent genome analysis and sequencing that an average one out of thousand nucleotide sequences are different in the genome of two individuals. Single Nucleotide Polymorphism or SNPs are an important variation observed between two individuals.

A **Single Nucleotide Polymorphism**, or SNP (snip), is a small genetic change, or variation, that can occur within a person's DNA sequence. Among the four **nucleotide** 'letters' A (adenine), C (cytosine), T (thymine), and G (guanine),which specify the gene sequence SNP variation occurs when a single nucleotide, such as an A, replaces one of the other three nucleotide letters—C, G, or T. For example, a SNP is the alteration of the DNA segment GTATTAGGTTA to GTATTACGTTA, where the seventh 'G' in the first snippet is replaced with a 'C'. On average, SNPs occur in the human population more than 1 per cent of the time. Snips can be in the coding and non-coding regions of the genome. But since only about 3 to 5 per cent of a person's DNA sequence codes for proteins, most SNPs are found in the non-coding sequences. Interestingly those SNPs found within a coding sequence are of very important because they are more likely to alter the biological function of a protein. Because of the recent advances in technology, coupled with the unique ability of these genetic variations to facilitate gene identification, there has been a recent flurry of SNP discovery and detection. This comparative genomics is very important in various fields particularly in pharmaceutical industry and in disease diagnosis. For example, a single base difference in the ApoEi gene is associated with the Alzheimer disease. A simple deletion within the chemokine-receptor gene CCR5 leads to resistance to AIDS. Therefore, SNPs are very important in diagnosis and prescription of medicines.

Each person's genetic material contains a unique SNP pattern that is made up of many different genetic variations. Researchers have found that most SNPs are not responsible for a disease state. Instead, they serve as biological markers for pinpointing a disease on the human genome map, because they are usually located near a gene found to be associated with a certain disease. Occasionally, a SNP may actually cause a disease and, therefore, can be used to search for and isolate the disease-causing gene. It will only be a matter of time before physicians can screen patients for susceptibility to a disease by analyzing their DNA for specific SNP profiles.

6.15 PHARMACOGENOMICS

The way a person responds to a drug (both positive and negative reactions) is a complex trait that is influenced by many different genes. Without knowing all of the genes involved in drug response, scientists have found it difficult to develop genetic tests that could predict a person's response to a particular drug. Once scientists discovered that people's genes show small **variations** (or changes) in their **nucleotide** (DNA base) content, all of that changed—genetic testing for predicting drug response is now possible.

Pharmacogenomics is a science that examines the inherited variations in genes that dictate drug response and explores the ways these variations can be used to predict whether a patient will have a good response to a drug, a bad response to a drug, or no response at all.

6.16 FUNCTIONAL GENOMICS AND MICROARRAYS

DNA microarrays are perfectly suited for comparing gene expression in different populations of cells. The hows and whys of such an experiment provide insight into the power of microarrays, their limitations, and the kinds of biological questions, which they can help to answer. Figure 6.16 shows the steps that make up a comparative cDNA hybridization experiment. Click on any of the steps in the image to jump to an explanation of how and why it is performed.

The major steps of a comparative cDNA hybridization experiment are :

1. Choosing a cell population
2. mRNA extraction and cDNA synthesis
3. Fluorescent labelling of cDNAs
4. Hybridization to a DNA microarray
5. Scanning the hybridized array
6. Interpreting the scanned image.

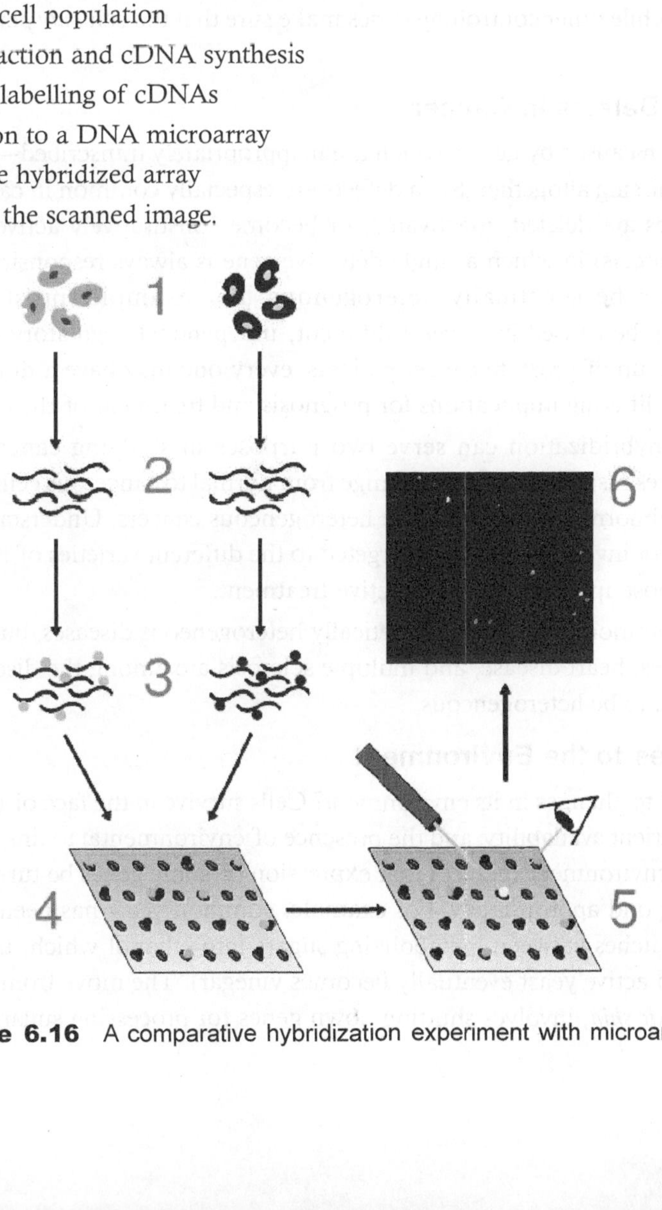

Figure 6.16 A comparative hybridization experiment with microarrays.

1. Choosing Cell Populations

The goal of comparative cDNA hybridization is to compare gene transcription in two or more different kinds of cells. We will describe some experiments of particular interest; however, the possibilities for informative comparative transcription studies are limited only by the investigator's imagination.

Tissue-specific Genes

Cells from two different tissues (say, cardiac muscle and prostate epithelium) are specialized for performing different functions in an organism. Although we can recognize cells from different tissues by their phenotypes, it is not known just what makes one cell function as smooth muscle, another as a neuron, and still another as prostate. Ultimately, a cell's role is determined by the proteins it produces, which in turn depend on its expressed genes. Comparative hybridization experiments can reveal genes, which are preferentially expressed, in specific tissues. Some of these genes implement the behaviours that distinguish the cell's tissue type, while other controlling genes make sure that the cell *only* performs the functions for its type.

Regulatory Gene Defects in Cancer

Genetic disease is often caused by genes, which are inappropriately transcribed—either too much or too little—or which are missing altogether. Such defects are especially common in cancers, which can occur when regulatory genes are deleted, inactivated, or become constitutively active. Unlike some genetic diseases (*e.g.* cystic fibrosis) in which a single defective gene is always responsible, cancers that appear clinically similar can be genetically heterogenous. For example, prostate cancer (prostatic adenocarcinoma) may be caused by several different, independent regulatory gene defects even in a single patient. In a group of prostate cancer patients, every one may have a different set of missing or damaged genes, with differing implications for prognosis and treatment of the disease.

Comparative hybridization can serve two purposes in studying cancer: it can pinpoint the transcription differences responsible for the change from normal to cancerous cells, and it can distinguish different patterns of abnormal transcription in heterogeneous cancers. Understanding the diverse basis of a cancer is crucial for inventing therapies targeted to the different varieties of the disease, so that each patient receives the most appropriate and effective treatment.

Cancers are common examples of genetically heterogeneous diseases, but they are by no means the only ones. Diabetes, heart disease, and multiple sclerosis are among the diseases for which genetic risk factors are known to be heterogeneous.

Cellular Responses to the Environment

How does a cell adapt to changes in its environment? Cells survive in the face of changes in temperature and pH, changing nutrient availability, and the presence of environmental toxins and ionizing radiation. Usually, a change in environment requires that expression of some genes be turned up or down so that the organism can respond appropriately. For example, common yeast has been extensively studied to understand how it switches between metabolizing sugars into ethanol which, in turn, into acetic acid (this is why wine with active yeast eventually becomes vinegar). The move from one metabolic state to the other, called *diauxic shift*, involves shutting down genes for processing sugars and activating others for processing ethanol, as well as a general stress response due to the greater difficulty of deriving energy from ethanol.

Comparative hybridization experiments can point out genes whose transcription changes in response to an environmental stimulus. In the simplest experiment, a population of cells is subjected to the stimulus and allowed to reach a steady state of transcription. Transcription levels in the altered cells can then be compared to those in a control population. A more informative experiment subjects cells to a change, then takes samples of the cell population at successive points in time. In this way, the experimenter can watch as the gene transcription patterns change from the old to the new steady state. Temporal studies can identify not only genes whose transcription changes but also the *order* of the changes, providing evidence about which genes control the response directly and which are only indirectly affected by it.

Environmental changes of interest also include the introduction of signalling molecules, such as hormones, interleukins, and interferons, as well as the actions of drugs. All these molecules stimulate a change in a cell's behaviour (including possibly its death). While some of the changes may be mediated purely at the protein level, others require new transcription, which can be detected by comparative hybridization.

Cell Cycle Variations

Even in a stable environment, cells undergo DNA replication, mitosis, and eventually death. These activities require quite different gene products, such as DNA polymerases for genome replication or microtubule spindle proteins for mitosis. A cell's genes encode the 'programs' for these activities, and gene transcription is required to execute those programs. Comparative hybridization can be used to distinguish genes that are expressed at different times in the cell cycle. In this way, the pathways responsible for controlling basic life processes can be uncovered.

2. mRNA Extraction and cDNA Synthesis

Genes that code for protein are transcribed into messenger RNAs (mRNAs) in the cell nucleus. The mRNAs in turn are translated into proteins by ribosomes in the cytoplasm. The transcription level of a gene is taken to be the amount of its corresponding mRNA present in the cell. Comparative hybridization experiments compare the amounts of many different mRNAs in two cell populations.

To prepare mRNA for use in a microarray assay, it must be purified from total cellular contents. mRNA accounts for only about 3% of all RNA in a cell, so isolating it in sufficient quantity for an experiment (1-2 micrograms) can be a challenge. Common mRNA isolation methods take advantage of the fact that most mRNAs have a poly-adenine (poly(A)) tail. These poly(A)$^{+}$ mRNAs can be purified by capturing them using complementary oligodeoxythymidine (oligodT) molecules bound to a solid support, such as a chromatographic column or a collection of magnetic beads.

Captured mRNAs are still difficult to work with because they are prone to being destroyed. The environment is full of RNA-digesting enzymes (there are some on your fingers, your keyboard, your mouse, and every other exposed surface around you right now), so free RNA is quickly degraded. To prevent the experimental samples from being lost, they are reverse transcribed back into more stable

DNA form. The products of this reaction are called complementary DNA or cDNAs, because their sequences are the complements of the original mRNA sequences. The reverse transcription reaction usually starts from the poly(A) tail of the mRNA and moves toward its head; such a reaction is called *oligo(dT)-primed.*

A problem with cDNA production is that not all mRNAs are reverse-transcribed with the same efficiency. This fact leads to *reverse transcription bias*, which can change the relative amounts of different cDNAs measured by the microarray assay. Reverse transcription bias is not a problem when comparing the same mRNA across two cell populations unless it causes the mRNA not to be transcribed at all. However, the bias does prohibit quantitative comparison between different mRNAs on one array. Another problem caused by bias is that some mRNAs may be reverse-transcribed for only part of their lengths, making them less likely to bind and stay bound to their complements on the array. One way of getting around this problem is to prime reverse transcription from random starting positions on the mRNAs, rather than always starting from their tails. This method can reduce bias, but it also makes useless cDNA from any remaining ribosomal and transfer RNAs (tRNAs) in the sample.

3. Fluorescent Labelling of cDNAs

In order to detect cDNAs bound to the microarray, we must label them with a reporter molecule that identifies their presence. The reporters currently used in comparative hybridization to microarrays are fluorescent dyes (fluors), represented by the red and green circles attached to the cDNAs in figure 6.16. A different coloured fluor is used for each sample so that we can tell the two samples apart on the array. The labelled cDNA samples are called probes because they are used to probe the collection of spots on the array.

The colours of the fluors in the diagram are just for illustrations. The actual fluors do not show their colours unless stimulated with a specific frequency of light by a laser. Even then, the colours are not directly observed; rather, the wavelength of the emitted light is used to tune a detector, which measures the fluorescence. The choice of red and green colours is reminiscent of the emission wavelengths of commonly used fluors, such as rhodamine and fluorescein or Cy3 and Cy5.

The number of fluor molecules that label each cDNA, depends on its length and possibly its sequence composition, both of which are often unknown. This is one more reason that fluorescent intensities for different cDNA's cannot be quantitatively compared. However, identical cDNA's from the two probes are still comparable as long as the same number of label molecules is added to the same DNA sequence in each probe.

To equalize the total concentrations of the two cDNA probes before applying them to the array, the probe solutions are diluted to have the same overall fluorescent intensity. This procedure makes two possibly unjustified assumptions: first, that the total amount of mRNA in each cell type being tested is identical; and second, that each fluor emits the same amount of light relative to its concentration. The second assumption can be eliminated by suitable calibration, but the first one is difficult to check. It may, therefore, be that cells with more abundant mRNA are made into a probe with artificially low mRNA concentrations.

4. Hybridization to a DNA Microarray

The two cDNA probes are tested by hybridizing them to a DNA microarray. The array holds hundreds or thousands of spots, each of which contains a different DNA sequence. If a probe contains a cDNA whose sequence is complementary to the DNA on a given spot, that cDNA will hybridize to the spot, where it will be detectable by its fluorescence. In this way, every spot on an array is an independent assay for the presence of a different cDNA. There is enough DNA on each spot that both probes can hybridize to it at once without interference.

Microarrays are made from a collection of purified DNA's. A drop of each type of DNA in solution is placed onto a specially prepared glass microscope slide by an arraying machine. The arraying machine can quickly produce a regular grid of thousands of spots in a square about 2 cm on a side, small enough to fit under a standard slide coverslip. The DNA in the spots is bonded to the glass to keep it from washing off during the hybridization reaction.

The choice of DNA's to be used in the spots on a microarray determines which genes can be detected in a comparative hybridization assay. For organisms whose genomes have been completely sequenced, including several bacteria and the yeast *Saccharomyces cerevisciae*, it is possible to array genomic DNA from every known gene or suspected open reading frame (ORF) in the organism. Each gene or ORF is amplified from total genomic DNA by PCR, producing enough DNA to make unlimited numbers of arrays. The Pat Brown Lab at Stanford University has arrayed all known or suspected genes of *S. cerevisciae* (roughly 6100) on a single microarray.

Because the human genome has not been completely sequenced, we cannot yet produce a comprehensive array for all its genes. The number of human genes has been estimated at somewhere between 30,000 and 50,000, so several arrays will probably be required to hold them all. Despite these limitations, several strategies can be used today to make arrays for studying human genes. We do know the location and sequence of quite a few human genes now, so the same method used to array yeast genes will produce at least a partial human genome array. There are two other ways to produce arrayable DNA even for unknown genes. Amplify clone inserts from human cDNA libraries, or synthesize oligonucleotides directly from known expressed sequence information such as ESTs. While neither of these methods will produce DNAs for every human gene, both can yield enough different expressed sequences to make substantial arrays. Both types of DNA have been used before in array-like applications: cDNA libraries were used for comparative hybridization before the advent of fluorescent microarrays, while oligonucleotide arrays are available commercially today from Affymetrix Corporation for rapid resequencing of a few genes important to AIDS and some cancers.

5. Scanning the Hybridized Array

Once the cDNA probes have been hybridized to the array and any loose probe has been washed off, the array must be scanned to determine how much of each probe is bound to each spot. The probes are tagged with fluorescent reporter molecules, which emit detectable light when stimulated by a laser. The emitted light is captured by a detector, either a *charge-coupled device (CCD)* or a *confocal microscope*, which records its intensity. Spots with more bound probe will have more reporters and will therefore fluoresce more intensely.

Each of the two fluorescent reporters (fluors) used has a characteristic excitation wavelength; only light of this wavelength will cause the molecule to fluoresce. The emitted light has a characteristic emission wavelength, which is different from the excitation wavelength. The detector for the emitted fluorescence from the array is sensitive to the emission wavelength but filters out the excitation wavelength; in this way, the fluorescent light of interest can be separated from the laser light scattered off the slide.

A good pair of fluors for a comparative hybridization experiment should have very different emission or excitation wavelengths. If the emission wavelengths are different, light emitted from the two fluors can be selectively filtered to measure the amount emitted by each fluor separately. If the excitation wavelengths are different, the two fluors can be stimulated and scanned one at a time. If one of these conditions is not met, the scanned intensities can be contaminated by crosstalk between the two fluorescent channels.

Although the purpose of the scanner is to pick up light emitted by probe cDNA's bound to their complementary spots, it also records light from a few molecules that hybridized either to the wrong spot or nonspecifically to the glass slide. This extra light becomes the background of the scanned array image. One advantage of current microarray technology over its predecessors is that the background is extremely low; consequently, the signal-to-noise ratio of the scanned data can be quite high.

6. Interpreting the Scanned Image

The end product of a comparative hybridization experiment is a scanned array image. A small piece of such an image is shown above. The measured intensities from the two fluorescent reporters have been false-coloured red and green and overlaid. Yellow spots have roughly equal amounts of bound cDNA from each cell population and so have equal intensity in the red and green channels (red + green = yellow). Spots whose mRNAs are present at a higher level in one or the other cell population show up as predominantly red or green. The intensities provided by the array image can be quantified by measuring the average or integrated intensities of the spots. The ratio of fluorescent intensities for a spot is interpreted as the ratio of concentrations for its corresponding mRNA in the two cell populations. Schena *et al.*, in 1996 have demonstrated the ability to detect quantitative changes of as little as a factor of two, with reasonable agreement between expression ratios measured on the array and ratios measured by an alternate form of RNA blotting.

Interpreting the data from a microarray experiment can be challenging. Quantisation of the intensities on each spot is subject to noise from irregular spots, dust on the slide, and nonspecific hybridization. Deciding the intensity threshold between spots and background can be difficult, especially when the spots fade gradually around their edges. Detection efficiency might not be uniform across the slide, leading to excessive red intensity on one side of the array and excessive green on the other side. Even after overcoming detection and calibration problems, the measured intensities for each spot only represent the ratio of cDNA's in each cell population. Low levels of cDNA due to reverse transcription bias, sample loss, or an inherently rare mRNA can cause large uncertainties in these ratios.

Numerous software packages, both free and commercial, exist for quantitating microarray data. The interpreted array data will highlight a relatively small number of spots representing very differentially expressed mRNAs whose genes deserve further investigation. Alternatively, the overall pattern of expression can be used as a "fingerprint" to characterize specific cell types (e.g. different classes of tumors), even if not all the differentially expressed genes on an array have been identified.

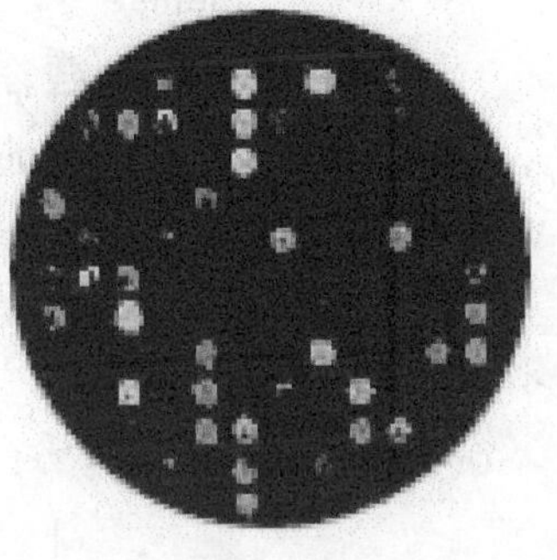

Figure 6.17 Colour scheme in a Microarray.

Green spots represents control DNA where either DNA or cDNA derived from normal tissue is hybridized to the target DNA.

Red represents sample DNA where either DNA or cDNA is derived from diseased tissue hybridized to the target DNA.

Yellow represents a combination of control and sample DNA where both hybridized equally to the target DNA.

Black represents areas where neither the control nor sample DNA hybridized to the target DNA.

Each spot on an array is associated with a particular gene. Each colour in an array represents either healthy (control) or diseased (sample) tissue. Depending on the type of array used, the location and intensity of a colour will tell us whether the gene, or mutation, is present in either the control and/or sample DNA. It will also provide an estimate of the expression level of the gene(s) in the sample and control DNA.

6.17 MOLECULAR PHYLOGENY

Phylogenetic systematics is that field of biology that does deal with identifying and understanding the evolutionary relationships among the many different kinds of life on earth, both living (**extant**) and dead (**extinct**). Evolutionary theory states that similarity among individuals or species is attributable to common descent, or inheritance from a common ancestor. Thus, the relationships established by phylogenetic systematics often describe a species' evolutionary history and, hence, its **phylogeny**, the historical relationships among lineages or organisms or their parts, such as their genes.

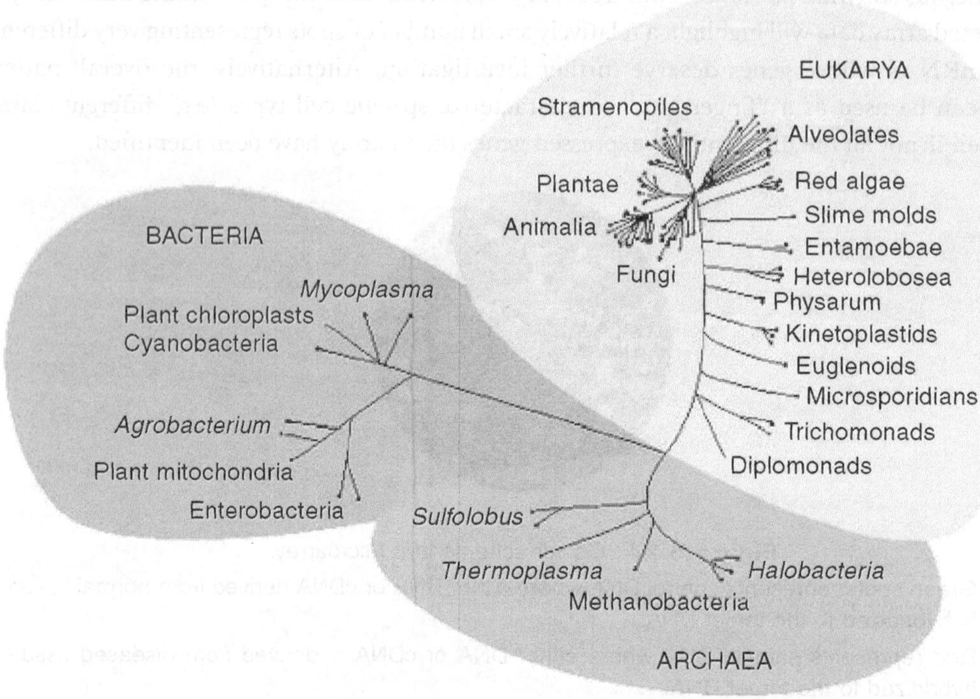

Figure 6.18 Molecular phylogeny.

Evolution is not always discrete with clearly defined boundaries that pinpoint the origin of a new species, nor is it a steady continuum. Evolution requires genetic variation and genetic variation, results from changes within a **gene pool**, the genetic make-up of a specific population. A gene pool is the combination of all the **alleles** —alternative forms of a genetic locus—for all traits that population may exhibit. Changes in a gene pool can result from **mutation**—variation within a particular gene—or from changes in **gene frequency**—a measure of the proportion of an allele in a given population.

Macromolecular data, meaning gene (DNA) and protein sequences, are accumulating at an increasing rate because of recent advances in molecular biology. For the evolutionary biologist, the rapid accumulation of sequence data from whole genomes has been a major advance, because the very nature of DNA allows it to be used as a 'document' of evolutionary history. Comparisons of the DNA sequences of various genes between different organisms can tell a scientist a lot about the relationships of organisms that cannot otherwise be inferred from **morphology**, or an organism's outer form and inner structure. Because genomes evolve by the gradual accumulation of mutations, the amount of nucleotide sequence difference between a pair of genomes from different organisms should indicate how recently those two genomes shared a common ancestor. Two genomes that diverged in the recent past should have fewer differences than two genomes whose common ancestor is more ancient. Therefore, by comparing different genomes with each other, it should be possible to derive evolutionary relationships between them, the major objective of molecular phylogenetics.

Molecular phylogenetics attempts to determine the rates and patterns of change occurring in DNA and proteins and to reconstruct the evolutionary history of genes and organisms. Two general approaches may be taken to obtain this information. In the first approach, scientists use DNA to study the evolution of an organism. In the second approach, different organisms are used to study the evolution of DNA. Whatever the approach, the general goal is to **infer process from pattern**: the processes of organismal evolution deduced from patterns of DNA variation and processes of molecular evolution inferred from the patterns of variations in the DNA itself.

As we just discussed, macromolecules, especially gene and protein sequences, have surpassed morphological and other organismal characters as the most popular forms of data for phylogenetic analyses. Therefore, this next section will concentrate only on molecular data.

It is important to point out that a single, all-purpose recipe does not exist for phylogenetic analysis of this type of data. Although numerous algorithms, procedures, and computer programs have been developed, their reliability and practicality are, in all cases, dependent upon the size and structure of the data set under analysis. The merits and shortfalls of these various methods are subject to much scientific debate, because the danger of generating incorrect results is greater in computational molecular phylogenetics than in many other fields of science. Occasionally, the limiting factor in such analyses is not so much the computational method used, but the users' understanding of what the method is actually doing with the data. Therefore, the goal of this section is to demonstrate to the reader that practical analysis should be thought of both as a search for a correct model (analysis) as well as a search for the correct tree (outcome).

Phylogenetic tree-building models presume particular evolutionary models. For any given set of data, these models may be violated because of various occurrences, such as the transfer of genetic material between organisms. Therefore, when interpreting a given analysis, a person should always consider the model used and entertain possible explanations for the results obtained. For example, models used in molecular phylogenetic analysis methods make 'default' assumptions, including:

The sequence is correct and originates from the specified source. The sequences are homologous—are all descended in some way from a shared ancestral sequence. Each position in a sequence alignment is homologous with every other in that alignment. Each of the multiple sequences included in a common analysis has a common phylogenetic history with the other sequences. The sampling of taxa is adequate to resolve the problem under study. Sequence variation among the samples is representative of the broader group. The sequence variability in the sample contains phylogenetic signal adequate to resolve the problem under study. A straightforward phylogenetic analysis consists of four steps :

- Alignment—building the data model and extracting a dataset.
- Determining the substitution model—consider sequence variation.
- Tree building.
- Tree evaluation.

Studies of gene and protein evolution often involve the comparison of **homologs**, sequences that have common origins but may or may not have common activity. Sequences that share an arbitrary level of similarity determined by alignment of matching bases are **homologous**. These sequences are inherited from a common ancestor that possessed similar structure, although the ancestor may be difficult to determine because it has been modified through descent.

The field of molecular phylogenetics has grown, both in size and in importance, since its inception in the early 1990s, attributable mostly to advances in molecular biology and more rigorous methods for phylogenetic tree building. The importance of phylogenetics has also been greatly enhanced by the successful application of tree reconstruction, as well as other phylogenetic techniques, to more diverse and perplexing issues in biology. Today, a survey of the scientific literature will show that molecular biology, genetics, evolution, development, behaviour, epidemiology, ecology, systematics, conservation biology, and forensics are but a few examples of the many disparate fields conceptually united by the methods and theories of molecular phylogenetics. Phylogenies are used essentially the same way in all of these fields, either by drawing inferences from the structure of the tree or from the way the character states map onto the tree. Biologists can then use these clues to build hypotheses and models of important events in history. Broadly speaking, the relationships established by phylogenetic trees often describe a species' evolutionary history and, hence, its phylogeny—the historical relationships among lineages or organisms or their parts, such as their genes. Phylogenies may be thought of as a natural and meaningful way to order data, with an enormous amount of evolutionary information contained within their branches. Scientists working in these different areas can then use these phylogenies to study and elucidate the biological processes occurring at many levels of life's hierarchy.

• REVIEW QUESTIONS •

1. How the gene number influence the genome size and organization ?
2. Explain the possible benefits of human genome projects in the medical science and drug industry.
3. What does a polycistronic gene mean ?
4. What are the components of a typical gene ?
5. Define the following :
   (*a*) Operon
   (*b*) Nucleosomes
   (*c*) An allelic pair
   (*d*) Introns
   (*e*) Triplet codon
   (*f*) Translation
   (*g*) Pseudogenes
   (*h*) Natural selection
   (*i*) Oncogenes
   (*j*) MHC.
6. Differentiate between multigene family and operon.
7. What is the relevance of MHC in medical science ?
8. What are the tumour suppress gene ?
9. Explain the relation between cancer and virus.
10. What is the special feature of retrovirus ?
11. Differentiate between oncogenes and proto-oncogenes.
12. What are binary switch genes ?
13. Give the structure of a typical mRNA.
14. What are the types of RNAs needed for the process of translation ?
15. What is the need of DNA repair ? Explain.

16. Give the secondary and tertiary structure of tRNA.
17. Distinguish between promoter and operator.
18. Differentiate between the eukaryotic and prokaryotic RNA polymerase.
19. What are the Okasaki fragments ?
20. Differentiate between leading strand and lagging strand.
21. List the enzymes involved in the process of DNA replication.
22. What are the various types of DNA polymerases present in prokaryotes ? Which is the main polymerase involved in the process of polymerase ?
23. What are the various steps involved in the process of translation ?
24. Differentiate between *lac* operon and *trp* operon.
25. Why the replication of DNA is called semiconservative replication ?
26. Distinguish between induction and repression.
27. What are transcriptional factors ?
28. Differentiate between introns and exons.
29. Explain how genetic is linked with evolution.
30. Distinguish between sense and antisense strands.
31. What is meant by origin of replication ?
32. Explain the process of splicing and its requirement. Why this process is absent in prokaryotes ?
33. Explain the gene organization in prokaryotes and eukaryotes.
34. Give the structure of ribosomes and its role in translation.
35. Explain the process of charging of amino acids in protein synthesis.
36. Cytoplasm has got an important role in the expression many genes related to the process of development. Explain it with example.
37. Name the amino acids, which has got only one codon.
38. Explain the major mechanisms of gene regulation in eukaryotes.
39. What are snRNAs ? What is its role in the post-transcriptional modifications of RNA ?
40. Explain the post-transcriptional modifications in the maturation of mRNA.

• SUGGESTED READING •

1. Genes VII, Benjamin Lewin. Oxford University Press, New York, 2000.
2. Molecular Biology of the Gene, J. D. Watson, N. H. Hopkins, J. W. Roberts, J. A. Steitz and A. M. Weiner, The Benjamin/Cummings Publishing Company. Inc., Menlo Park, California.
3. Molecular Cell Biology, H. Lodish, A. Berk, S. L. Zipursky, P. Matsudaira, D. Baltimore and J. Darnell, W. H. Freeman and Company, New York.
4. Genetics: Analysis and Principles, R. J. Brooker, Addison Wesley Longman. Inc., New York.
5. Genetics: A Molecular Approach, T. A. Brown, Nelson Thrones Ltd., Cheltenham, London.
6. The Cell: A Molecular Approach, G. M. Coopper, ASN Press, Washington,1997.

UNIT 3

7 DEVELOPMENT OF BIOTECHNOLOGY

7.1 NATURE AND SCOPE OF BIOTECHNOLOGY

1. Introduction and Definition

Man has been employing living organisms knowingly or unknowingly from pre-historic times for the production of various things like alcoholic beverages, curds, cheese etc. There are evidences to show that Sumerians and Babylonians used to produce various types of wines and beers. In India, there is the description about 'somarasa', an intoxicating juice, probably a fermented juice, from a plant named 'Soma' in Vedic texts. The ancient Indian system of medicine Ayurveda used the self-generated alcohol as the medium in the medicines. From these it is very evident that the use of biotechnology is not a recent phenomenon but a very old technology, as old as human race.

The term 'biotechnology' was used before the 20th century for traditional activities like making dairy products cheese and curd, bread, wine, beer etc. But none of these could be considered biotechnology in the modern sense. Genetic alteration of organisms through selective breeding, plant cloning by grafting etc. do not come under biotechnology. The process of fermentation for the preparation and manufacture of products like alcohol, beer, wine, dairy products, various types of organic acids like vinegar, citric acid, amino acids and vitamins can be termed as the classical biotechnology or the traditional biotechnology. Fermentation is the process by which living organisms like yeast or bacteria are employed to produce useful compounds or products.

Modern biotechnology is similar to classical biotechnology in utilizing living organisms. Then what is it that makes modern biotechnology modern ? It is not modern in the sense of using various living organisms, but in the techniques for doing so. The introduction of a large number of new techniques changed the face of classical biotechnology forever. These modern techniques, applied mainly to cells and molecules make it possible to take advantage of the biological process in a very precise and exact way. For example, genetic engineering has allowed us to transfer the property of a single gene from one organism to another. But before going into the details of biotechnology and the techniques that made it possible, let us see how we can define biotechnology.

2. Definition of Biotechnology

There are several definitions for biotechnology. One simple definition is that it is commercialization of cell and molecular biology. According to United States National Science Academy, Biotechnology is the "controlled use of biological agents like cells or cellular components for beneficial use". It covers both classical as well as modern biotechnology. More generally, biotechnology can be defined as "the use of living organisms, cells or cellular components for the production of compounds or precise genetic improvement of living things for the benefit of man".

Even though biotechnology has been practiced for thousands of years, the technological explosion of the 20th century, in the various branches of sciences—physics, chemistry, engineering, computer application and information technology revolutionized the development of life sciences, which ultimately resulted in the evolution of modern biotechnology.

Supported by an array of biochemical, biophysical and molecular techniques in addition to engineering and information technology, life scientists were able to develop new drugs, diagnostics, vaccines, food products, cosmetics, and industrially useful chemicals. Genetically altered crop plants, which can resist the stress of pests, diseases and environmental extremes, new tools and techniques to extend the studies on genomics and proteomics, not only of man but other organisms also were developed. The involvement of information technology and internet in biotechnology particularly genomics and proteomics has given birth to a new branch in biotechnology—the science of bioinformatics and computational biology. The skills of biotechnology, like any other modern science is founded on the previous knowledge acquired through the ages. If one wants to understand biotechnology thoroughly one should know the history of its development also. Otherwise, it will be like watching a movie half way or reading a story from its middle chapter. The history of biotechnology is nothing but the history of life sciences.

7.2 HISTORIC PERSPECTIVES

Biotechnology as a science is very new, about 200 years old, but as a technology it is very old. The word **biotechnology** used in 1917 refers to large-scale fermentation process for the production of various types of industrial chemicals. But the roots of biotechnology can be traced back to pre-historic civilizations like Egyptian and Indus valley civilization, when man learned to practice agriculture and animal domestication. Even before knowing about the existence of microorganism, they learned to practice biotechnology.

Biotechnology in Pre-historic Time

The primitive man became domesticated enough to breed plants and animals; gather and process herbs for medicine; make bread and wine and beer and create many fermented food products including yogurt, cheese and various soy products; create septic systems to deal with digestive and excretory waste products, and to create vaccines to immunize themselves against diseases. Archeologists keep discovering earlier examples of the uses of microorganisms by man. Examples of most of these processes dates back to 5000

BC. Ancient Indus people, for example, prepared and used various types of fermented foods, beverages and medicines. The ancient Egyptians and Sumerians used yeast to brew wine and to bake bread as early as 4000 BC. People in Mesopotamia used bacteria to convert wine into vinegar. Many ancient civilizations exploited tiny organisms that live in the earth by rotating crops in the field to increase crop yields. For example, the Greeks used crop rotation to maximize crop yield. They also practiced various methods of food preservation such as drying, smoking, curing, salting etc. All these techniques and processes were practiced in Middle East and South East of Asia including ancient India. The Egyptian art of mummification of dead bodies used the technique of dehydration using a mixture of salts.

Use of Genetic Resources

The ancient people were also aware of the role of natural genetic resources such as plants in the economic growth of a land. The rulers at those time used to sent plant-collectors to gather prized exotic species of plants that produced valuable spices and medicines. Indians, even before 3000 BC, realized the medicinal properties of herbs and systematically used to prepare herbal medicines for the treatment of various diseases. That has given birth to the oldest of the medical systems—The Ayurveda. Likewise, in modern times, colonial powers mounted huge plant-collecting expeditions across Latin America, Asia and Africa, installing their findings in botanic gardens. These early 'gene banks' helped the colonial powers to establish agricultural monocultures around the globe. A single coffee plant brought to Kerala from Arabia in the17th century spread to different parts of India. And the British started growing Chinese tea plants in the foothills of the Himalayas in their first big attempt to establish an economic base using an exotic plant in the colonial India. By the late 19th century, India had displaced China as the main exporter of tea to Britain.

Microorganisms and Fermentation

Although baking bread, brewing beer and making cheese have been going on for centuries, the scientific study of these biochemical processes is less than 200 years old. Clues to understanding fermentation emerged in the 17th century when Dutch experimentalist Anton van Leeuwenhoek discovered microorganisms using his microscope. He unravelled the chemical basis of the process of fermentation by using analytical techniques for the estimation of carbon dioxide. Two centuries later, in 1857, a French scientist Louis Pasteur published his first report on lactic acid formation from sugar by fermentation. He published a detailed report on alcohol fermentation later in 1860. In this, he revealed some of the complex physiological processes that happens during fermentation. He proved that fermentation is the consequence of anaerobic life and three types of fermentation processes were identified.

1. Fermentation, which generates gas,
2. Fermentation that results in alcohol and
3. Fermentation, which results in acids.

At the end of 19th century, Eduard Buchner observed the formation of ethanol and carbon dioxide, when cell free extract of yeast was added to an aqueous solution of sugars. Thus, he proved that cells are not essential for the fermentation process and the components responsible for the process are dissolved in the extract. He named that substance as Zymase. The fermentation process was modified in Germany during First World War, to produce glycerine for making the explosive nitroglycerine. Similarly, military armament programmes discovered new technologies in food and chemical industries, which helped them, win battles in the First World War. For example, they used the bacteria that convert corn or molasses into acetone for making the explosive, cordite. While biotechnology helped kill soldiers, it also cured them and Sir Alexander Fleming's discovery of penicillin, the first antibiotic, proved highly successful in treating wounded soldiers.

In India our ancestors developed a number of traditional fermented food in addition to curd by using yeasts and bacteria.

The Origin of Genetics

In 1906, biotechnology took a leap forward when Gregor John Mendel announced the findings of his experiments as the 'laws of genetics'. He predicted the presence of 'units of heredity'—later called genes—which did not change their identity from generation to generation but only recombined. The science of genetics, derived from the term 'genesis', which relates to the origin of a thing, tried to explain how organisms both resemble their parents and differ from them. It was believed that every gene corresponded to a specific trait directly. By the 1920s genetics was helping plant breeders improve their crops. By the 1940s, genetics had transformed the agriculture sector, which led to the Green Revolution in the 1960s. The green revolution has changed India into a food surplus state.

DNA and Genetic Engineering—The Beginning of Modern Biotechnology

The science of genetics was transformed by the discovery of DNA (or deoxyribonucleic acid), which carries the hereditary information in the cells. The chemical DNA had already been found in 1869 by Friederich Miescher but was not taken seriously as the chemical basis of genes until the early 1950s. Two scientists, Francis Crick and James Watson along with Rosalind Franklin, in 1953, discovered that the DNA structure was a double helix: two strands twisted around each other like a spiral staircase with bars across like rings. The structure, function and composition of DNA are virtually identical in all living organisms—from a blade of grass to an elephant. What differs—and makes each creature unique—is the precise ordering of the chemical base in the DNA molecule. This gave scientists the idea that they might change this ordering and so modify life forms. Marshall Nirenberg and H. Gobind Khorana carried out the deciphering of the genetic code in 1961.

Soon scientists and industrialists were seeking to alter the genetic make-up of living things by transferring specific genes from one organism to another. They can now modify life forms by altering the hereditary material at molecular level. Walter Gilbert carried out the first recombinant DNA experiments in 1973 and the creation of the first hybridomas was done in 1975. The production of monoclonal antibodies for diagnostics was carried out in 1982, and the first recombinant human therapeutic protein, insulin (humulin) was produced in 1982. In 1976, the US company Genentech became the first biotech company to develop technologies to rearrange DNA. Commercial uses of recombinant DNA assisted biotechnology include the development of interferon, insulin and a number of genetically modified crop plants like the high-solids-processing tomato that has 20% less water. Transgenic animals have been created like the unfortunate onco-mouse designed to develop cancer ten months after birth to study about cancer.

Companies have been assisted and encouraged in their research by the 1980 ruling of the US Supreme Court allowing genetically engineered microorganisms to be patented. This means that virtually any life form on this planet can theoretically become the private property of the company or person who 'creates' it. One of the greatest threats of the new biosciences is that life will become the monopoly of a few giant companies.

An estimated 600 pharmaceutical companies worldwide are conducting research and development into genetically engineered products. According to industry by the year 2,005 over 5000 genetically engineered new products will be on the market. Mistakes are bound to happen. And with something so powerful as genetic engineering, one mistake could have profound and wide-ranging effects. The whole gene revolution is on the verge of becoming the private property of a few multinationals. We must impose tough controls on the genetics supply industry. We must work to make sure that the new techniques are in the service of the global community.

7.3 SCOPE OF BIOTECHNOLOGY

In the past, biotechnology had concentrated on the production of food and medicine. It also tried to solve environmental problems. The process of fermentation was used to produce alcohol, beer, wine, and other types of chemicals important for industry. In the 19th century, industries linked to the fermentation technology had grown tremendously because of the high demand for various chemicals like ethanol, butanol, glycerine acetone etc. The advancement in fermentation process by its interaction with chemical engineering has given rise to a new area—the bioprocess technology. Large-scale production of proteins and enzymes can be carried out by applying bioprocess technology in fermentation. Applying the principles of biology, chemistry and engineering sciences, processes are developed to create large quantities of chemicals, antibiotics, proteins and enzymes in an economical manner. The bioprocess technology includes media and buffer preparation, upstream processing and downstream processing. Upstream processing is to provide the microorganism the media, substrate and the correct chemical environment to carry out the required biochemical reactions to produce the product. Downstream process is the separation method to harvest the pure product from the fermentation medium. Thus, the fermentation technology changed into biotechnology, now known as the classical biotechnology. Now if we look at biotechnology, we find its application in various fields like food, agriculture, medicine and in solving environmental problems. This has lead to the division of biotechnology into different areas like agricultural biotechnology, medical or pharmaceutical biotechnology, industrial biotechnology and environmental biotechnology.

The modern biotechnology is mainly based on recombinant DNA (rDNA) and hybridoma technology in addition to the bioprocess technology. The rDNA technology is the main tool not only to produce genetically modified organisms, including plants, animals and microbes but also to address the fundamental questions in life sciences. In fact the modern biotechnology began when recombinant human insulin was produced and marketed in the United States in 1982. The effort leading up to this landmark event began in the early 1970's when research scientists developed protocols to construct vectors, by cutting out and pasting pieces of DNA together to create a new piece of DNA (recombinant DNA), that could be inserted into the bacterium, *Escherichia coli* (transformation). If one of the pieces of the new DNA included a gene for insulin or any other therapeutic protein or enzyme, the bacterium would be able to produce that protein or enzyme in large quantities by applying bioprocess technology.

Another way of preparing human therapeutic proteins, vaccine and diagnostic proteins is by hybridoma technology. The first hybridoma experiments were carried out in 1975. In hybridoma technology, a B-lymphocyte secreting antibodies against a specific antigen is fused with a myeloma cell (a cancerous B-lymphocyte). The resulting cell if injected in a mouse's abdomen or if cultured in a bioreactor by applying bioprocess technology will grow and divide, indefinitely, producing large quantities of antibody, which can then be harvested. The resulting proteins are called monoclonal antibodies (MAb) and are most often used in diagnostic kits. The most famous MAb containing diagnostic kit is the pregnancy testing kit.

In agriculture, rDNA technology can be used to produce new varieties of crop plants having improved agricultural and nutritive qualities. Transgenic plants, which are resistant to biotic and abiotic stresses like salinity, drought and diseases have been produced. Recently, the Government of India has given permission to cultivate genetically engineered cotton plants, which are resistant to attack by insects. By cultivating this cotton plant farmers will be able to increase the productivity by protecting the plants against insects. And indirectly this will protect the environment from the use of hazardous insecticides. The bacterium *Bacillus thuringiensis* is a natural enemy of some insects, which attack cotton and other agricultural crops. A protenacious toxin—Bt. toxin is responsible for the insecticidal activity of the bacteria. The insect resistant transgenic cotton plant has been produced by transferring the gene of Bt. toxin, isolated from the bacterium *Bacillus thuringiensis* to a single cell of the cotton plant and regenerating a number complete plants from that genetically transformed plant cell. Similarly, the nutritive qualities of food plants also can be increased to meet specific nutritive requirements. For example, production of golden rice by incorporating three genes required for the production of vitamin A in rice.

Recombinant microorganisms, plant cells and animal cells can be cultivated in large-scale and can be used for the large-scale production of industrially important enzymes and chemicals. Examples of such enzymes are protease, amylase, lipase, glucose isomerase, invertase etc. Amylase is used in the starch industry. Glucose isomerase is used in fructose formation from glucose syrup. Proteases and lipases are incorporated into detergent products to take out stains. Protease is also used in the meat and leather industries to remove hair and soften meat and leather. A list of enzymes and its industrial use is given in table 7.1.

Ever since the manufacture of human insulin using recombinant *Escherichia coli* began in 1982, many other proteins (for human and veterinary therapeutics, vaccines and diagnostics) have been manufactured. Today, there are a large number of human therapeutic proteins or vaccines made by modern biotechnology methods, approved by the government and marketed in the country. There are more than 200 other human therapeutic and vaccine proteins at clinical trial stage. Products are being tested to target the diseases like cancer, AIDS, heart disease, multiple sclerosis, rheumatoid arthritis and viral diseases. Products are also being developed to reduce bleeding from surgical procedures, aid in wound healing and prevent organ transplant rejection. Vaccines are also being developed to prevent lyme disease, herpes, AIDS, rheumatoid arthritis, and cancer.

It is difficult to predict the future of this exciting new field of modern biotechnology using recombinant DNA technology. There is no doubt in its ability to improve the human life and the economy of the country. But along with the advancement in the research and development of life science and biotechnology, there arise a lot of social, environmental and ethical problems. A lot of organizations are looking into various issues and addressing the general concerns.

TABLE 7.1 Some of the major Industrial enzymes—sources and uses.

Enzymes	Sources	Uses
Amylases	*Aspergillus niger* *A. oryzae, Bacillus licheniformis* *B. subtilis*, germinating cereals germinating barley	Hydrolyze starch to glucose, detergents, baked goods, milk cheese, Fruit juice, digestive medicines, dental care
Invertases	*Saccharomyces cerevisiae*	Production of invert sugar, Confectionery
Glucose isomerase	*Arthrobacterglobiformis* *Actomoplanes missouriensis* *Streptomycesolivaceus und E. coli*	Conversion of glucose to fructose, production of high fructose syrup, other beverages and food
α D-Galactosidase	*Mortierella vinaccase*	Raffinose hydrolysis
β D-Galacosidase	*Aspergillus niger*	Lactose hydrolysis
Papain	Papaya	Meat, beer, leather, textiles, pharmacuticals, meat industry, digestive aid, dental hygiene etc.
Protcases	*Bacillus subtillis* *B. licheniformis*	Detergents, meat tenderizer, beer, cheese, flavour production.
Pepsin	Hog (pig) stomachs	Cereals, pharmaceuticals
Trypsin	Hog and calf pancreases	Meat, pharmaceuticals
11-β-Hydroxylase	*Curvularia lunata*	Steroid conversion, bioconversion of organic chemicals
Ficin	Figs	Leather, meat, pharmaceuticals
Bromelain	Pineapple	Meat, beer, pharmaceuticals

7.4 COMMERCIAL POTENTIAL

Biotechnology is considered to be the commercialization of life sciences. It did have a significant impact on various applied sciences, manufacturing processes, on medicine and health, agriculture and environmental sciences. By monitoring and diagnostic systems it made giant strides in the field of health and medicine. It can play an important role in monitoring the use of both traditional and non-conventional energy resources.

Commercial biotechnology products are already available and include for instance new diagnostics, recombinant vaccines and therapeutic proteins, biochips or DNA chips. The biochips or the DNA micro arrays currently being produced are revolutionizing the design and output of gene analysis in the fields of molecular medicine. Bioremediation technologies for the elimination of toxic factory effluents with the help of genetically altered microorganisms, purification of rivers, fresh water ecosystems and drinking water are now carried out commercially.

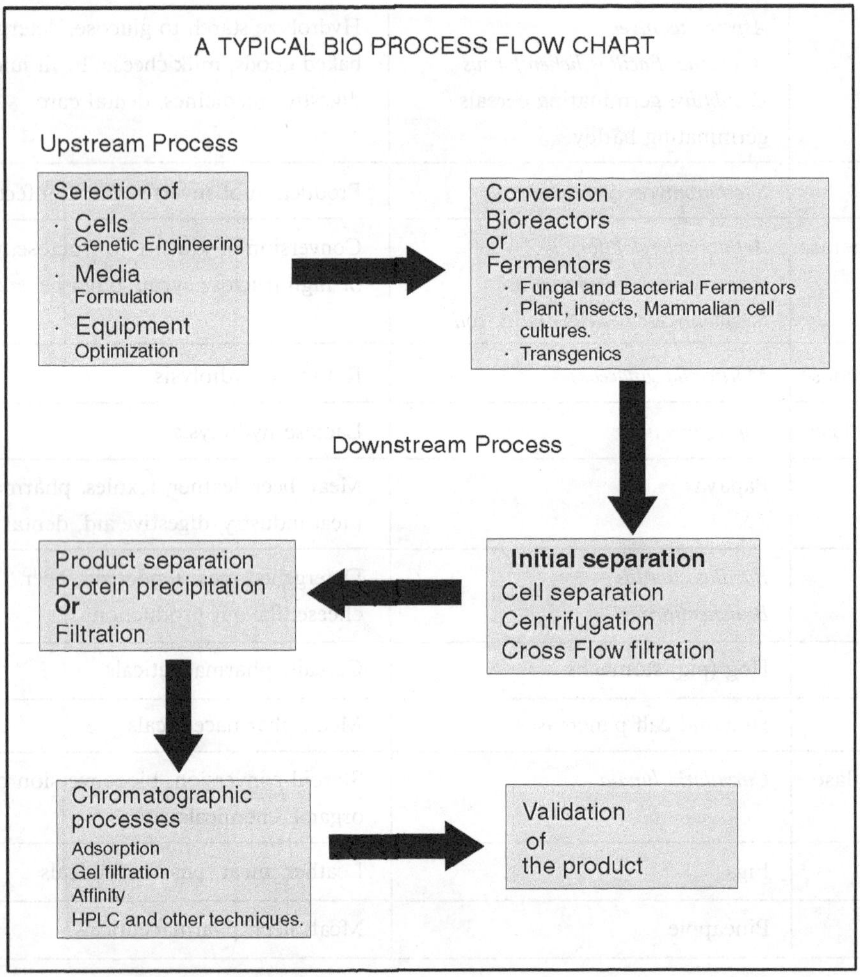

Figure 7.1 An overview of modern bioprocess technology.

In its economic potential, biotechnology runs parallel with computer industry. After the birth of genetic engineering, the biotech industry is waiting to explode in the consumer market. Consumers are going to see scores of new biotech products, such as foods that contain vaccines or super-nutritious foods that will change the way people view agriculture.

In addition to the similarities in their economical potential, there is a resemblance of the biotechnology industry and the computer industry in the technical side also. There is a parallel between genetic code and computer code. Computers and living organisms both organize their essential information in a similar fashion. Computers are directed by a series of ones and zeroes, which is known as the binary code. Living organisms all use a code made up of four parts, a quaternary code. Instead of ones and zeroes, the information is conveyed by a series of four chemicals, adenine, thymine, guanine and cytosine, which geneticists simply call as A, T, G and C. Like a computer code, the arrangement of these four chemicals strings together to form genes, which contain the information that tells the cells whether you are to be a linebacker-sized human or a lemming!

Scientists first learned that they could manipulate these four chemicals to form new genes in the mid-1970s. Recombinant DNA technique was first developed in 1974, and today, even high school kids can cut and stitch genes together. The development of this science has been mind-boggling, and so has the rise of biotech industries all over the world.

7.5 CLASSICAL AND MODERN CONCEPTS OF BIOTECHNOLOGY

Though there are no differences in the principles, the technological advancement in the mode of utilizing living cells for the benefit of man differentiates between the classical and modern biotechnology. The classical biotechnology that emerged during the early 20th century was, basically a microbial-based fermentation process, in which the principles of biochemical engineering have been applied to change it into an industrial process. In short it is a hybrid of fermentation and biochemical engineering. The modern biotechnology is based on the ability of recombinant DNA and Hybridoma technology to genetically alter the cells and organisms—microbes, plants and animals and to use them for different purposes.

Another important aspect of modern biotechnology is the ownership of the bioprocess technology, transgenic organisms and the socio-political, ethical and economical consequences that accompany these experiments.

In plant biotechnology the transgenic crop plants are used in agriculture to enhance productivity. By the classical plant breeding techniques it is possible to transfer a trait from one plant to another plant. It has got its limitations. It is highly nonspecific in the sense that the hybrid plant produced may not be with the desired trait. Gene transfer through hybridization is only possible between related plants. But genetic engineering made it possible to transfer a specific gene to a plant from any organism. A major use of transgenic plants is in its application in non-agricultural sectors. For example, bioremediation of toxic wastes having heavy metals like arsenic, mercury etc. The genetically altered organisms, both plant and animals, also may be used as bioreactors for the production of vaccines, hormones and other therapeutic proteins. Recombinant DNA technology or genetic engineering, in addition to these, is a powerful tool in molecular biology. Along with genetic engineering, hybridoma technology has also made the modern biotechnology powerful in medical and pharmaceutical industries. Isolation and characterisation of new genes, proteins and genome sequencing has led to the formation of a huge volume of computerized data and a lot of associated problems, giving birth to the new science of bioinformatics and computational biology.

The completion of a "working draft" of the human genome—an important landmark in the history of biotechnology—was announced in June, 2000 at a press conference at the White House and was published in the February 15, 2001 issue of the journal *Nature.* According to this, there are only 32,000 genes, which is just about 2 to 3% of the total genome sequence. The remaining 97% of the sequence does not code for any gene. It is not even known whether these sequences do have any function.

But with these 32,000 genes a lot of things can be done by geneticist, molecular biologists and pharmacologists. For example, with all these 32,000 genes on a microchip, it is possible to do a large throughput screening of new drugs for a specific group of population and develop a population specific or a tribe specific designer drug. A newly developed drug to treat breast cancer is good for the treatment of those patients in whom one specific gene, Her-2 Neu, is over expressed. The development of gene chips *i.e.* complete genome or specific DNA sequences, immobilized on a microscopic silica, glass or nylon chip, facilitate rapid screening of genomes and proteomes for various purposes such as drug design and development, toxicological and pharmacological trials of drugs. Similarly, there are protein chips, where protein molecules are immobilized on microscopic chips. The emergence of these protein and DNA chips has changed the proteomics and genomic researches and new fields such as Pharmaco genomics and toxico genomics have emerged.

The excitement and optimism in biotechnology has made both public and private sectors to make huge investments in research and technology developments. Biotech industries started as university supported private enterprises, have undergone many changes. Many small industries closed or merged with bigger pharmaceutical or chemical companies. The market response to biotechnology products was not encouraging in the beginning. The main reason was the regulatory policies of the concerned governments and the negative approach of the public towards genetically engineered products.

A lot of other factors too hindered the growth of biotechnology industries, as against the great expectations. In agricultural sector, the biotech products, particularly, the edible materials have to compete with the products from the conventional sources, that people prefer. But in the case of life saving medicines, vaccines, alcohol, pesticides, weedcides etc.—the products of conventional methods has been taken over by the biotech products.

The biotech industries also have the potential to replace several polluting industries, using environment-friendly manufacturing process. The vaccines produced by genetic engineering are supposed to be safer than the vaccines made out of conventional methods. The vaccine produced conventionally carries the inactive virus, which may become active at any time. However, the optimism on biotechnology products has to be supported by hard scientific evidences. Only the hard scientific proofs can boost the morale of the public and industry. For example, there is no evidence yet to show that the rDNA product (a protein) has the correct folding of polypeptide chain to form the 3-dimensional structure by the post-translational modifications as it happens in the actual system. Today most of the drugs that are coming to the market or those in the process of development are from modern biotechnology. This is because the pharma industries are mainly focusing on discovering new biotechnology based drugs.

7.6 BIOTECHNOLOGY IN INDIA AND GLOBAL TRENDS

Indian Scenario

In India the prospects of biotechnology is very favourable right from the beginning, in the early 1980s. Government of India has constituted a separate department under the Ministry of Science and Technology—the Department of Biotechnology (DBT), to facilitate the development of biotechnology in India. DBT is the main funding agency and the body, which regulates research and developments in biotechnology. We have other scientific bodies like Council of Scientific and Industrial Research, Indian

Council of Medical Research, Indian Council of Agricultural Research, Department of Environment and several other agencies under government and international levels which promote the research and development. In addition to the funding, they also monitor and conduct manufacturing, clinical trials, and marketing of products in India. Government of India has set up another body, the Technology Development Board to promote product development with universities, industries and national institutes. The Technology Information Forecasting and Assessment Council (TIFAC) prepared a document—Vision 2020 that also includes biotechnology. Several states have created biotechnology parks for promoting the growth of biotechnology industries in India. A number of rDNA vaccines like that of Hepatitis B indigenously developed has been manufactured and marketed since 1998. There are nearly 50 companies devoted to the development of modern biotechnology products in India. The big companies like Reliance Industries and Wipro entered the biotechnology segment and India is expected to be a significant player in coming years. The major national institutes and universities involved in the Biotechnology Research and Development in addition to the numerous universities under state government are listed in the table below.

1. Indian Institute of Sciences
2. Madurai-Kamaraj University
3. Jawaharlal Nehru University
4. Poona University
5. University of Kerala, Thiruvananthapuram
6. Centre for Cellular and Molecular Biology (CCMB)
7. National Institute of Immunology (NII)
8. Centre for Biochemical Technology (CBT)
9. International Centre for Biotechnology and Genetic Engineering (ICBGE)
10. Indian Agricultural Research Institute (IARI)
11. All India Institute of Medical Sciences (AIIMS)
12. Bose Institute, Calcutta
13. Institute of Microbial Technology (IMTECH)
14. Tata Institute of Fundamental Research (TIFR)

In addition to these there are a number of private institute, research foundations and biotech companies involved in the research activities and development of biotech products.

Global Scenario

When you view the global picture of biotechnology, you can see that United States of America is the forerunner in biotechnology research and development. Eversince the start of biotechnology research, USA is promoting biotechnology enterprise by shaping the policies, which can promote biotechnology innovations as far as possible. The Federal Government has supported the basic research in National Institute of Health, the prime institute of biotechnology research in USA, and other scientific agencies, by funding, accelerating the administrative processes for approving new medicines; encouraging private-sector research investment and small business development through tax incentives; promoting intellectual property protection and open international markets for biotechnology inventions and products. They developed public databases and Internet facilities that allowed the scientists to coordinate their efforts and research findings for easy and systematic studies. In addition to all these, the government also gave more concern to improve the science education and promote the freedom of scientific enquiries. The administrations has provided guidelines to protect patients from misuse or abuse of new drugs and sensitive medical information and provide the federal regulatory agencies with sufficient resources to maintain sound, science based review and regulations of biotechnology products. It has also ensured that science based regulatory programs worldwide promote public safety; earn public confidence and guarantee fair and open international markets. More or less the same types of guidelines are also available in other developed counties like UK, Germany, France, Australia, Japan and Canada.

• REVIEW QUESTIONS •

1. Define Biotechnologies.
2. Compare the modern biotechnology with classical biotechnology.
3. What is your justification to consider biotechnology as an old technology ?
4. What are the different types of fermentation ?
5. What is lactic acid bacillus? Explain its role in the formation of curd.
6. Name some of the fermented food products of India.
7. Which is the first recombinant product that came into Indian market ?
8. Name four recombinant DNA products available in the market.
9. How is the modern biotechnology differ from the classical biotechnology ?

• SUGGESTED READING •

1. Physics in Biology and Medicine, Paul Davidovitis (Academic Press, San Diego), 1999.
2. Biotechnology unzipped—Promises & Realities, Eric & Grace (Joseph Henry Press), 1997.
3. Microbes and Man, John Postgate, Cambridge University Press, Cambridge, 2000.

8 APPLICATIONS OF BIOTECHNOLOGY

8.1 BIOPROCESS AND FERMENTATION TECHNOLOGY

Introduction

Biotechnology is the commercial application of all types of biological reactions and processes. We are all aware of the opportunities created by advances in molecular biology. Living cells and its components can be used to produce a large number of useful compounds like therapeutics and other products. But to obtain the significant benefits as a commercial operation, it needs the support of biochemical engineering. The vital area of biotechnology that is concerned with practical application of biological agents (whole cell systems and biocatalysts), the methodologies and processes associated with it on an industrial scale is the biochemical engineering. It is applicable in different areas of biotechnology like biochemical reactions, enzyme technology, environmental biotechnology, microbial manipulations, bioseparation technology, plant and animal cell cultures and food technology. It consists of the development of new process technology, designing bioreactors, developing efficient, economically feasible extraction and purification procedures (downstream processing).

A detailed understanding about the behaviour of the biological agents—living cells or its components, and an expertise in the principles of biochemical engineering is a prerequisite to develop a bioprocess technology to beneficially exploit the potential of the biological system. Thus biochemical engineering has an important role in transforming a biological laboratory process into an economically feasible industrial process.

It is very difficult to handle living cells in a bioprocess development. Because the cells are very fragile, require a very specific type of chemical and physical environment, including nutritional requirements, pH and temperature. The product formed in the cells may not be secreted into the medium. In such cases the cells has to be ruptured to extract the products. In most cases the product may be combined with other similar cellular products, from which the actual product has to be separated and purified economically.

The basic problems that can be associated with the development of a cost effective bioprocess technology are :

1. The low concentration of substrate in the culture media.
2. The reaction should be conducted in the optimal temperature and pH, that facilitates both culture growth and maximum product output.
3. The pressure built up in the bioreactor due to the generation gas by the growing cells should be effectively controlled.
4. The cost of the process can be minimized when adopting special methods for the recycling of the biocatalysts, nutrients and byproducts.

When a laboratory process is converted into an industrial scale a large number of similar problems can arise. These problems have to be addressed properly for the success of the technology.

The main areas of studies in biochemical engineering are :

(*a*) Operating considerations for bioreactors for different types of cell cultures, suspension and immobilized cell cultures of plant animal, microbes and genetically engineered organisms.

(*b*) Selection, scale-up, operation, and control of bioreactors (fermentors)

(*c*) Methods for recovery and purification of products

(*d*) Medical applications of bioprocess engineering

(*e*) Mixed cultures

(*f*) Stoichiometry of microbial growth and product formation and

(*g*) Metabolic pathways and pathway engineering.

Applications of Biochemical Engineering include novel technologies for cheaper and cleaner alternatives for energy, chemical feedstocks. New biochemical engineering techniques are also available for cost effective pollution control.

The bioprocess is the industrial application of biological pathways or reactions mediated by living whole cells of animals, plants and microorganisms or enzymes under controlled conditions for the biotransformation of raw material into products. The product may be directly useful as food, medicines or industrial compounds or the bioprocess may be without any direct products. In such cases the bioprocess may be used as a means to detoxify the industrial wastes or treatment of factory effluents with or without byproduct. The conversion of a laboratory scale biochemical reaction into an industrial bioprocess technology involves the application of biochemical engineering.

Biochemical engineering also includes the allied subjects like Biochemistry, Molecular Biology, Microbiology, Genetics, Cell Biology, Immunology, Biomedical and Environmental Sciences in addition to engineering principles. The need for biochemical engineering specifically developed because traditional chemical engineering deals primarily with petroleum and chemical industries.

The bioprocess technology involves mainly three parts.

1. Upstream process.

2. Bioreaction and bioreactors.

3. Downstream processing.

1. Upstream processing. This part of biochemical engineering involves the development of process for ascetic treatment of the substrate or the raw material with the microorganism or the biocatalyst. It includes the media preparation its sterilization, feeding into the bioreactor, regulation of temperature, pH and pressure, inoculation of the cell culture into the medium aseptically etc.

2. Bioreactor. The designing of the appropriate type of bioreactor or fermentor is one of the important steps in the development of a bioprocess technology. **Bioreactors** are vessels in which raw materials are biologically converted into specific products, using microorganisms, plant, animal or human cells or individual enzymes. A bioreactor supports the natural process of cells by trying to maintain their environment to provide optimum growth conditions—by providing appropriate temperature, pH, substrate, salts, vitamins, and oxygen. In most of the bioreaction processes the substrate of the biotransformation and the carbon source of the organisms will be the same. Table 8.1 gives some of the commonly used carbohydrates used in the various fermentation process as the carbon source and substrate for the reaction.

Bioreactors can be classified according to the type of biocatalysts and the type of bioreaction. The first classification is based on the type of biological agent used :

- Microbial fermentors
- Enzyme (cell-free) reactors.

Further classification is possible based on biochemical reaction and process requirements.

3. Downstream processing. The recovery and purification of the required product from the growth medium through asset of separation and purification technique is called downstream processing. Each stage in the overall separation procedure is strongly dependent on the history and quality of the biological production process. Maximization of production can lead to great difficulties in downstreaming and recently more attention is being paid to overall process optimization. It includes the techniques like filtration, centrifugation, sedimentation various types of chromatographic techniques, electrophoresis etc. Some fermentation parameter factors affecting the DSP are listed below.

- The properties of organisms or cells used in the fermentation process (safety, classification, morphology, thermal stability, tendency to flocculate, size, cell wall rigidity,....) influence the filtration, sedimentation and homogenization performance.
- The location of the product (intracellular, deposited in vacuoles or inclusion bodies or excreted into the broth or biomass itself) will define the initial separation steps and purification strategy.
- The stability of the product defines the need and kind of pretreatment for inactivation or stabilization.
- The product, by-products and impurities as well any additions to the broth (antifoam) may form an interfacial layer in extraction steps, give peaks in chromatography, block membranes in ultra filtration and analytical equipment; also salts and trace elements often have to be removed prior to pharmaceutical use.
- Nutrient medium residues (pesticides, herbicides, etc.) present in the product.

TABLE 8.1 Substrates or carbon sources frequently used in fermentation process

Substrate/Carbon source	Type of Biocatalysts	Product
Sucrose Molasses Hydrolysed starch	Saccharomyces cerviceae	Ethyl alcohol
Sucrose Molasses Hydrolysed starch	Aspergillus niger	Citric acid and allied products.
Sucrose Molasses Hydrolysed starch	Pencillium sps.	Penicillin
Sucrose/Glucose	Plant cells	Various secondary metabolites
Glucose	Escherichia coli	Various recombinant products.
Glucose/whey	Lacobacillus	Lactic acid
Glucose	Hybridoma	Monoclonal antibodies
Glucose	Animal cells	Various products.

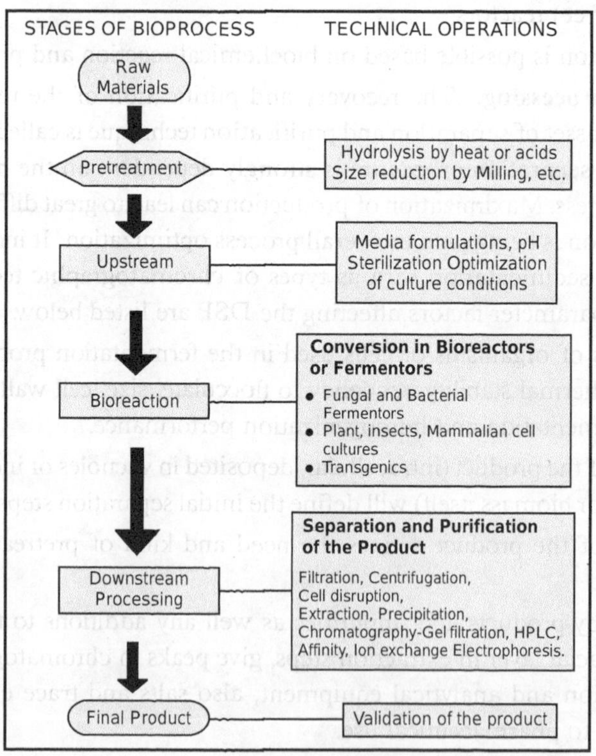

Figure 8.1 Flow diagram of a bioprocess.

Material and Energy Balances

Material Balance

In bioprocess technology developments stoichiometric equations indicating the both material balance as well as energy balance is very important to determine the exact molar concentration of the substrates that facilitate the high rate of formation of the products and energy conservation in the biochemical process.

Consider the growth of microorganism in a culture, which produces only carbon dioxide and water as the final products with increase in the biomass formation. This aerobic growth can be represented by the following equation.

$$aC_wH_xO_y + bO_2 + cN_iH_gO_h \longrightarrow C_\alpha H_\beta N_\gamma O_\delta + dH_2O + eCO_2 \quad \text{...(1)}$$

In the above equation $aC_wH_xO_y$ represents chemical formula for the substrate or the carbon source such as glucose, $cN_iH_gO_h$ represents the formula of nitrogen source and $C_\alpha H_\beta N_\gamma O_\delta$ indicates the dry biomass.

$$aC_wH_xO_yN_z + bO_2 + bH_gO_hN_i \longrightarrow cCH_\alpha O_\beta N_\delta + dCO_2 + eH_2O + fC_jH_kO_lN_m \quad \text{...(2)}$$

In the above biochemical equation in addition to the biomass ($CH_\alpha O_\beta N_\delta$) formation, there is the product $fC_jH_kO_lN_m$. Note that compounds like ATP and NADH have been ignored. They may be part of metabolism, but are not exchanged with the environment. In addition, other compounds like vitamins, which are needed in very small quantities, have been neglected. When the formula for biomass is rewritten to have one carbon it is called a unit carbon formula (UCF). The letters a, b, c, d and e are stoichiometric coefficients. The chemical equations are complete only when these stoichiometric coefficients are known. This equation can be generalized further by including the formation of a byproduct like ethanol. In equation (2) the product in addition to the biomass, CO_2 and water is $fC_jH_kO_lN_m$.

Once the stoichiometric coefficients are known, the biochemical formula can be evaluated by using normal procedures for balancing the equation. For equation (1), the elemental balances are (assuming that no byproduct forms, *i.e.* f = 0)

$C : w = c + d$

$N : z + bi = c\,\delta$

$H : x + bg = c\,\alpha + 2e$

$O : y + 2a + bh = c\,\beta + 2d + e$

For a given system of substrates of known composition and with biomass composition known, there will be 5 variables (a, b, c, d and e) and 4 balances. A charge balance could provide additional data if ions are present in the balance, but is not relevant here. The solution is to use an experimental measurement of respiratory quotient (RQ).

$$\text{Respiratory Quotient (RQ)} = \frac{[CO_2 \text{ produced}]}{[O_2 \text{ consumed}]} = \frac{d}{b}$$

Hence the elemental balances can now be solved exactly. The mass balance of the cell growth and product formation gives an idea about the substrate utilization and product formation during the industrial scale fermentation process. This will be very helpful in designing the bioreactors assessing the overall cost of production of the material. The mass balance analysis also gives the wastage of nutrients due to the build up inside the bioreactors, which can interfere with overall reaction procedure of the bioprocess.

Energy Balances

Biological reactions, like any other phenomenon always occur with a change in energy content and free energy, which can be explained by the principles of thermodynamics. The reaction can be either endergonic or exergonic. The product formed may be with comparatively with high-energy state or low-energy state, which is proportional to the degree of reduction or oxidation that the substrate molecule has undergone.

The energy state of molecule is given by the molar heat of combustion at standard conditions (Table 8.2) for that molecule. The heat of combustion of a molecule at standard conditions (normal temperature, pressure and pH) is directly proportional to the degree of reduction. If the product formed is relatively in a higher reduced state, the reaction have taken place with an input of energy in the form of ATP consumption and if the product is in the oxidized state there is the formation of heat or ATPs since it is exergonic reaction.

ΔG, Gibb's free energy and ΔH, the heat change, are the terms commonly used to show the energy balance of a biochemical reaction. Gibb's free energy is the usable form of energy in the biological system. Heat is not a usable form of energy as far as the biological system is concerned. Knowledge about the total energy change—energy consumption or heat formation in the biochemical reaction is very important in designing the bioreactor for the bioprocess development.

TABLE 8.2 Heat of combustion for various compounds at standard conditions (298 °K and 1 atm) and pH 7

Compound	Formula	γ	ΔH °C (kJ/mole)	ΔH °C (kJ/C-mole)	ΔH °C/γ (kJ/C-mole)
Formic acid	CH_2O_2	2	255	255	127.5
Acetic acid	$C_2H_4O_2$	4	875	437	109.5
Propionic acid	$C_3H_6O_2$	4.67	1527	509	109.0
Butyric acid	$C_4H_8O_2$	5.00	2184	546	109.2
Valeric acid	$C_5H_{10}O_2$	5.2	2841	568	109.2
Palmitic acid	$C_{16}H_{32}O_2$	5.75	9978	624	108.5
Lactic acid	$C_3H_6O_3$	4	1367	456	114.0
Gluconic acid	$C_6H_{12}O_7$	3.67			
Pyruvic acid	$C_3H_4O_3$	3.33			
Oxalic acid	$C_2H_2O_4$	1	246	123	123.0
Succinic acid	$C_4H_6O_4$	3.50	1491	373	106.6
Fumaric acid	$C_4H_4O_4$	3	1335	334	111.3
Malic acid	$C_4H_6O_5$	3	1328	332	110.7
Citric acid	$C_6H_8O_7$	3	1961	327	109.0
Glucose	$C_6H_{12}O_6$	4	2803	467	116.8
Fructose	$C_6H_{12}O_6$	4	2813	469	117.2

Galactose	$C_6H_{12}O_6$	4	2805	468	117.0
Sucrose	$C_{12}H_{22}O_{11}$	4	5644	470	117.5
Lactose	$C_{12}H_{22}O_{11}$	4	5651	471	117.8
Methane	CH_4	8	890	890	111.3
Ethane	C_2H_6	7	1560	780	111.4
Propane	C_3H_8	6.67	2220	740	110.9
Methanol	CH_4O	6	727	727	121.2
Ethanol	C_2H_6O	6	1367	683	113.8
iso-Propanol	C_3H_8O	6	2020	673	112.2
n-Butanol	$C_4H_{10}O$	6	2676	669	111.5
Ethylene glycol	$C_2H_6O_2$	5	1179	590	118.0
Glycerol	$C_3H_8O_3$	4.67	1661	554	118.6
Acetone	C_3H_6O	5.33	1790	597	112.0
Formaldehyde	CH_2O	4	571	571	142.8
Acetaldehyde	C_2H_4O	5	1166	583	116.6
Urea	CH_4ON_2	6	632	632	105.3
Ammonia	NH_3	3	383		127.7
Biomass	$CH_{1.8}O_{0.5}N_{0.2}$	4.8	560	560	116.7

Fluid Flow and Mixing

The proper mixing of the reactants is very important because it maintains the uniformity of the physical and chemical factors of the reactants and the reaction conditions in the bioreactor. It provides the correct uniform biological environment. The bioprocess has to be carried out at the precise range of physiological temperature and pH that facilitate the maximum reaction rate and maximum product output. Continuous supply of oxygen is also an important factor in the cell growth and biomass formation to increase the reaction rate. Therefore, fluid mixing is very important in maintaining homogeneous chemical and physical environment in the bioreactor.

The media that is used in fermentation process may be Newtonian fluids bacterial broth media or a non Newtonian fluids like semisolid media of solid state fermentation of fungus for the production of certain enzymes and antibiotics like penicillin.

Viscosity and fluidity of the culture media in large scale fermentation process is an important factor which interfere with the mass transfer in terms of nutrient mixing, oxygen supply and marinating the proper temperature and pH.

There are different types of fluid mixing devices in fermentors that allows the proper mixing of the media components depending on the viscosity of the fluid. There are different types of mixing arrangements in bioreactors. Depending on the type of mixing devices there are different types of fermentors such as stirred tank reactors, air-lift reactors, and bubble column reactors for aerobic fermentations.

Mass Transfer

Mass transfer refers to transfer of oxygen from bulk gas into the cells. The transfer of oxygen from the gas phase to the liquid phase through gas/liquid film is one of the rate limiting steps in large-scale fermentation. The oxygen transfer rate (OTR) is depended on a number of factors. There are some methods given below, by which the OTR of reaction system can be increased and maintained for smooth operation and better product output.

- Increasing the pressure
- O_2 enrichment of the inlet air
- Increasing agitation
- Increasing airflow rate
- Reducing foaming and removing bubbles

Heat Transfer

This is another major factor, which can affect the overall performance of a biochemical reaction in a bioprocess technology. The reaction that is taking place in a fermentor can be exothermic or endothermic. Therefore, most of the fermentors need careful temperature control. The increase in temperature of the fermentor can be by the exothermic reaction as well as by the agitation and by the respiration of the growing organisms. Once the details about the heat transfer—the evolution of heat by fermentation and heat generated due to agitation and aeration—are known, the heat transfer area needed to attain desired temperature can be calculated. Based on the overall heat production during the process of fermentation, heating and cooling devices can be designed to provide the optimal temperature needed for the smooth operation of the process always. A well designed jacket around the bioreactor and circulation of steam, hot water or chilled water through it can very well guard the fluctuation from the optimal temperature of the bioreactor. This facility also can be used for the sterilization of the media and fermentor before starting the process and autoclaving and cleaning of the fermentor after the process.

Scaling of a Fermentation Process to a Bioprocess

The major points that have to be considered during scaling up of a microbial or fermentation process are the following :

(*a*) The biomass of cells should be of sufficient quantity and should be uniformly distributed in the culture medium. So that the cells will be able to multiply and grow freely.

(*b*) The substrate molecules present in the culture medium has to be in contact with the cell surface.

(*c*) The substrate has to be transported to the site of action within the cell.

(*d*) The concentration of the substrate molecules within the cell should be sufficient enough to get the maximum efficiency of the process. The transport of substrate and product molecules across the cell membrane is known as Mass Transfer.

(*e*) The byproducts formed during the process of reactions have to be removed from the site of action and should be transported out of cells.

(*f*) There should not be any localized holding up of products or localized accumulation products within the reacting system. This can lead to the inhibition of cell growth and product formation.

(*g*) If aeration is a factor in the growth or product formation, there should be sufficient availability of oxygen in the culture medium. In many microbial systems mass transfer and diffusion of oxygen to the microorganisms are important rate limiting factors.

(*h*) The quality and efficiency of the process depends on the sterility maintained in the reacting system. Contamination by unwanted microorganisms can reduce the quality of the product as well as the efficiency of the process.

(*i*) The chemical and physical environment of the culture with respect to the pH and temperature has to be maintained optimum.

All these parameters can be maintained in specially designed culture vessels called fermentors or bioreactors, which are used extensively in microbial process studies.

The Fermentor

Fermentor, in general terms, is something that, as its name would suggest, ferments. The process of fermentation has been known for thousands of years, but has been mainly used over this time to convert the glucose found in various fruits, seeds and tubers into alcohol, later used for human consumption. In recent times, however, with increased knowledge of bacteria and fungi, fermentors have been put to a more productive use. More scientific and recent uses of fermentors in biotechnology would include those used to grow large quantities of genetically engineered organisms like bacteria and yeast, plant and animal cells. These bacteria, having had the genes that code for various proteins (human insulin, for example) spliced into them, will grow and reproduce, and will express the inserted gene. This will result in the desired protein being released into the growth medium, from where it can be harvested, purified, and then sold and used.

The general idea behind the fermentor is to provide a stable and optimal environment for microorganism, in which they can reproduce and do whatever they do. It is a specialized container where all the culture conditions that are optimized for the growth and product synthesis can be maintained continuously throughout the process of fermentation under sterile condition. Fermenters are specially designed to suit the industrial scale fermentation for the production of antibiotics, hormones, vaccines, enzymes and special chemicals. But now fermentation experiments can be conducted in the laboratories using laboratory fermentors, which are also available in a range of volumes. A fermentor is a closed vessel made of glass or steel, with an air inlet attached with an air filter to maintain sterility and an agitator or stirrer to maintain uniformity in the culture medium and other features that can monitor and adjust temperature, pH etc. automatically by micro process controller. There are various types of fermentors designed for cultivating different types of cells, like plant and animal cells. For example, an airlift fermentor is a non-mechanical system for cultivating many types of cells (microbial, plant and animal). Mixing is achieved by forcing air through a porous plate at the bottom of the up riser column. The gas then escapes at the top and the denser fluid returns in an external downcomer column.

Figure 8.2 shows an ordinary laboratory fermentor with microprocess controller. The main components and features of an ordinary laboratory fermentor are explained below.

Type of Vessel

A closed vessel made up of glass or steel, attached with an air inlet having filter to maintain sterility inside.

1. **Agitator.** An agitator or metallic blades to stirrer the medium to properly mixing the medium and cells to improve the material and oxygen transfer between the cells and the medium.
2. **Baffles.** This is a rectangular strip of metal attached to the vessel wall. It can improve the oxygen transfer by increasing the turbulence of the culture medium.

Figure 8.2 A Laboratory Fermentor (BioG-Micom) is a precise and unique fermentation system with microprocessor controller and display system. This system is designed for various types fermentation process.

3. **Sensors.** The main sensors in the fermentor are the pH sensor, temperature sensor and antifoam sensor.

   (*a*) ***pH regulator.*** When the cells grow and multiply by metabolizing the media constituents, the pH will change continuously. It can change the physiological state of the cells and thereby the efficiency of the product formation can be affected negatively. The sensor constantly monitors the pH change of the media and its specific pH will be maintained constant by adding acid or alkali.

   (*b*) ***Temperature regulator.*** Microbial culture under active metabolic conditions will produce heat and that will increase the temperature of the medium beyond its optimum temperature. To counter this there is a heat exchange coil inside the vessel or a jacket around the vessel through which hot water or cold water circulates to maintain the temperature.

   (*c*) ***Antifoam monitor.*** Agitation and aeration of the culture medium can result in excessive foaming. Foaming will increase at high cell density. There will be a sensor placed just above the culture medium, which can detect the formation of foam when it touches the sensor. When the foam touches the sensor the electrical signals activates the pump attached, to supply the antifoam agent into the medium.

5. **Addition ports.** This is the inlet for the adding the culture medium and inoculum (microbial culture) required for the fermentation process.
6. **Outlet port.** There may be a separate outlet for removing samples intermittently during the process and also at the end after the completion of the fermentation.
7. **Sterilization.** The entire process of fermentation depends on the sterility of the culture medium and other equipments involved in the process. Therefore, the fermentor vessel and its accessories that come in contact with the culture media should be sterilized. Usually the fermentor vessel is sterilized by circulating boiled water followed by filling it with steam for a specific period of time. For carrying out this process there is a steam inlet and outlet. Fermentor vessels of small volumes can also be sterilized by autoclaving.

Aseptic Operation of Fermentor

Fermentor is a small bioreactor specially designed to cultivate microorganisms in aseptic condition. There should not be any contamination by other microorganisms. This is very essential for the successful completion of fermentation. The entire fermentor and all other accessories of fermentor and solutions such as growth medium used in the experiment are to be sterile. The air used in the fermentation process should be free of microbes. In short the entire operation of fermentation has to be conducted under aseptic condition.

The fermentor vessel in which the cells are growing has to be washed with hot water and then it should be sterilized by conducting steam through it. Care should be taken to see that steam is reaching all parts of the fermentor assembly. If it is a small volume fermentor, it can be separated from the assembly and can be autoclaved separately. The medium normally steam sterilized separately in autoclave, but also can be sterilized in the fermentor vessel itself by passing steam through the jacket or cooling coil around the vessel. In addition to the media components, the other additives such as antifoam agents also should be sterilized. But there is no need of sterilizing the acids or alkalis used to maintain the pH, if it is strong. The air used in the fermentor should be filtered using a bacterial filter or a better filter. By using ordinary bacterial filters it is possible to make the air free of bacteria and fungal spores. But it is not possible to avoid the presence of bacteriophages in the incoming air and this can cause serious damage to the culture. By choosing the right type of air filters, compressors and by regulating the speed of airflow, the contamination of cultures in the fermentor can be minimized. For example, some type of compressor can generate sufficient heat to kill the bacteriophages present in the air. In certain special cases, if cultivating a pathogenic microorganism or a genetically engineered organism, the air leaving the fermentor also need to be sterilized as a safety measure. This will also help in protecting a new strain of microorganism from freely available to others.

Bioreactor Designing

The designing of a bioreactor depends on a number of factors like type cells, details about the type metabolic reaction, information about the mass transfer and heat transfer, fluid mixing etc.

Types of Bioprocesses

Bioreactors or fermentors can be classified in different ways. For example, classification is based on the type of biological agent, the type of metabolism or product formation, classification based on modes of operation and another classification based on the mode of stirring or mixing.

First, classification is based on the type of biological agent used. They are of two types :

1. Microbial fermentors
2. Enzyme (cell-free) reactors

Further classification is possible based on other process requirements.

For example, on the basis of mode of fermentation bioreactors can be designed for :

1. Aerobic and anaerobic type or
2. Submerged or surface type of fermentation etc.

Based on the type of metabolism there are three types of bioreactors :

TYPE 1 : *Production of primary metabolites*—Growth-associated production. For example, ethanol production.

TYPE 2 : *Overproduction of primary metabolites*—The product may be an intermediate (and not the end product of a pathway). It has something to do with catabolism/energy metabolism. The production phase can be distinguished from the growth phase. For example, citric acid production, production of amino acids.

TYPE 3 : *Non-growth associated production*—The product is not connecting to catabolism or energy metabolism. For example, production of secondary metabolites like antibiotics. There are distinct growth on production phases. Sometimes production only begins when the main carbon source is exhausted and a secondary carbon source is used.

Classification based on the mode of operation—There are two types of reactors :

1. **Batch mode**: No exchange of liquid medium. On addition of inoculum, culture grows uncontrolled until some nutrient is exhausted. This is easy to set up.
2. **Fed-batch**: Additions of media, on demand. Some medium components may be continuously fed without taking out anything. Hence reaction volume will increase. The substrate conc. must be kept low in the beginning to minimize byproduct formation. As biomass increases, the feed rate might have to be correspondingly increased. This requires that a feeding strategy be devised based on a predetermined rate or by a feedback control.

In another type of classification, the mechanism of mixing has been taken into consideration.

The type of mixing that is provided in the fermentor is an important factor in fermentors that facilitate proper mass transfer and heat transfer. The mechanism of mixing should not give a negative impact in the growth and metabolism of the cells. For example, plant cells, animal cells and hybridoma, which grows slowly and are fragile compared to bacterial cells needs less amount of oxygen for growth and metabolism. A mechanical stirring can cause damage to the cells. At the same time mixing needed to maintain uniformity in the distribution of nutrients and other components in the medium. The following are some of the bioreactors having different modes of agitation or mixing.

1. *Stirred tank reactors*: In this type impellers provide agitation.
2. *Bubble column reactors*: Mixing is provided by airflow through the medium at constant velocity.
3. *Air lift fermentors*: Mixing is provided by forced circulation of nutrients inside the reactor by air.

Units of Bioprocess Operation

The large-scale fermentation or the bioprocess operation involves mainly three steps—Upstream processing, bioreaction and finally the downstream processing and product recovery. Each of these three steps involves further individual steps and each individual step of operation is called unit operation.

The unit operations coming under upstream processing are :

1. Milling
2. Mixing
3. Media preparation
4. Sterilization
5. Cooling
6. Heating.

Unit operation coming under bioreaction are :

1. Mixing
2. Handling of microorganism, plant or animal cells
3. Inoculation of the cells
4. Heating and cooling.

Unit operation under downstream processing and product recovery are :

1. Filtration
2. Centrifugation
3. Dialysis
4. Crystallization
5. Membrane separation
6. Chromatography
7. Solvent extraction
8. Evaporation
9. Precipitation.

Homogenous Reactions

Microbial Growth Kinetics and Fermentation

In large-scale fermentation, information about the nature of growth and metabolism of the organism or cells is very important. In microorganism usually the growth (increase in biomass) take place by binary fission. Yeasts normally reproduce by budding. Fission, mycelial growth and branching are also possible in the case of yeast.

After inoculation, the microbial culture starts growing. The growth of the microorganism follows a typical pattern of growth. A microbial batch culture has four distinct phases of growth. The growth cycle shows a (1) lag phase, (2) exponential phase or log (logarithmic) phase, (3) stationary phase and (4) death phase.

1. The lag phase. Inoculation of cells into the growth medium results in a period where there is no increase in cell number. This forms a preparation for the cell multiplication and growth. This period is known as lag phase. The duration of the lag phase depends on the state of the inoculum and its growth history. The lag phase is not an inactive period, but it is a very active period preparing for a rapid growth phase.

2. The log phase. This is the logarithmic phase or the exponential growth phase, which follows the lag phase. During this phase the biomass or the number of cells increase exponentially with time. Cells undergo doubling several times and the specific growth rate of the culture remains constant. Assuming growth rate depends on the available biomass, and if μ = specific growth rate, the rate of increase in biomass (the increase in cell mass with time), dX/dt is the product of specific growth rate μ and total biomass concentration X.

$$dX/dt = \mu X$$

Similarly, if N is the total number of cells, then the increase in cell number dN/dt is the product of specific growth rate and cell number N.

$$dN/dt = \mu N$$

The cell mass concentration X_0 at the beginning of the exponential phase of growth, the time required for doubling the biomass could be calculated by the above formula

$$dX/dt = \mu t$$

$$\text{Ln } X/X_0 = \mu t$$

$$\text{Ln } 2 = \mu\, t, \text{ thus } t = 1\, 2/\mu$$

The generation time of a culture is the time taken by the cells in the culture to double the number of the biomass under defined conditions. In the case of commonly used industrial microorganisms have a specific growth rate, μ varies from 2.1–0.8 h^{-1}. It corresponds to the doubling time ranging from 20 min to 8 hrs. The doubling time and specific growth rate of the organisms are very important in the large scale fermentation process, because it determines the requirements of media and other growth related factors such as fermentation batch time for the production of the required biochemicals.

The specific growth rate of an organism is very characteristic of that organism and is depended on the growth environment including ionic concentration, pH, temperature, media composition and dissolved oxygen levels. The doubling time of some the microorganisms are given the table below.

TABLE 8.3 Doubling time of some microorganisms under normal culture conditions

Microorganisms	T_d (minutes)
E. coli	20
Bacteria	45
Yeasts	90
Moulds	160
Animal cells	630—1260
Plant cells	3600—6600

3. Stationary phase. The logarithmic phase or the exponential phase of growth is followed by a stationary phase. During this phase the growth rate and death rate of cells are almost equal. Depletion of a growth-limiting nutrient in the medium or accumulation of toxic byproduct or secondary metabolite may the cause for the stationary phase.

4. Death phase. Following the stationary phase, finally the culture enters the death phase. Here the cells lose its viability and the death rate overtakes the growth rate. The damage for the cells may be due to cell lysis or by some other mechanisms.

Effect of Temperature on Growth

Growth of organisms is influenced by a number of factors and temperature is one among them. All biochemical reactions of the cells are controlled by the state of temperature and thus it influences the growth rate. Microorganisms are classified into three groups based on its optimum temperature for growth. They are :

- Psychrophiles—grows around 15°C
- Mesophiles—grows around 37°C and
- Thermopiles—grows around or above 55°C

Each organism has a specific range of optimum temperatures and exact value depends on the species or strains of the organisms and other growth conditions. The ability of the organism to utilize the carbon source for the production of biomass is temperature depended and it declines with increase in temperature in the case of normal organisms. Therefore, optimization of growth temperature for microbial cultures is essential to get maximum product output in bioprocess technology.

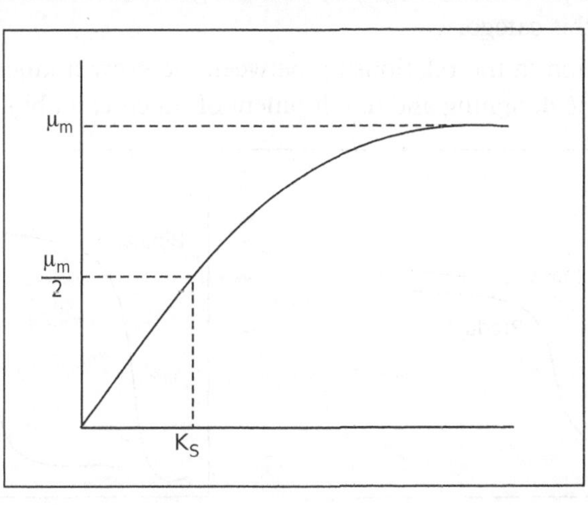

Figure 8.3 Effect of substrate concentration on specific growth rate K_s is the substrate concentration at half μ_m.

Effect of Substrate Concentration on Growth

Substrate concentration is another important factor that controls growth and rate of product formation in a microbial process. Usually the carbon source is the rate limiting component of the culture media and influence of such substrates can be described by the equation $\mu = \mu_m S/(K_s + S)$, where S is the concentration of the limiting nutrient, K_s is the substrate specific constant and μ_m is the maximum specific growth rate. The value K_s indicates the affinity of the organism for the substrate or the nutrient. The effect of substrate concentration on growth of organism is represented in Figure 8.3. The specific growth rate μ increases with substrate concentrations until it becomes a rate-limiting factor. However, there can be inhibition to growth and product formation due to high substrate concentration in the culture media. K_s is the substrate concentration corresponding to the half of μ_m ($\mu_m/2$). Thus, the growth on a given substrate can be described by two constants—μ_m and K_s. Therefore, it is essential to know and evaluate the values of K_s, μ_m and μ of the organism with respect to the respective substrate to get an idea about the suitability of that material as substrate for growth and product formation.

Relation between growth and product formation is very important in the designing and development of a bioprocess technology. It is also very important in the designing of the bioreactor or fermentor. Based on this relationship fermentation process can be classified into growth associated and non-growth associated fermentation.

In growth associated fermentation, product formation occurs along with the biomass increase. Both take place simultaneously. In such cases, the rate of product formation is directly proportional to rate of biomass formation (Figure 8.4). The product will be a primary metabolite and its production takes place primarily during the logarithmic phase of the growth curve. Most of the alcohol and acid fermentation belong to this class. In non-growth associated fermentation, the product formation is not directly related to the rate of growth of the cells. But it is proportional to the concentration of biomass. The product formation takes place at the stationary phase of growth. Production of secondary metabolites like antibiotics belongs to this category.

Therefore, information in the relationship between the growth kinetics and product formation kinetics is very crucial in the designing and development of an efficient bioprocess technology.

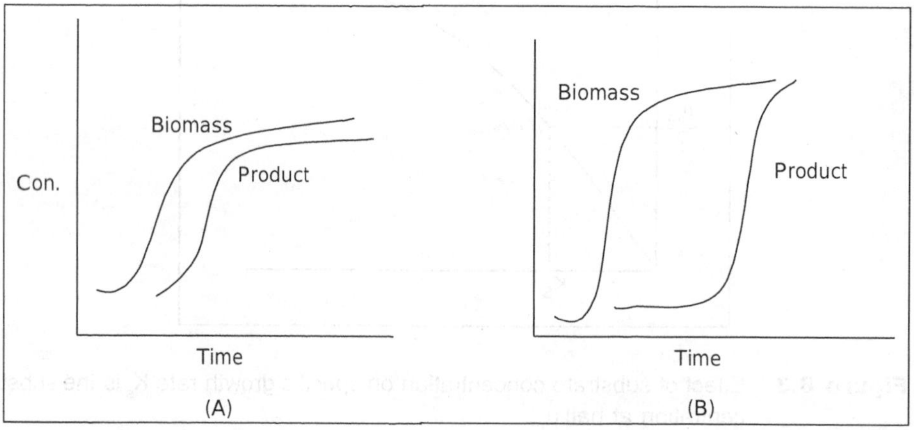

Figure 8.4 Relationship between growth and product formation
(A) Growth associated (B) Non-growth associated product formation.

Therefore, selection and composition of the media should provide all the requirements for increasing the growth rate and product formation and also it should minimize or simplify the process operations. In certain cases of fermentation there may be two types of media particularly in the cases of plant cell culturing.

Growth media. The media that facilitate growth of the cells and the product formation will be minimum.

Production media. The media that is meant for product synthesis is called production media.

The process is designed in such a way that when the cells were grown to a sufficient amount of biomass in the growth media, it will be transferred to the production media where it starts synthesizing the product.

Reactor Engineering

The reactor engineering includes all the three stages of bioprocess engineering—pretreatment or upstream process, bioprocess or the biochemical process and the downstream processing. While designing all these stages of the bioprocess special care should be taken to bring down the cost of production.

The upstream processing equipment consists of a pretreatment vessel. This vessel is specially designed to prepare the media of a specific volume by putting and mixing the correct components in correct quantity and volume, followed by sterilization.

The bioreactor or the fermentor is designed to carry out the specific biological reaction with the help of the biological agent—living cells or immobilized enzyme. The type of bioreactor depends on the type of cells nature of growth and the biochemical reactions. Huge bioreactors are needed for carrying out the bioreaction in controlled environment with large volumes of the media and substrate.

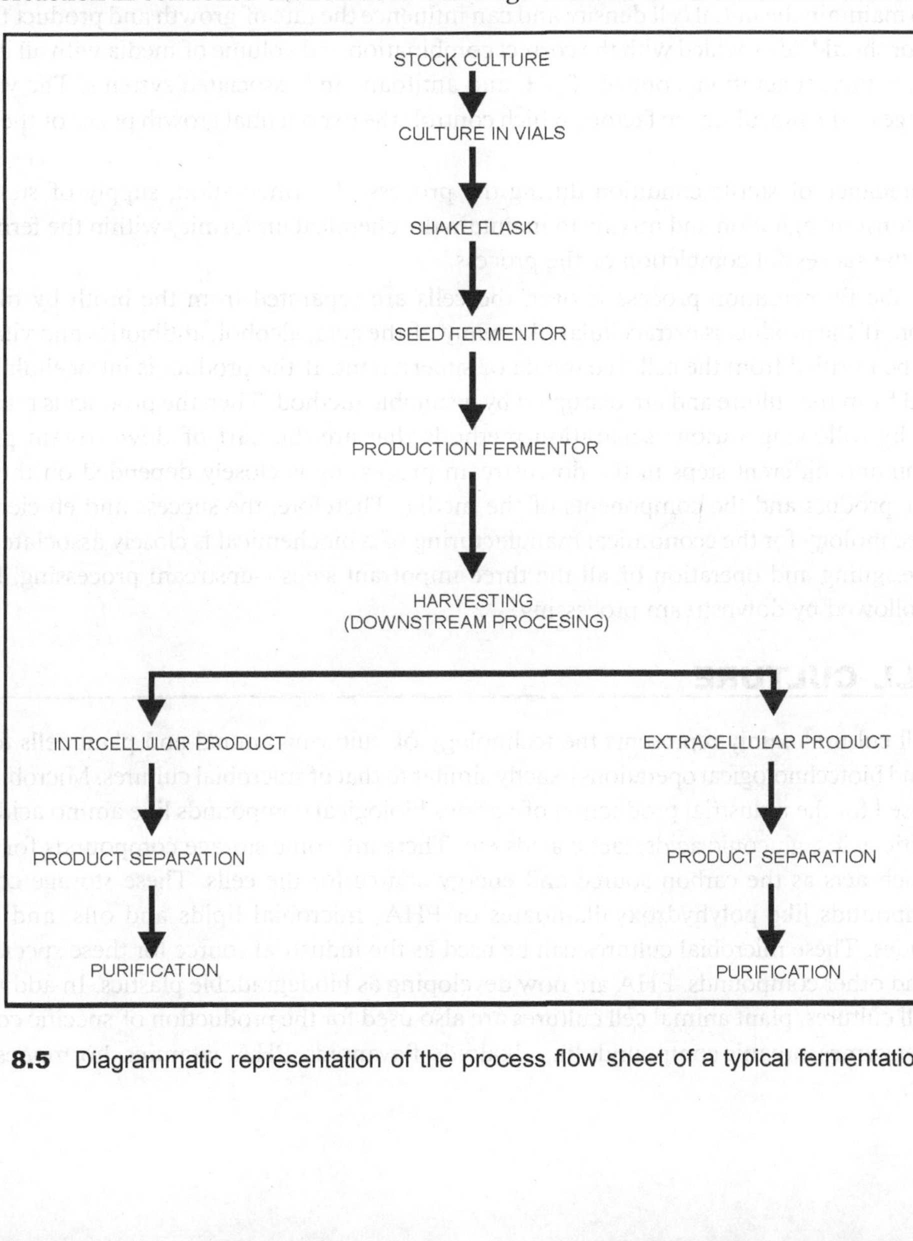

Figure 8.5 Diagrammatic representation of the process flow sheet of a typical fermentation process.

Downstream processing includes the operations needed for the recovery of the product in a very simple and economic way. It includes a serried of different filtration, separation and purification methods in series. In general, large-scale fermentation and purification are carried out in a stepwise manner. A diagrammatic representation of the process flow sheet of the fermentation process development is given in Figure 8.5.

The procedure of the entire operation starts with the media preparation and its sterilization, sterilization of the bioreactor or the fermentor with or without media. The microbial cultures are usually maintained in petri plates or in slants of culture tubes as stock cultures. These cells are in the dormant stage and should have to be activated. First the cells are inoculated in small culture volumes (5–10 ml) and grow overnight. Then it is inoculated in to larger volumes (100–1000 ml) and then these are either transferred into the bioreactor directly or through a seed fermentor of 10–100 l (Figure 8.5). This is very important to maintain the initial cell density and can influence the rate of growth and product formation. The fermentor should be provided with the correct combination and volume of media with all its required components, sufficient aeration, control of pH, and antifoam and associated systems. The volume cell density and age of the inoculum are factors, which controls the exponential growth phase of the microbial growth.

Maintenance of sterile condition during the process of fermentation, supply of sterile air for respiration, constant agitation and mixing to maintain the chemical uniformity within the fermentor are essential for the successful completion of the process.

Once the fermentation process is over, the cells are separated from the broth by filtration or centrifugation. If the product is extracellular like many of the acid, alcohol, antibiotics and vitamins, the product can be purified from the cell free media or supernatant. If the product is intracellular, the cells are separated from the culture and are disrupted by a suitable method. Then the product is purified from the mixture by following various separation methods that are the part of downstream processing. Oraganisation and different steps in the downstream processing is closely depended on the type and nature of the product and the components of the media. Therefore, the success and efficiency of the bioprocess technology for the economical manufacturing of a biochemical is closely associated with the integrated designing and operation of all the three important steps—upstream processing, bioreactor operations followed by downstream processing.

8.2 CELL CULTURE

The term cell culture mainly represents the technology of culturing animal and plant cells for various bioprocess and biotechnological operations exactly similar to that of microbial cultures. Microbial cultures are mainly used for the industrial production of various biological compounds like amino acids, organic acids like citric acids, gluconic acids, lactic acids etc. There are some storage compounds for microbial systems, which acts as the carbon source and energy source for the cells. These storage compounds include compounds like polyhydroxyalkanoates or PHA, microbial lipids and oils, and microbial polysaccharides. These microbial cultures can be used as the industrial source for these special types of polymeric and other compounds. PHA are now developing as biodegradable plastics. In addition to the microbial cell cultures, plant animal cell cultures are also used for the production of specific compounds like various enzymes, organic compounds like alkaloids, flavanoids, PHA, vitamins, hormones, steroids, pigments etc.

The use of crop plants and plant cell cultures for the production of biodegradable plastics (PHA) is attracting biotechnologists because of the high cost of production with microbial systems. Cultivation of transgenic plants capable of producing PHA with the help of photosynthesis to provide the carbon source and energy required for the polymer synthesis.

Microbial polysaccharides are special polysaccharides; most of them are not detected in other living systems like animals and plants. These polysaccharides accumulate within the bacterial and fungal cells and act as reserve food materials. Glycogen is a very good example of storage polysaccharides present in bacterial and fungal cells. In eukaryotic systems, it is the storage polysaccharides of animals. There are the exopolysaccharides in microbial systems, which are excreted by the cells and are commercially important microbial polysaccharides. Some of the important microbial polysaccharides of commercial interests are the following :

Xanthan produced by *Xanthomonas campestris*. It is a branched polymer having a β (1 $\rightarrow$ 4) linked glucan (polymer of glucose) backbone with a trisaccharides side chain on alternate glucose residues. It is an important and widely used microbial polysaccharide in food industry for various purposes like stabilization, gelling agent, viscosity control, suspension etc.

Dextran produced by a wide variety of gram positive and gram-negative bacteria like *Leuconostoc mesenteroides*, *Streptococcus* species etc. It is a polymer containing glucan having various types of branching, which depends on the type of organism. Dextran is not produced within the cell, but by an extracellular enzyme, dextransucrase. This enzyme acts on the sucrose, polymerizing the glucose units liberating the fructose into the medium. Dextrans have various uses including clinical applications. It was first used as a plasma extender. It is used for the prevention of thrombosis and in wound dressing to absorb fluids. Dextrans now have various laboratory uses and also in food industries. Sephadex is used as an efficient Gel filtration medium.

Gellan is an unbranched heteropolysaccharides. Its repeating units contain two glucose one gluconic acid and one rhamnose residue. It is produced by *Pseudomonas elodea* and it is an acidic gel forming polysaccharide. It can be used instead of agar in microbiology and as a gelling material in food industries.

Scleroglucan is a natural polysaccharide with a β (1 $\rightarrow$ 3) backbone and branches consisting of single glucose residues attached to the every third glucose of the main polymer chain. It is fungal exopolysaccharide produced by various species of *Sclerotium*.

Alginate is another linear polymer composed of mannuronic acid glucoronic acid. It is produced by *Azotobacter vinelandii* and various species of *Pseudomonas*. There are some seaweeds that producing various types of alginate which are extensively used in industries including food industries.

Pullulan is another α-glucan with a trisaccharides repeating unit. It is commercially produced by the fungus *Aureobasideum pullulans*. Pullulan forms a strong resilient film and fibres. The film has a lower permeability to oxygen and is biodegradable. Therefore, these films and fibres are better than cellophane polypropylene.

Curdlan is another exopolysaccharide composed of β (1 $\rightarrow$ 3) glucan. It is produced by *Alcaligenes faecalis* var. myxogenes. Similar types of polymers are also produced by *Agrobacterium radiobacter* and *A. rhizogenes*. It is insoluble in water and it gels above 55°C and the process is irreversible. It can be used in food industries as a gelling agent in cooked foods and as a support for immobilized enzymes.

Various types of commercial enzymes like alpha amylase, single cell oils and single cell protein are some other products that can be obtained by culturing microbial cells.

Animal and Plant Cell Culture

Eukaryotic cells, both animal and plant cells can be cultured *in vitro* like microbial systems and has been started using for various purposes. The cultivation of plant cells is mainly for the biosynthesis of certain plant-derived chemicals and for basic studies like genetic transformation, somaclonal variation etc. Cultivation of animal cells particularly mammalian cells has been started since 1945. Media components and culture conditions that support cell multiplication and proliferation has been developed over time. Various types of cell lines of the same organism as well as of different organisms were established. Cell lines of mammals like human, rat, hamsters, cats, dogs, mice, rabbit, monkeys, horse, cattle, sheep etc., and cell lines of individual organs such as skin, heart, liver, spleen, kidneys, ovaries, lymph nodes, thymus and muscles. Cell lines were also established from various cancer cells and from other organisms like fish reptiles, insects etc. The purposes of establishing the cell lines were to study the basic aspects of developmental biology, cell differentiation and other cellular activities. But the major practical applications that drived the art and technique of cell culture was its use as the source of vaccines like polio vaccines.

Today production of vaccines and other therapeutic proteins are the major applications of mammalian cell culture technology. Certain complex proteins like diagnostic proteins and therapeutic proteins like antibodies and hormones, when produced by prokaryotic systems through genetic engineering may not be biological active and therefore become useless. This is mainly because the microbial system may not have the correct machinery to carry out the posttranslational modification of proteins like glycosilation, phosphorylation acetylation etc. The folding pattern of the polypeptide in prokaryotes may not be proper resulting in a totally new 3D structure, which may not be biologically active. In addition to this cell cultures are used for the toxicological and pharmaceutical experiments to test effect of various chemicals as potential drugs. This has tremendously reduced the use of animals as experimental models. Now the cell culture particularly the stem cell research has started using for the production of artificial organs. For example, layers of cultured keratinocytes can provide artificial skin. Similarly, efforts are being made for the development of artificial liver and kidneys.

Another example for the use of mammalian cell culture for the industrial use is the hybridoma technology. Fusion of lymphocytes and myeloma cells (cancer cells) results in hybridoma cells. In hybridoma cells characters of both lymphocytes and cancer cells are combined—the capacity for indefinite multiplication and the ability to produce antibodies. The construction of hybridoma cells were first demonstrated by Kohler and Milstein to produce monoclonal; antibodies. They have the high potential for the production of glycosylated proteins. Myeloma cells are also used as the host cells for the production of recombinant proteins, which needs specific posttranslational modifications as in the case of antibodies and hormones. Baby hamster kidney cells have been used for a long time for vaccine production in pharmaceutical industry. Chinese Hamster Ovary cell lines (CHO cell lines) are another type of cell cultures favoured for the industrial production of recombinant eukaryotic proteins.

8.3 ENZYME TECHNOLOGY

Like microorganisms, enzymes also have been used by mankind for thousands of years for various purposes directly and indirectly. But the uses of enzymes, in most cases were indirectly by using microorganisms. For example, in leavening of bread, fermentation of grape and other types of fruit juices, conversion of milk to cheese, bating of leather etc. Today crude as well as purified enzymes have various industrial including medical uses. Approximately 75 % of the industrial enzymes are used for the hydrolysis and depolymerization of complex natural substances. Proteases form the highest category of enzymes, which is used in detergent manufacturing and diary industries. Applications of enzymes in food industries form the largest among the uses of enzymes. These applications include conversion of starch to glucose and fructose, cheese wine, beer, flavouring compounds, fruit juices and animal and cattle feeds. The production of enzymes have increased tremendously by the widespread use of fermentation technology and bioprocess technology along with the introduction of genetic engineering. Genetically modified microorganisms are becoming the dominant sources of industrial enzymes and this trend will increase in future. Many enzymes produced from plant and animal sources are now produced in transgenic microorganisms.

Site Directed Mutagenesis

Even though microbes form the main sources of enzymes animal and plant tissues also form the sources of certain specific enzymes like lipases, esterase, and peroxidases and very efficient proteases like pepsin, papain, trypsin, rennets etc. Molecular biology techniques like site directed mutagenesis can be used for producing enzymes having altered properties and efficiency. Enzymes, which have known amino acid sequences can be altered by site, directed mutation by changing specific amino acids in the active centres of the molecules very selectively. Specific amino acids in the sleeted region of the protein molecule can be modified to other amino acids. This is easily performed by altering the codons of the respective amino acids in the enzyme gene and that altered gene is expressed in a host cell by conventional genetic engineering and cloning techniques.

Immobilised Enzymes

The enzymes that are used in industries are in solutions on a single use basis. This is applicable mostly in food and beverage industries. The enzymes may not be reusable. But in certain cases the enzyme can be used repeatedly over a period of time without lose in its efficiency. In such cases the enzymes can be used in an immobilised form. This allows the reuse of enzymes to carry out the reaction or biotransformation and is very easy to recover the product without damaging the enzyme. This immobilization of enzymes is particularly applicable in pharmaceutical industries where some reaction steps can be carried out by specific enzymes to introduce some chiral centres in the molecule or adding some specific functional groups to certain positions. The immobilization techniques help in the reuse of enzyme for a very long time and it help in the easy recovery of product with minimum carry over of the residual enzyme protein.

The techniques used for the immobilization of enzymes depends on various factors such as the type of reaction the final product and the sensitivity of the enzymes to the physical and chemical factors, and finally the medium of reaction—aqueous or organic solvents. Adsorption of the enzyme to an inert support is the simplest of the techniques. Adsorbed enzymes can be permanently fixed onto the support by cross-linking reactions with glutaraldehyde treatment. This can make the enzymes reusable in aqueous solutions. The nature of the support and its selection is very important in the final appearance of the immobilised enzyme and the selection of the bioreactor to carry out the biotransformation reaction.

For example, the supports like cellulose or diatomaceous earth produce an enzyme immobilised product that can be stirred in a bioreactor and the product and enzyme can be separated by filtration. Enzymes that are immobilised on beaded supports are suitable for column bioreactors with a reactant or substrate inlet and product outlet with a sufficient retention time to allow the proper reaction, retaining the enzyme in the bioreactor. A large number of supports are available for enzyme immobilization, which include both inert and activated. Resins, porous glass beads, cellulose, gelatin, chitosans and alginate require chemical activation to act as a support for enzyme immobilization. Enzymes can be immobilised by entrapping in some supports, where the enzyme is mixed with the soluble support and is then solidified. Example—cellulose acetate precipitation from an organic solvent to aqueous medium, entrapment in calcium alginate, chemical polymerization with polyacrylamide etc. the resulting entrapped enzyme can be packed in a suitable bioreactors and can be used for the biotransformation for long time.

8.4 BIOLOGICAL FUEL GENERATION

We are slowly depleting our fossil fuel energy resulting in the need to seek out alternative sources of energy. So far, these have included the harnessing of hydro, tidal, wave and wind power, the capture of solar and geothermal energy supplies, and nuclear power. There is now a growing appreciation of biological solar energy systems and biotechnological advances in this area will soon bring economic reality to selected processes. As fossil fuel resources are depleted and become increasingly more expensive, conversion of organic residues to liquid fuels will become a more economically attractive consideration.

There are three main directions that can be followed to achieve biomass supplies: (1) cultivation of so-called energy crops, (2) harvesting of natural vegetation, and (3) utilization of agricultural and other organic wastes. The conversion of the resulting biomass to usable fuels can be accomplished by either biological or chemical means or by a combination of both. The two main end products that will be formed will be either methane or ethanol although other products may arise depending on initial biomass and the processes utilized, for example, solid fuels, hydrogen, low-energy gases, methanol and longer-chain hydrocarbons.

Although biomass may ultimately only supply a relatively small amount of the world's energy requirements, it will nevertheless be of immense overall value. In some parts of the world, such as Brazil and countries of similar climatic conditions, biomass will surely attain wider exploitation and utilization. There may still be some disadvantages when comparing it with coal or oil, but the very fact that it is renewable and they are not must be the spur to further research. In time, biomass will become much more easily and economically used as a source of energy for mankind.

8.5 SINGLE CELL PROTEIN

One of the biggest problems facing the world today is population growth, especially in developing nations. At present there are 4 billion mouths to be fed and if the growth rate continues, by the year 2000, this number will exceed 5 billion. Conventional agriculture may not be able to a sufficient supply of food and, in particular, protein. The search for protein is recently pursued. Today, new agricultural practices are widespread, high protein cereals have been developed, the cultivation of soybeans and groundnuts is ever expanding, etc. After all this, the use of microbes as protein producers has gained wide experimental success. This field of study has become known as single cell protein production (SCP) and refers to the fact that most of the microorganisms used as producers grow as single or filamentous individuals rather than as complex multicellular organisms like plants or animals.

There are many reasons why microbes are the prime candidate for SCP production, some of which include: (1) microorganisms can grow at remarkably rapid rates under optimum conditions: some microbes can double their mass every 0.5–1 hour, (2) microorganisms are more easily modified genetically than plants and animals: they are more amenable to large-scale screening programmes to select for higher growth rate, improved amino acid content etc. and can be more easily subjected to gene transfer technology, (3) microorganisms have relatively high protein content and the nutritional value of the protein is good, (4) microorganisms can be grown in vast numbers in relatively small continuous fermentation processes using relatively small land area and are also independent of climate, and (5) microorganisms can grow on a wide range of raw materials, in particular low value wastes, and can also use plant-derived cellulose.

The acceptability of SCP, when presented as human food, depends not only on its safety and nutritional value, but also on other factors. People do not usually fancy the idea of eating food derived from microbes. In many cultures, there are guidelines of what you can and cannot eat. Also, odour, colour, taste, and texture need to be taken into consideration when dealing with peoples' desires. Thus, if SCP is to be used as direct food for man, then the skills of the food technologist will be greatly challenged.

8.6 SEWAGE TREATMENT

Waste can be considered as any material or energy form that cannot be economically used, recovered or recycled at a given time and place. Growth in human populations has generally been matched by a greater formation of a wider range of waste products, many of which cause serious environmental pollution if they are allowed to accumulate in the ecosystem. In rural communities recycling of human, animal and vegetable waste has been practiced by man for centuries, providing in many cases valuable fertilizers or fuel. In urban communities where most of the deleterious wastes accumulate efficient waste collection and specific treatment processes have been developed since it is impractical to discharge high volumes of waste into natural land and waters. The development of these practices in the last century was one of the main reasons for the spectacular improvement in health and well being of the community.

Mainly by empirical means a variety of biological treatment systems have been developed, ranging from cesspits, septic tanks and sewage farms to gravel beds, percolating filters and activated sludge processes coupled with anaerobic digestion. The primary aims of all of these systems or biotreaters is to alleviate health hazards and to reduce the amount of oxidizable organic compounds and thus produce a final effluent or outflow, which can be discharged into the natural environment without producing any adverse affects. Bioreactors rely on the metabolic versatility of mixed microbial populations for their efficiency. The fundamental feature of biotreaters is that they should contain a range of microorganisms with the overall metabolic capacity to degrade any compound entering the system. Controlled use of microorganisms has lead to the virtual elimination of such water-borne disease as typhoid, cholera, and dysentery in industrialized communities.

8.7 ENVIRONMENTAL BIOTECHNOLOGY

In industrialized countries, there is increasing public concern over the impact of human activities on the environment and the legacy we leave for future generations. Attention is being given to minimizing environmental damage, and to cleaning up past environmental damage, while trying to ensure that the standard of living to which we have been accustomed can be maintained.

As an example, US $100 billion worth of crops are destroyed annually by soil-dwelling nematodes damaging crop plants. Chemical nematicides are the only current option for crop protection, but these are among the most toxic and environmentally damaging pesticides that are in widespread use.

It is obviously far better not to pollute the environment in the first place, rather than having to develop ingenious ways of cleaning up the environment. For this reason, many industries and agriculture are now investigating the potential of plants and microorganisms to provide cleaner processes and products.

Benefits that agriculture could have on the environment are :

- An increase in the productivity of crops, without an increase in the dependency on environmentally damaging agrochemicals.
- As a result of increased productivity, a reduced pressure to exploit the remaining uncultivated habitats.
- As a result of increased productivity, a reduction in energy inputs (mostly from reduced agrochemical manufacture).
- The creation of alternative, renewable, sources of energy (*e.g.* biodiesel).
- The creation of new more environmentally-friendly raw materials for industry (*e.g.* biodegradable plastics from plant starches, or high-value speciality chemicals).
- As a result of the development of genetically modified crops (if properly used), a reduction in the amounts of agrochemicals (*e.g.* pesticides and herbicides) released into the environment.

Environmental concerns over the use of genetically modified crops

Many environmental groups, and some biotechnologists, have concerns about the use of genetically modified crops with respect to their effect on the environment. They warn that the environmental impact of their use might not become apparent until after their large-scale commercial use.

Concerns voiced include :

Herbicide use. The use of herbicide-resistant crops will increase rather than decrease herbicide use.

Genetic pollution and superweeds. Genes that have been copied from one species and inserted into another might 'escape' and spread to other organisms, thus causing 'genetic pollution'. For example, herbicide-resistant crops might cross-breed with related weeds and produce herbicide-resistant 'superweeds'.

Antibiotic resistance. Genes that code for antibiotic resistance are sometimes inserted into plant cells together with the 'useful' gene; these 'marker' genes enable scientists to select cells that have been successfully modified. Such antibiotic-resistance markers in crops might spread to animals or humans, rendering medical or veterinary use of the antibiotic ineffective.

Unexpected effects. Some genes that are inserted into genetically modified plants might unstable (there is a high probability that they will be lost from the plant cells), or might show unexpected effects because it is difficult to predict where in the plant genome the genes will insert.

Pest resistance. Genetically modified crops that are designed to be resistant to a particular pest might have unintentional (and unpredictable) effects on harmless or beneficial organisms (e.g. ladybirds and bees).

Persistence and weediness. Genetically modified plants might, intentionally or unintentionally, be more vigorous than their non-modified relatives. They can therefore effectively become 'weeds'. If plants are more 'persistent' (e.g. survive over winter better), they could rapidly dominate ecosystems at the expense of other plants. If they show 'weediness' characteristics, they could also spread to new habitats.

Damage to wildlife and biodiversity. Monocultues of pest-resistant crops also harm beneficial insects and non-target organisms, with knock-on effects on the food chain.

Addressing Environmental Concerns

Because it is difficult to assess the risks associated with the potential environmental impact of using genetically modified crops, precautionary environmental legislation is in place in many countries. Before allowing experimental or commercial release of genetically modified organisms, most countries use risk assessment procedures that are based on estimating the impact and interactions of genetically modified crops and their environment. The following counter-vicws might also be considered :

Genetic modification of plants achieves essentially the same result as conventional plant breeding

The ability to move single genes, or sets of genes, from one species to another allows 'step changes' in genetic composition to be made quickly. However, in some cases, the end product is effectively the same as might be achieved eventually by conventional breeding.

Wide-crossing

'Wide-crossing' is regarded as a conventional breeding technique, but is used to create new varieties of crops between plants that would not normally interbreed. For example, some varieties of wheat that are used to make bread contain some genes from rye. This is the result of a plant-breeding experiment that introduced to the wheat the rye genes that confer disease resistance.

In nature, genes are exchanged between species

Bacteria that are not related can exchange genes quite readily (this is one mechanism for antibiotic resistance spreading). *Agrobacterium tumefaciens* is a bacterium known as 'Nature's own genetic engineer', which routinely transfers some of its genes into plant cells.

Gene instability in conventional crops

The genes in new varieties of crops that have been developed by conventional breeding can also be unstable, particularly when they are produced as a result of wide-crosses. As with GM crops, selective breeding and variety evaluation 'weeds out' undesirable gene combinations.

Conventional breeding is not 'natural' either

Plant breeders have for many years used techniques to introduce more variation than nature produces. Bombardment of a plant with chemicals or radiation causes mutations randomly throughout its genes.

Among the plants that survive might be tens or thousands of mutations—perhaps including ones that plant breeders want. Why is it that chemical or radiation mutagenesis, which generates such a vast array of unknown genetic changes, is acceptable as part of conventional breeding, while changes to one or a few genes by GM—much more easily tracked through variety evaluation—are not ?

Gene transfer from crops to their 'wild' relatives

Little is known about the transfer of genes from any crops to the environment. 'Escaped' genes are most likely to spread to the crops' nearest relatives. Many, but not all, crops are grown in environments in which their 'wild' relatives do not exist. For example, in Europe, genes are much less likely to transfer from genetically modified potatoes than from those grown in Latin American countries where there are wild relatives of potato. There is no evidence to suggest the genes in genetically modified plants 'escape' any more easily than genes in other plants. Herbicide-tolerant crops—produced by conventional breeding—already exist, as does the possibility of gene transfer and 'superweeds'. To date this has not been a problem.

Novel, sustainable, agricultural practices needed

There are concerns that pest-resistant plants will remove important food sources for wildlife and affect non-target organisms. The ecological impact of such crops must be evaluated, particularly for large-scale use. But we must also remember that many pesticides currently in use are not highly selective for pests, harming non-target organisms. Agriculture as currently practiced is the apparent cause of some disastrous effects on wildlife biodiversity—we must look for new options.

Application of biotechnology that we discussed in biological fuel generation, single cell protein single cell oils, sewage treatment etc. can also be covered under environmental biotechnology and can be used for the above purposes while cleaning the environment using biotechnological methods.

8.8 MEDICAL BIOTECHNOLOGY

The use of biotechnology in medicine is growing rapidly and is providing us with opportunities to develop new, more effective drugs and other therapeutics. Studying the genetics of humans is allowing us to understand what happens when genes go wrong in inherited diseases, or cancers, and to start to develop new therapies that treat the genetic cause, not the symptoms. By studying the genetics of viruses, fungi or bacteria, we can understand how they cause disease and develop better drugs and antibiotics that target them more specifically.

Safer, Cheaper Medicines from Biotechnology

Many medicines have traditionally been extracted from plants, microorganisms or animals. However, in many cases, it is not possible to produce the medicines in large enough amounts, safely, or cheaply enough to treat large numbers of people. Valuable drugs may be available only from endangered plant species (*e.g.* the anti-cancer drug tactual from the Pacific yew), or produced naturally in such small amounts that extraction is just too expensive. Biotechnology offers ways to produce valuable medicines in larger amounts by 'cell culture', or using genetic modification to increase the amounts produced, or to make entirely new medicines.

Medicines from 'Cultured' Cells

Many different types of plant, animal and human cells can be grown 'in culture', that is, as a suspension of cells in a large fermentor vessel of liquid that contains all of the nutrients they need, just as though they were yeasts or other microorganisms. These cells can be treated in various ways to make them produce large amounts of valuable compounds: microorganisms grown in this way can produce antibiotics. Plant cells grown in this way have been used to produce the anti-leukemia drugs vinblastine and vincristine: these are natural products of the pink periwinkle. Sedatives and heart drugs (*e.g.* digoxin from foxglove) have also been produced in plant culture.

Genetic Modification for Medicine Production

Organisms can be genetically modified either to increase the amount of medically useful products, such as antibiotics, that they produce or to produce entirely new medicines. Microorganisms and plants are both being genetically modified to produce large amounts of medicines safely and cheaply: genetically modified microorganisms containing the gene for human insulin produce a safe and plentiful supply of insulin for diabetics; while genetically modified plants are starting to be used to produce vaccines for a variety of human and animal diseases.

New Antibiotics through Genetic Modification

Research into finding new antibiotics is probably more urgent now than at any time since the discovery of penicillin some 50 years ago. Why? Many antibiotics that were once highly effective are now proving to be useless as the disease-causing microorganisms develop resistance.

Antibiotics and Antibiotic Resistance

Antibiotics are chemicals produced naturally by microorganisms in the environment that kill, or prevent the growth of other microorganisms. Antibiotic resistance is just as natural, and resistance to natural antibiotics is believed to have evolved hundreds of millions of years ago in soil bacteria. The widespread use of antibiotics enables resistant organisms to survive and multiply: use of antibiotics in livestock to promote growth by suppressing disease also encourages antibiotic resistance.

New Antibiotics

A new approach is being used to try to make entirely new antibiotics. A soil bacterium—Streptomyces—is a source of antibiotics, immunosuppressants, anti-cancer and anti-parasitic agents, and natural herbicides. Many of these valuable products are made by the bacterium from smaller building blocks according to the 'blueprint' determined by the bacterium's genes. It is possible to generate new compounds by changing the blueprint—switching some of the genes, and thus some of the units, around. A whole range of new compounds can be made using this approach, potentially leading to new antibiotics. This area of research is extremely important, considering the rising number of diseases caused by antibiotic-resistant microorganisms.

8.9 AGRICULTURE AND FOREST BIOTECHNOLOGY

Genetic engineering can be used to modify the genetic compositions of plants, animals, and microorganisms. The number of genes that have been isolated and are available for transfer is growing daily. Currently, the technology is used primarily to modify crops, although a number of other applications are in the wings.

Like other products, genetically engineered products undergo a period of research and development before they are ready for commercial release. Many products never emerge from the research and development pipeline. While this is true for almost any technology, genetic engineering is turning out to be more difficult and more expensive than early proponents expected. Although in the early 1980s biotechnology was touted as a miracle technology that was going to usher in a new era of agricultural abundance with minimal harm to the environment, the initial set of products has proved modest. Some of the most important commercial applications of biotechnology are discussed below :

- Crops that are resistant to pests or herbicides, which will allow reduced or more selective use of agrochemicals.
- Crops that are better able to tolerate hostile environmental conditions such as frost or drought.
- New and more efficient crop-breeding systems that allow production of hybrids that cannot be produced by conventional plant breeding.
- New crop varieties that are qualitatively different (*e.g.* plants that produce seeds or tubers that have an altered starch composition, and thus represent new valuable products).
- New crop varieties that have different growth characteristics (*e.g.* altered flowering time or rate of growth) to increase yield.

These tools fall into two major categories: improving crop performance in the field and developing new products with enhanced values.

Genetic Improvement through Traditional Crop Breeding

The cultivation and improvement of food crops began centuries ago, when seeds from the most successful plants of one season were used to sow the next year's crop. Improved crop varieties have been developed

Figure 8.6 Many of our modern cereals bear little resemblance to their 'wild' grass ancestors. Through breeding and selection, the characteristics of these crops have been greatly altered. Left to right: pearl millet, barley, wheat, maize, rice and sorghum.

by selection and breeding, and few of today's crops resemble their wild ancestors. Some crops were developed by these conventional plant-breeding processes to be disease resistant or to grow faster for example, others to make the edible parts of the plant larger or tastier. However, developing new varieties of crops by traditional methods is not only slow but also limits plant breeders to transferring genes between crops that normally cross-fertilize.

Plant Genetics

Over the past few years, scientists have been able to identify the individual genes that determine particular characteristics in a plant. Just as importantly, we can now transfer these genes between plants, which means that novel features can be introduced into crop species. It is unlikely that biotechnology will solve every agronomic problem, but it can offer farmers new options.

Much of our understanding of plant genetics, including the inheritance of particular characteristics, comes from experience with plant breeding. The idea of genes being the unit of inheritance originated more than a century ago when Gregor Mendel started to study peas to understand the principles of inheritance.

Plant Physiology and Biochemistry

Studies of plant physiology, and biochemistry too, have given us an understanding of how plants sense their environment, and respond by germinating, growing, flowering or setting seed. All these processes are determined by their genes. How they photosynthesize, and make proteins, oils or carbohydrates is now well understood. How they recognize insect pests or disease-causing fungi, bacteria or viruses (pathogens) and mount their own defence responses is also much better understood.

Genetic Modification of Crop Plants

The discovery that genes are naturally transferred from the bacterial pathogen *Agrobacterium tumefaciens* when it infects plants allowed a breakthrough in introducing new genes and desirable traits into crop plants. The subsequent development of other genetic modification methods means that it is now quite straightforward to introduce a gene, or set of genes, into most of the world's major crops.

Advanced breeding programmes mean that genes introduced by genetic modification can often be moved into 'elite varieties' (highly productive, and commercially successful varieties) within economically viable timeframes. A better understanding of how genes work is allowing the introduced genes to be controlled in quite sophisticated ways: switched 'on' or 'off' at a particular time during plant development or in response to environmental signals. For example, research is underway to make plants switch from using their energy supplies for 'growing tall' to storing energy in the form of starch and sugars. Other studies are aimed at getting plants to switch on the genes for defence mechanisms in response to attack by pests or diseases, or at stages of growth when they are most likely to be susceptible to attack.

Engineered Crops

The most widespread application of genetic engineering in agriculture by far is in engineered crops. Thousands of such products have been field-tested and over a dozen have been approved for commercial use. The traits most commonly introduced into crops are herbicide tolerance, insect tolerance, and virus tolerance.

Herbicide Tolerance

Herbicide tolerance allows crops to withstand otherwise lethal doses of herbicides, which are chemicals that kill plants. Some herbicides kill virtually all plants and cannot be used on crops. By offering crops tolerant to herbicides, chemical companies can expand the market for their products. And indeed, the major developers of herbicide-tolerant plants are companies with herbicides to sell. The current set of commercially available herbicide-tolerant crops is tolerant to three herbicides based on three active ingredients: bromoxynil, glyphosate, and glufosinate.

Insect Tolerance

All of the commercially available insect-tolerant plants contain a version of the toxin *Bacillus thuringiensis* (Bt), which is found in nature in soil bacteria. Bt toxins are highly effective for many pest organisms, like beetles and moth larva, but not toxic to mammals and most other non-target organisms. A major concern among farmers and environmentalists is that wide use of Bt crops will lead to the rapid development (over the course of perhaps as few as three to five years) of resistance to the toxin. If resistance develops, the Bt toxin will be useless as a pesticide. In this case, the environmental benefits of the product will be short lived.

Loss of Bt will affect those who currently use the engineered Bt crops, but also many other farmers who use Bt in its natural bacterial form, usually as a spray. These other farmers include those who grow food organically and those who use Bt as part of integrated pest management (IPM). Natural Bt sprays are a valuable mode of pest control for these farmers. Organic farmers and others who rely on Bt question whether the companies who sell the Bt crops have the right to use up this resource guided only by commercial calculations. UCS considers Bt a public good and is working to save Bt for everyone.

Virus Tolerance

The third major application of biotechnology to crops is virus tolerance. These crops contain a gene taken from a virus. By a process that is not well understood, plants that produce certain viral proteins are able to fend off infections by the viruses from which the proteins were taken. Two virus-tolerant crops are currently approved for commercial use: papaya and squash. The squash, which is resistant to two viruses, is currently off the market. Although it is difficult to get information on why products are not on the market, it is possible that the squash did not perform well enough in the field to capture market share.

Other Engineered Products

Many other genetically engineered products have been envisioned, but only a few have so far come to market.

Several tomatoes engineered to delay ripening have been approved for commercial use. In some cases, delayed ripening just prolongs shelf life. But for the FlavrSavr^{TM}, the objective was to increase the time on the vine without softening, producing a transportable, tasty winter tomato. After a highly publicized rollout, the FlavrSavr^{TM} too is off the market. The problem appears to have been with transportability rather than taste.

On the livestock side, a drug has been produced for dairy cows—Bovine Growth Hormone (BGH) or Bovine Somatotropin (BST)—by engineering a bacterium to contain the gene for the hormone. The drug is administered to cows to increase milk production, despite the chronic oversupply of milk in the United States. A highly controversial product when it was first introduced, BGH is currently used on about 10 percent of the US dairy herd.

An interesting product not related to agriculture is a rabies vaccine intended for use on wild raccoons. In this case, genetic engineering was used to construct a 'hybrid' virus made up of a component of the rabies virus inserted into an unrelated 'carrier' virus. The resulting virus confers immunity to rabies but poses no danger of causing the disease. Baits laced with the vaccine have been distributed in many parts of the eastern United States in attempts to combat rabies in wild raccoon populations. The vaccine has been approved by the USDA, despite suspicions that it has been only marginally, if at all, effective. Early studies on efficacy failed to demonstrate that the product could control rabies in wild raccoon populations. Data from more recent studies are being withheld from the public as confidential business information.

Agricultural Biotechnology Research Projects

The following lists some of the organisms being engineered by agricultural researchers and indicates which products are commercially available and which are not. For up-to-date information about what is on the market, see the most recent Gene Exchange and check the commercialization chart.

Genetically Engineered Livestock and Poultry

Animals Engineered for Leaner Meat

No livestock engineered for leaner meat is currently near commercialization. Research done early in the 1980s to genetically engineer leaner pigs failed because of unacceptable side effects, including low fertility, arthritis, and impaired immune systems. Some low level of research activity may still be underway.

Animals Engineered as Drug-Production Facilities

Status: Goats and sheep have been engineered to secrete bioactive molecules into their blood, urine, or milk. Companies are in the process developing commercial enterprises based on these animals. So far, none of the drugs is on the market. It is likely that producers will want to slaughter the animals for food after they are no longer useful for drug production.

Animals Engineered as Sources of Transplant Organs

Status: Commercial entities are engineering pigs so that their organs will not be rejected by human transplant recipients. So far, the organs are not commercially available. It is likely that producers will want to use the carcasses of donor pigs as food.

Animals Engineered for Disease Resistance

Status: Chickens and turkeys have been engineered to resist avian diseases. None have been commercialized.

Genetically Engineered Fish and Shellfish

Status: Fish and shellfish have been engineered to cause changes in hormones that accelerate growth in several laboratories. So far, none have been commercialized in the United States.

Genetically Engineered Plants Eaten Whole as Food

Status: Many plants have been commercialized, including tomatoes and squash and commodity crops like corn and soybeans. Most have been engineered for one of three traits: herbicide tolerance, insect resistance, or virus tolerance.

Genetically Engineered Fibre Plants

Status: Genetically engineered cotton has been approved for commercial use.

Engineered Insects Used in Agricultural Systems

Status: No engineered insects have been approved for commercial use. An engineered predatory mite has been field tested in Florida. Researchers have produced honeybees and other beneficial insects engineered to tolerate pesticides.

Engineered Microorganisms Used as Pesticides

Status: Several bacteria engineered to enhance their ability to kill or repel pests have been approved for commercial use. These products are used as pesticides in agricultural fields and gardens.

Food Processing Aids Made from Engineered Bacteria

Status: Bacteria have been genetically engineered to produce rennet, an enzyme important in making cheese. Genetically engineered rennet (chymosin) is approved for commercial use and widely used by US cheese processors.

Animal Drugs Made by Engineered Bacteria

Status: Bovine Growth Hormone (BGH) made from engineered bacteria is used commercially to stimulate milk production in cows. Currently this hormone is being used in about 10 percent of the nation's dairy herds.

8.10 FOOD AND BEVERAGE BIOTECHNOLOGY

Why do we need Food ?

The human body is like a machine, and requires fuel and materials to enable it to grow, repair and protect body tissues against wear and tear, and function normally. Food is composed of various chemical elements, arranged in a variety of ways to form molecules. The body uses some of the molecules (nutrients) in food to help it function correctly and stay healthy. There are many different nutrients, which have are variety of functions in the body and are vital to health. Most foods are mixtures of nutrients. No one food provides all of the nutrients our bodies need and in the correct balance, which is why we must eat a variety of foods to meet our nutritional needs.

Energy, Nutrients and Fibre

Like all machines, the human body requires a source of energy. This energy is supplied by food, and is released for use in the body after digestion and absorption. Energy can be supplied by the three main nutrients carbohydrate, fat and protein, although this is not the primary role of protein. Some nutrients, such as vitamins, minerals and trace elements, are very important to health but are required only in tiny amounts—and are thus known as micronutrients. Although not a nutrient, roughage (or dietary fibre) is another important component of our diets. It acts as a laxative, helping to keep our bowels in good working order.

How Healthy is our Diet ?

There is general consensus about the type of diet that will maximize health and minimize disease risk. The average diet in most Westernized societies is too high in fat, particularly saturated fat. Too much fat in the diet, along with other factors such as smoking and lack of exercise, increases the risk of heart disease. However, some fats contain vitamins or are essential to our health (*e.g.* the polyunsaturated fatty acids contained in oily fish). We also tend to consume too much salt (sodium) and sugar, both of which are often added to our food, either by food processors or ourselves. Too much salt can lead to high blood pressure, which in turn can cause strokes and heart attacks, whereas eating too many sugary foods is a major cause of tooth decay and can lead to obesity (and sometimes diabetes). Furthermore, sugars contain calories and no other nutrients. Many people consume excess food and, therefore, energy (calories). The excess is stored as fat in the body, and can lead to obesity. Few of us eat enough fibre and starchy carbohydrates, and possibly not enough of some vitamins and minerals, which are all essential for maintaining good health.

When we say food—what comes to mind ? Apples, carrots, bread, pizzas, chocolate? All these—whether raw, semi-processed or highly processed—are food. Because we all have to eat to survive, and because food is so important to health, what can be sold for human consumption is closely regulated. The word 'food' as used in the food industry therefore has strict meaning—perhaps surprisingly, it also includes drinks, any substance used in the production of food, and even chewing gum.

Biotechnology has already benefited the food industry; it has given us high-quality foods that are tasty, nutritious, convenient and safe. As research continues, it seems inevitable that biotechnology will have an increasing impact on the food we eat. It offers huge potential for increasing the range and quality of food available to us, particularly more nutritious, palatable and stable food. It also seems likely that it will continue to bring advantages to the processing and safety monitoring of our food supply.

Food Biotechnology : Old and New

Food Processing

There is an increasing desire among many consumers for food to be as 'natural' as possible, because such food is perceived by consumers to be more nutritious and safer to eat. However, food processing preserves food that would otherwise perish and gives us an interesting and varied choice of products. Most food nowadays undergoes some form of processing. In fact, it has been estimated that more than 80% of food sold in the UK is processed. What's more, food processing has been going on for centuries—and so too has the use of biotechnology in food processing.

Traditional Biotechnology

There cannot be many people who have not consumed bread, wine, yoghurt or cheese. These products, like many others including salami, pickles, soy sauce, vinegar, beer and cider, are made using the natural processes of living organisms (*e.g.* fermentation), in other words, using biotechnology.

Such 'traditional' biotechnology has played an important role in the development of our food supplies for thousands of years. Biotechnology was practised by our ancestors in an effort to find new ways of producing foods. Through trial and error, they learned how to breed and domesticate animals, and developed a variety of food crops. They also developed techniques, such as baking, brewing and fermentation, to process and preserve raw foods. Some traditional biotechnology techniques (such as the production of mycoprotein Quorn™ as an alternative to meat) have been developed much more recently. Nevertheless, such conventional techniques are widely accepted and generally do not cause public concern.

Modern Biotechnology

'Modern' biotechnology, in contrast with 'traditional' biotechnology, also uses the techniques of genetic modification—the joining of pieces of genetic material (DNA) in new combinations. This allows characteristics to be transferred between organisms, to give new combinations of genes and improved varieties of plants or microorganisms for use in agriculture and industry. Some consumers are concerned about the safety of using the techniques of modern biotechnology.

Application of Modern Biotechnology being used in the Food Industry

The techniques of modern biotechnology are becoming an increasingly important part of the overall effort to improve methods of food production and to increase the variety and quality of foods. Modern biotechnology has potential applications in the production of food, food processing, and also in the assurance of food quality and safety.

Although many genetically modified food crops are under development, very few have so far reached the supermarket shelf. The first genetically modified whole food to reach the shops was the FLAVR SAVR™ tomato in the USA, and the first product in the UK was tomato puree. The recent introduction of genetically modified soybean has caused controversy. Up to 60% of processed foods contain soy or soy derivatives, such as starch, protein or oils.

Modern biotechnology is also being used to improve microorganisms, processing aids (*e.g.*, enzymes) or ingredients for use in the production and processing of food. The enzyme chymosin, which is used in the manufacture of 'vegetarian cheese', is an example of this type of genetically modified product.

The first whole-food product of modern biotechnology that gone to market (in the USA in 1994) was a genetically modified tomato, the FLAVR SAVR™ tomato produced by the US company Calgene Inc. The first product to arrive in UK supermarkets, in 1996, was a tomato puree manufactured from genetically modified tomatoes and produced by Zeneca Plant Science.

Tomato Ripening

Like all soft fruit, tomatoes are usually picked while they are still under-ripe (*i.e.*, green), and then treated with ethylene (a natural ripening gas) to turn them red. This approach ensures that they remain firm during transportation and storage, thus avoiding softening and loss of product. Tomatoes turn soft, and ultimately rot, when an enzyme (polygalacturonase or pectinase) that is naturally found in the fruit breaks down the cell walls.

Genetic Modification Slows Ripening

Both of these genetically modified tomato products have been produced by switching off most of the polygalacturonase enzyme production, allowing pectin levels to remain high, and the tomatoes to soften more slowly. To achieve this, researchers removed the gene that codes for polygalacturonase, copied it and then reinserted the copy back into the plant, but backwards (a genetic modification technique termed 'anti-sense' technology), which 'switches off' the polygalacturonase enzyme. Because the rate of softening has been slowed, the modified tomatoes can remain longer on the vine to develop their full flavour (and colour), but stay firm enough to be transported to the market.

Less Waste and Improved Processing

The major advantage of the modified tomatoes is less wastage. In the case of the tomato puree, the higher level of pectin in the tomatoes means that waste is reduced, energy is saved and the need for the addition of thickeners is minimized during its manufacture. The manufacturers also claim that this product is cheaper to produce than conventional tomato purees produced from non-modified tomatoes.

'Vegetarian' Cheese

Another product of modern biotechnology to be sold in the UK is 'vegetarian' cheese. Although it does not contain a genetically modified organism, it is produced with the aid of a product of genetic modification.

Chymosin/Rennin

Cheesemaking requires the use of the enzyme chymosin (rennin) to make milk clot (by causing the protein particles in the milk to coagulate) and turn it into curds and whey. Traditionally, this enzyme is contained in rennet, which is extracted from the stomachs of calves after slaughter.

Calf Gene in Yeast

The calf gene for chymosin has been copied and inserted into yeast cells (*Kluyveromyces lactis*), which produce pure chymosin. This yeast-derived enzyme is identical to the animal enzyme, and is now routinely used in cheese production in the UK. In addition to being more acceptable to vegetarians, cheese produced using the genetically modified enzyme is of a more uniform quality; furthermore, its development has counteracted a worldwide shortage of calf rennet.

8.11 SAFETY IN BIOTECHNOLOGY

Based on international guidelines, governments in each country frame rules to ensure not only the safety of the product but also for those who conduct the Genetic Engineering experiments. In India these guidelines for carrying out the rDNA experiments were framed by the Department of Biotechnology.

When the first rDNA experiments were carried out, scientists themselves imposed a moratorium on further genetic experiments until proper guidelines were structured. In 1977, the Asilomar Conference formulated some guidelines and the major points are the following :

1. Recombinant DNA experiments should be carried out in laminar flow-chamber kept in a clean room, so that the recombinant organisms do not escape.
2. The host organism used to carry out the gene cloning or the cloning of rDNA should be specially made for such experiments. Even if the organisms escape the laboratory it will not survive. And the recombinant DNA molecule will be lost. This led to the development of special safe vectors that can be used for rDNA experiment with out fear. There are also non-pathogenic *E. coli* developed to use in the rDNA experiments as host cells.

The product in most cases of modern biotechnology will be therapeutic proteins like vaccines, hormones and enzymes. These proteins that are to be used as a drug should meet all the safety requirements and regulations imposed by the regulatory authorities. They are very strict about the data regarding the experiments on animal models, terminal patience, non-terminal patience and independent clinical trials by separate agencies. If a transgenic organism is released into the environment, in addition to the ordinary safety norms it should also monitor its environmental impact. The clinical trials of drugs produced by rDNA method have to be carried out at least in three phases.

Phase I. In the first phase of study, the side effects and dose of tolerance by the patients is systematically carried out on selected patients.

Phase II. In the second phase, collecting data from pharmacological, pharmacokinetic, metabolic and toxicological studies on a selected number of patients optimizes the use of the drug.

Phase III. In the last phase, the studies focus on the safety aspects of the drugs. During these studies the harmful effects of the drug if any is monitored in addition to its effectiveness as a drug. In dosage range, the interactions with other drugs are also investigated.

A new biotechnology product, whether it is a drug or a food material has to under go trial studies, and only after the regulatory agencies are fully satisfied by the data, they give permission or license for its mass production and marketing. If the product is a transgenic organism like plants, animals or microorganisms, it has to be cleared by the biosafety committee of the Department of Biotechnology, Ministry of Environment and Forestry, Ministry of Agriculture, and Ministry of Health before it is released to the environment or to the market.

8.12 QUALITY CONTROL IN MANUFACTURING

The ultimate aim of any good manufacturing processes is to bring a product of superior quality into the market. Therefore, quality control is of great importance in any manufacturing process.

A product has to undergo stringent quality control tests and procedures. Quality Control (QC) is included to ensure that the product is released after carrying out the relevant and necessary tests. It ensures that a product is not released for use until its quality has been judged satisfactorily. Quality control is concerned with sampling, specification, and testing as well as organization, documentation, and release procedures. Each manufacturing unit has a separate quality control unit to ensure this, which operates independently. Quality control has to take care of all parameters, which directly or indirectly affect the quality of the product.

The parameters of quality control depend on the type and the final use of the product. For example, if the product is a therapeutic protein like vaccine or hormones, it should be biologically active, pure and chemically stable. The product has to be examined for its

1. Biological activity,
2. Purity, chemical and physical properties and finally its
3. Shelf-life and stability.

The following are some of the major points to be taken into consideration to successfully incorporate quality control in a manufacturing unit.

1. The procedure of the manufacturing process should be standardized and should be approved by a research laboratory.
2. Enough trained manpower to carry out manufacturing, sampling, inspecting and testing of the materials at the different stages of the process—starting materials, intermediates and final products.
3. Proper packing with correct labelling, which includes some information about the property and handling procedures.
4. Samples should be taken in approved manner to test its qualities.
5. Records of sampling, inspection and test procedures carried out should be maintained properly and should be made available for inspection by the authorities on demand.
6. The final product should contain the active ingredients complying with the qualitative and quantitative composition of the stated formulation.
7. The products should be certified by a qualified person or by an approved committee before it is released for sale.
8. Sufficient quantity of the starting materials, intermediates and final products should be retained for future examinations if required.
9. The packaging materials should also be of the right type, which would guard the product from any degradation or change. The impact of physical parameters on the property of the product should be recorded on the label.

8.13 GOOD MANUFACTURING PRACTICES (GMP)

A standard industry is the one having a good manufacturing practices. Quality control actually forms a part of that. Since most of the products of biotechnology are directly used for human consumption, it is very important to realize these two aspects of manufacturing very stringently. A good manufacturing practice follows prescribed guidelines including the procedure for the manufacturing of a particular product intended for a specific use. It should follow a well-defined manufacturing process and should be provided with all the necessary facilities. This includes the space, suitable building and premises, equipments and instruments, in addition to the facilities for routine supply of the raw materials, storage and transport of the products. Trained manpower is an essential requirement of GMP to carry out all the processes very promptly and accurately and to keep the records of the manufacturing processes. Finally the GMP will reflect in the quality of the product. The use of the correct type of raw materials, its quality and quantity etc. very much matters the GMP, which finally determines the quality of the product. Therefore, it is very important to monitor the manufacturing processes in all its different steps very stringently to ensure the superior quality of the product.

The following are some of the important points to be observed to achieve good manufacturing practice :

(*a*) The manufacturing unit should be housed in a compound that is well protected. It should be protected from insects, animals, and migration of extraneous materials from outside to the building or from one region of the building to another.

(*b*) The building should be very strong to accommodate the machinery required for the manufacturing. It should meet all the required criteria with enough space, ventilation and provisions to meet any urgent situations like fire.

(*c*) The building should be constructed in such a way that it should guard the machinery, raw materials and products from contamination.

(*d*) Temperature and humidity should be controlled to protect both raw materials and products.

(*e*) There should be good sanitary practices. The toilet, resting room, dress-changing room etc. should be separated from the premises of manufacturing areas.

(*f*) The production area should be well maintained in terms of sterility and hygienity. The flow of personnel in the production area should be controlled to minimize contamination.

(*g*) The sophisticated instruments and analytical equipments should be well maintained and should be protected from vibrations, electrical impacts, moisture, heat etc. If necessary separate rooms should be provided in such situations.

(*h*) Utilities and support systems like sterile water, compressed gas, nitrogen and liquid nitrogen; etc. should be of standard quality and should be supplied promptly.

(*i*) Storage and distribution of the product should be well organized. Well-maintained storage and packing facility is essential to prevent contamination and mixing of products.

(*j*) To carry out all these, the factory and its premises should be well maintained in a very good working state.

8.14 GOOD LABORATORY PRACTICES (GLP)

To carry out research projects, experiments including trial experiments, some basic guidelines and criteria are to be followed by laboratories. This is applicable to manufacturing establishments also. These guidelines are necessary to ensure the safety of the environment, the public and the researcher. Moreover it is a must for the success of the experiments. There should be committees constituted as per the directions of the administrative authorities (For example Department of Biotechnology, Government of India) within the institute or the research establishments to monitor the safety norms and ethics.

There are four components to Good Laboratory Practices :

1. *Management*: The administrative authority that coordinates the implementing and monitoring activity of different committees and organizations to watch the safety and ethical issues.
2. *Quality Assurance*: This is an internal body or committee ensuring the practice of GLP within the laboratory. Animal ethics committee, bioethics committee and biosafety committee are some examples.
3. *Study Director*: The person who is responsible for the overall conduct of the safety and study committees.
4. *National Compliance Monitoring Authority*: The approval of safety studies related to health and environment conducted on a new biotechnology product by this authority is necessary for international recognition and mutual acceptance. A guideline or framework is provided by International Organization of Economic Co-operation and Development (OECD) to the National Compliance Monitoring Institution regarding GLP for product development and marketing.

The GLP related to the product development and its health and environment safety are compulsory for testing any product before it goes into the market for human consumption. The studies and its data will be assessed by a National Regulatory Authority for the purpose of giving license or registration. These regulations in the trial studies can prevent scientific misconduct and fraud, which can affect the safety of man and environment.

Good Laboratory Practice for Students

Certain code of conduct is to be followed in a laboratory to ensure maximum safety. Every person has a responsibility towards the health and safety of himself/herself and of all other persons who may be affected by his/her acts or omissions. As all laboratories are potentially dangerous, it is essential that the students should follow the safety instructions strictly. Eating and drinking in laboratory areas should be prohibited at all times. Students are advised to wear laboratory coats whenever they are in the laboratory.

Use fume cupboards and personal protective equipments (gloves, goggles etc.) when advised to do so by a teacher or supervisor, if the work involves the use of hazardous materials like radioactive chemicals, pathogenic organisms or UV radiations.

If you are going to use unfamiliar equipment, try to understand how it works with the help of the instruction manual or seek the help of your supervisor or a technician. Look out for instructions or warnings labelled on the equipment. Students should plan their experiments in such a way that it can be carried out during normal working hours. Avoid working alone.

Keep your work within the bench area allotted. Do not leave equipment such as pH meters, laminar flow cabinets, fume cupboards etc. in an untidy state: Clean all equipments, including balances after use. You are responsible for keeping your laboratory clean.

All ways avoid mouth pipetting; use filling devices (rubber bulbs, Pi-pumps) or transfer pipettes. Rinse empty glassware as soon as possible after use and remove to the designated location for washing.

Pay particular attention to the removal of corrosive or toxic chemicals, plant material, microbial cultures, soil, agar etc. and their proper disposal. Make sure all chemicals and media are clearly and unambiguously labelled.

Clean up spillages immediately using the appropriate procedure. For example, if it is a liquid, mop with cloth or paper towel. Do not try to suck them up using pipettes etc. In the case of strong acids or alkalis, flush with water, neutralize and then mop it up. In the case of solid chemicals, wipe with damp cloth or paper towel. Never blow or brush the dusty spills or other solid chemicals. This can only increase the risk from hazardous materials.

Do not put paper towels used for cleaning up chemical spillages into waste paper bins where cleaning staff may come into contact with the harmful substances. Use black waste sacks for disposing such waste material.

Students will need to keep a lab notebook in order to write down procedures, data, results, observations, and any good ideas that pop into your mind about what you are doing in lab. Keep your notes organized because you will be able to use it later for reference.

8.15 MARKETING

Marketing high-technology products and innovations like that of biotechnology is not the same as marketing more traditional products and services. The marketing of a familiar product is very different than marketing products with which customers are unfamiliar. The marketing of Biotechnology products is also depended on a number of factors as that of the normal products. Therefore, the normal criteria of marketing are applicable in this case also. Introducing a new product particularly an unconventional products like that of modern biotechnology is very difficult than making that new product. Marketing is the process of planning and executing the conception, pricing, promotion, and distribution of ideas, goods, and services to create exchanges that satisfy individual and organizational objectives.

In 1995, sales of products derived from recombinant DNA-based technologies approached $8 billion, less than 4% of worldwide pharmaceutical sales. Modern biotechnology products start appearing in the market only in the late 1980s. Now there are a lot of products of modern biotechnology are available in the national as well as international market.

The industry has its own strategies for getting into and competing in the market. New technologies and its demonstration among the consumers like controlled drug delivery technologies, targeted drug delivery technologies etc. and specialized, more patient-friendly technologies to deliver protein-based drugs are vital to the future marketing success of protein based drugs and other biotechnology products. Because :

1. Proteins are erratically absorbed or are inactivated when swallowed, thus they tend to be injected.
2. Proteins are generally unstable and reactive.
3. Most proteins have a short life in plasma after injection.

TABLE 8.4

No.	Name of the drug	In market	Under trial/development
1.	Vaccines	8	68
2.	Monoclonal antibodies	16	70
3.	Gene therapy	0	44
4.	Growth factor	4	26
5.	Interferons	7	21
6.	Interleukins	1	18
7.	Recombinant human proteins	12	8
8.	Human growth hormomes	16	14
9.	Clotting factors	5	6
10.	Colony-stimulating factors	2	4
11.	Erythroproteins	2	4
12.	Others	20	48

A combined survey conducted by Biotechnology Industry Organization and Pharmaceutical and Manufacturers of America.

Some of the technologies that could appear in the future markets are the following. Insulin could be delivered by nasal sprays, or orally, self-regulated implants, artificial pancreases with monoclonal insulin-producing cells. Protein drug doses could be administered in very small, controlled delivery systems over days, months, or even years on a single re-filling. In the case of drug targeting, drug goes only to infected cells and not systemic blood stream. The use of liposomes as drug containers or vehicle to carry toxic solutions, labelling liposomes with antibodies or surface reactive chemicals will lead to better targeting and tissue-selective delivery systems. New technologies in the production of drugs by biotechnology can reduce the cost of production of drugs and other products, and then it will become more consumers-friendly. Table 8.4 gives some of the therapeutic proteins and vaccines that already striked the market by the 2002.

• REVIEW QUESTIONS •

1. What is a bioreactor ?
2. What are the different steps of bioprocess technology development ?
3. What are the factors that influence the designing of a fermentor ?
4. Explain the importance of biotechnology in food and agriculture.
5. Explain the contribution made by biotechnology and molecular biology in medicine.
6. Describe Good Laboratory Practice for Industry and Research Laboratories.
7. Which is the Government Organization that controls and promotes the Research Development (R & D) activities in Biotechnology in India ?
8. Explain the role of biotechnology in the production of fuel.

9. What is single cell protein ? Explain how can it be related to environmental biotechnology ?
10. Explain the use of cell culture in industrial biotechnology. Explain the use of plant cell and tissue culture in agriculture.

• SUGGESTED READING •

1. People's Perception of Biotechnology, Renato Schibeci, Ian Barns.
2. Biotechnology and Ethics: A Blueprint for the Future, Daniel Callahan President, Hastings Center, Center for Biotechnology, Northwestern University.
3. Social Issues in Science and Technology: An Encyclopedia, David E. Newton (ABC-CLIO, Santa Barbara), 1999.
4. Biotechnology: Issues, Ethics, and Regulations, Tina M. Prow, Communications Specialist, Office of Agricultural Communications and Education.
5. Biotechnologies and the Public: An International Study of Policy, Media Coverage and Public Attitudes from 1973 to 1996 (1995-1998), Helge Torqersen, Institute of Technology Assessment.
6. Patenting in Biotechnology—Part I, R. Stephen Crespi, Tibtech, Vol. 9, 117-122, 1991.
7. A Passion for DNA: Genes, Genomes, and Society, James D. Watson (Cold Spring Harbor Laboratory Press, New York).
8. Plant Biotechnology: Facts and Public Perception, D. Boulter, Department of Biological Sciences, University of Durham, South Road, Durham DH1 3LE, U.K. *'Phytochemistry' (Vol. 40, No. 1, pp. 1-9, 1995).*
9. Public Attitudes to Genetically Engineered Products, Wendy Ross, Katy Marsh, Alexi Jackson, Jaqui Skoyles, (1998), John Innes Centre, Norwich, U.K.

9 BIOTECHNOLOGY AND SOCIETY

9.1 PUBLIC PERCEPTION OF BIOTECHNOLOGY

Science and Society

Our perceptions or attitudes towards things are not always rational and are often culturally influenced. They are a combination of thoughts or the cognitive dimension, feelings or the affective dimension, and the way we react or the behavioural dimension. The cognitive dimension consists of things we know, the affective dimension of things we feel, and the behavioural dimension is how we will act on the attitudes we build. Attitudes help us to become socially acceptable, belonging to a group is very important, and it gives meaning to things we experience.

The advancement in science and technology made our life very simple and fast. At the same time some of them has caused great concern regarding their long-term impacts on environment and life. The World Commission on Environment and Development (WCED) also known as Brundtland Commission appointed by United Nations (UN) in 1985 recommended sustainable development preserving environment without any degradation. The Commission defined the sustainable development as "the development that meets the needs of the present without compromising the ability of future generations to meet their own needs". There are a lot of definitions and views regarding the sustainable development for different nations. In all these views the common point in which all nations agreeing is that science and technology is portrayed as a double-edged sword.

There are two opposing ideologies regarding the use of science and technology.

1. Holistic Ideology
2. Reductionistic Ideology

The holistic ideology recommends the use of traditional methods in all fronts of life from agriculture to industry. They argue that the modern society gives more importance to formal knowledge neglecting the informal and traditional types of knowledge. According to them there is no problem in the agriculture sector. Earth can produce sufficient food materials for the entire population, if we could return to consume some of the old or traditional grains, which can be easily grown. They also recommend the use of the chemical manures, pesticides, herbicides and use minimum tillage to conserve the land and to produce a crop. Whatever problem persists in the world is mainly due to the unequal distribution of food that is produced.

The reductionisms argue for the use of the new knowledge coming out of scientific discoveries and researches and put them in application to improve the quality of agriculture and crop plants. They recommend the continuous research and development studies to find out new solutions to the problems that are coming to the scene time to time.

Both these groups and their variant types have given different views regarding the sustainable development with minimum impact on environment.

Certain specific events can change people's perceptions, especially if it is transmitted through the mass media (television, radio, the press), in a more socially acceptable form or to reflect the biases of the mass media. Thus, the media may distort the facts. This distortion means the media image of anything from biotechnology to bee-keeping usually differs from the reality. This means that the image of biotechnology can be distorted. However, it is also important to remember that if any new information contradicts existing beliefs and practices, we are likely to disregard the new information rather than our beliefs.

The Impact of Biotechnology on Society

At first glance, biotechnology may seem like simply a science and technology matter. Yet, biotechnology is valued and feared because of its potential impact on environment and our lives. Because of its power to effect social values, social relations and environmental problems, biotechnology has become a lightening rod for the debate between competing socioethical perspectives.

There are significant differences in public perception of biotechnology according to country, sex, age, educational level, religious practices, and social groups. On the whole, attitudes are not nearly as positive as the industry would like them to be.

There are two things, which dominate the public's perception of biotechnology: lack of trust and ignorance of science. Ignorance has both advantage and disadvantages. It is difficult to feel you can argue effectively with someone about a topic they do not know about in the first place. Ignorance means that many people will, have anxieties about the technology, which are not focused but which are easily linked to much more sinister issues. Unless people can understand the issues, their perceptions will easily be influenced by arguments that say that all biotechnology is morally dangerous, simply because of their in-built fear of the unknown.

For example, it was observed in an old survey conducted that, only 34% of Briton's and 46% of Americans appeared to know that the earth goes around the sun once a year, and 23% of Britons and 25% of Americans knew that antibiotics are ineffective against viruses. If this is the state of the developed nations, what will be condition of India where more than 50% of the population is illiterate? How can we change levels of public ignorance ? In order to raise awareness amongst young people, food biotechnology and its implication could be given greater status in The National Curriculum, and could be discussed via General Studies programmes. It could also be offered as a broadening element at universities and colleges. In order to educate the present, older public, who are not in full-time education, an independent report should be written and published in regional languages. It could be communicated via the media, television, leaflets, information packs and advertising.

The development of modern technologies is often the subject of heated and polarized debate. At first glance, biotechnology may seem like simply a science and technology matter. Yet, biotechnology is valued and feared because of its potential impact on our lives. Because of its power to effect social values and social relations, biotechnology has become a lightening rod for the debate between competing socioethical perspectives. It can constantly change our social values and restructure the social relations. Because of this there are several religious and social groups who view the developments in biotechnology in a very different perceptions. For example, two important socially and ethically important issues are :

1. **Should man be allowed to play God** ? Should he be allowed to genetically alter organism by genetic engineering: clone animals and man ?
2. **Who is the owner of the technology and products** ? In the case of genetic engineering, who is the owner of the genetically altered new organism ? In the case of cloning, who are the parents of the cloned organism? Even in the case of test tube babies, who should be considered as the real mother—the women who is giving birth to the child or the contributor of ovum.

In debates the public, scientists, governments and religious groups should discuss the social implications of biotechnology as a science, technology and a fast developing industry. On one hand we want to address the obstacles to the successful development and commercialization of biotechnology products. They should develop a program that includes activities to harmonize and clarify biotechnology regulations, to understand consumer attitudes to biotechnology, and to understand the social and ethical concerns raised by this technology. They should analyze realize the competing social and ethical perspectives of biotechnology, and to discuss the competing pressures on governments to address socioethical and environmental issues.

Biotechnology is a method of production the applications of which will impact on virtually every aspect of human life. Both the method and its applications have socioethical implications. There is significant disagreement over whether biotechnology, and its many applications, will benefit or harm society. Biotechnology applications are so diverse that each instance must be judged individually. Each application of biotechnology should be seen as the actualization of a specific preference for how society develops.

There are mainly three central levels of disagreement. First, there is disagreement over forecasts of the social consequences of this technology—what could happen. Second, there is disagreement over the ethics that should direct the consequences—what should happen. And finally, there is disagreement over the mechanism, which will be used to develop the ethics—who decides what should and will happen? The debate over the development of biotechnology is, in fact, a debate over the way our social relations and social values will develop. Biotechnology is a subject of debate because, as a method and in its applications, it will affect our social values and our social relations.

The application of biotechnology in society raises many social, ethical and legal questions. To date, in many countries did not develop any system or mechanism to manage the social and ethical issues raised by biotechnology. Instead, government allocation of public moneys to research and development has determined the social consequences of biotechnology.

While public debate revolves around specific applications of the technology, such as BGH, new reproductive technologies, the patenting of higher life forms and the release of GMOs into the environment, under the debate on any specific application are still unanswered questions. The commercialization process for biotechnology products has been uncertain and prolonged as products are forced to confront socioethical issues in hearings and parliamentary committees. Industrial development depends on the creation of a publicly derived socioethical framework to direct biotechnology's development.

If the ultimate goal is the successful commercialization of biotechnology, public support is crucial. Public support will, in turn, rest on the ethical social applications of biotechnology. The ethics that will guide the social consequences of biotechnology must be representative of the public. To date, no organized investigation has been undertaken to explore the social and ethical guidelines the public would like to manage the development of biotechnology.

Public understanding of biotechnology as a science and technology is important because the products of biotechnology—and consequent benefits and risks—are going to affect everyone. Biotechnology holds great promise as a tool to preserve and enhance environmental quality, food problems and human health. It can create environmentally safe methods and technologies to address these problems that affect human and his environment. But without public understanding, acceptance, and support, the role that biotechnology could play in solving environmental and food production problems could be stymied.

9.2 PATENTING (INTELLECTUAL PROPERTY RIGHTS–IPR)

Intellectual property right is not a new concept in biotechnology and in science. It is a product of industrial world. It became famous among scientists and laymen equally due to the advancement in the scientific researches and marketing of its products. Some of the concerns about modern biotechnology have focused on the nature, impact and legitimacy of IP rights as they are applied to gene technology, biodiversity and to inventions that draw on genetic resources and associated traditional knowledge. Now most of the institutes and universities have their own expert people or groups to deal with their own IP problems.

Intellectual Property Rights is a collective term given to a number of different types of legal rights granted by each country. It can be considered as recognition of the contribution or the inventor by the country to the development of a new technology, process or product. It is protected against the unauthorized exploitation for the industrial or commercial purpose.

Types of Intellectual Property Rights

There different types of intellectual properties. It is mainly classified into five categories. Patents, trademarks, industrial designs, geographic indications of sources, plant species protection, and copyright.

These right are protected in each countries by the laws enacted by the parliament or the governing body of the country.

1. **Patents.** This gives protection for inventions, manufacturing process, or products. It gives its owner a monopoly right to the commercial utilization of the process or product for a limited period of time.

2. **Industrial design.** It relates to the shape and design of certain products like ornaments and instruments.
3. **Trademark.** This relates to the distinctive words or symbols applied to products, services or companies. Example—Agmark in food products, Tata symbol etc.
4. **Protection to plant varieties.** It is a type of IPR coming under biotechnology. It is known as plant variety right or plant breeder's right. It is given to new varieties of plant produced by hybridization procedures or by other genetic modifications.
5. **Copyright.** This IPR is for literary and artistic works and craftsmanship, engineering drawings and software.

9.3 PATENTS

This is a system of laws giving protection to inventors and investors involved in an invention to protect the creation of the inventor for a period of twenty years against use for profit without his /her consent.

The Criterion for Patents

For an invention of product or process to be patented, it should fulfill the following criteria :

- The invention or the information should have novelty. It should not be already be in the possession of public.
- It should be inventive. It should not be a traditional technique or product. The technique or process should not be obvious to person skilled in the particular art.
- It should be capable of industrial application.
- It should have process utility.

The economists have accepted the concept of patent as an encouragement for :

- Economic development through scientific discoveries.
- High investments in research and developments.
- Investment in the production and marketing of new products and processes.
- Revelation of knowledge and technological information on inventions as against secrecy.

Discovery *vs* Invention

There is a sharp difference between discovery and invention. Discovery primarily relates to the acquiring new knowledge by experiments, investigation and thinking. It is of intellectual in nature. The term itself indicates that it is the finding of a new principle, theory or knowledge, which is already existing or concealed in nature. It adds to the existing human knowledge. Example: Discovery of gravitation force by Sir Issac Newton, elucidation of genetic code by Hargobind Khorana etc.

Invention is actually the development of something new, which is not pre-existing. There is the application of pre-existing knowledge established by discovery. For example, electricity is a product of discovery, whereas electric bulb is an invention. Similarly, the knowledge about the power of steam is a discovery and the steam engine is an invention. The techniques of gene cloning, PCR amplification, etc. are inventions. Invention also gives out new knowledge or information. A new procedure which resulted in an old product, a known procedure resulted in a new product, a new result or a new process or a new combination of material to produce a known product etc.

Generally speaking an invention is patentable and discovery is not patentable. Now there arise confusion about materials that preexisted and new one. For example, a synthetic product is a new product, which was not existed previously. So it is definitely patentable. But when you consider a specific protein or gene or any other natural compound, which is isolated or purified. It is not a new compound, but it is patentable—why? The reason is that these proteins or natural products never existed in previous times in isolated form or purified form. Similarly, a new strain of microorganism isolated from nature will qualify for patenting if it exhibits a new property that is lacking in the previously known natural form of the bacteria. The patent offices of US, Europe and Japan have cleared these points through a joint statement of policy.

Product and Process Patent

A new product, which has been synthesized or isolated and purified from natural sources, is patentable according to the patent laws of many countries. Now the question is whether a process is patentable or not. A new method or procedure or process is used for the synthesis, isolation or purification of a known material—a compound or a microorganism is patentable. The normal procedure is not patentable. If the product is a new one and the procedure used is a known procedure, then the procedure or the process is not patentable, only the product. If both process and product are new one then both are patentable.

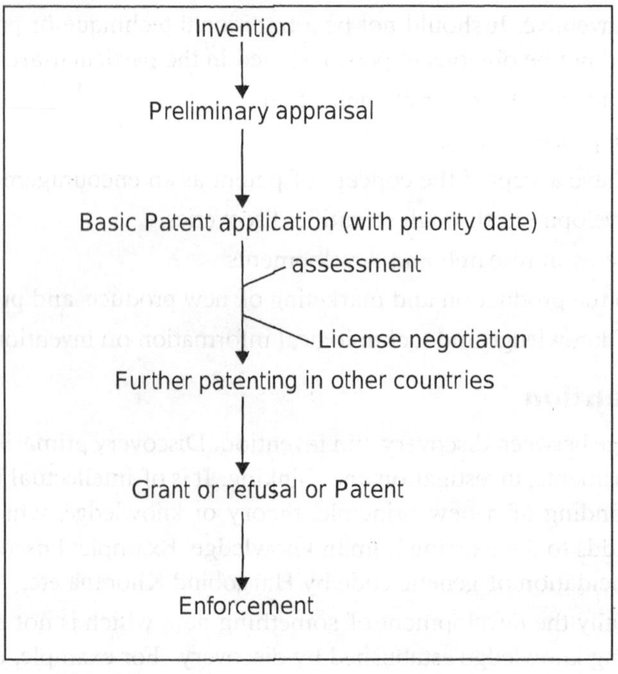

Figure 9.1 The procedure of patenting.

Reading a Patent

Patent document has a specific format as per the international guidelines. The patent structure has two parts :

1. Description and
2. Claims.

Description. This technical part of the patent contains the following :

1. Field of invention and its objective.
2. Previous attempts to meet the objective, if any, with the results.
3. The remaining part of the problem to be solved.
4. The inventive solution and how to apply it.

The solution consists of a general discussion of the different parameters of the process and products with possible variations and substitutions. It also will have the description of some worked examples.

Claims. This part is the legal component of the patent. This part defines the scope of legal protection required in terms of apparatus, process, products, use and any other suitable part or category.

Patent Strategy

When any body invents something, the inventor first apply for a patent for that product or process in his country of residence and then he can file an application for patent in other countries. The patent application should have a clear title. The novelty of the invention should be checked thoroughly. The filing of application and all other procedures has to be through a patent attorney.

9.4 NATIONAL AND INTERNATIONAL PATENT LAWS

International Patent Scenario

Each country have their own specific type of patent law as per an international agreement or treaty or individually. But it has got a tradition of a strong international cooperation over a century by means of international conventions. These conventions make agreements and consensus regarding the formal and substantive patent matters between member states. Some of these meetings that shaped international patent laws are described below.

Paris Convention. This convention was held in 1883. It was also known as 'The International Convention for the Protection of Industrial Property'. Now there are 151 states in the Paris Union, and include the great majority of industrial countries. Member countries must treat nationals of other member countries of the union on equal terms with their own nationals as regards the protection of industrial property. There are many understanding and agreements between the member states with regards to patent laws and patent application. The text of Paris convention has been modified several times and the last one was in Stockholm (1979).

Strasbourg Convention. This was held in 1963. This is the "Convention on the Unification of Certain Points of Substantive Law on Patents for Invention". This defines the common requirements for the patentability of an invention such as that it must be amenable to industrial application, must be novel, and must involve an inventive step etc. many of these features of this convention have been incorporated into the European Patent Convention. For example, the definition of 'the state of the art' against which the degree of novelty and inventiveness of the subject matter of a patent application must assessed and judged. The exclusion of plant varieties and animal varieties from patent protection also decided in this convention.

Patent Co-operation Treaty (PCT). This is considered as the international patent body because it is administered by the World Intellectual Property Organization (WIPO) based in Geneva. This treaty was actually signed in 1970 and came into force in 1978 along with the European patent convention. It is the broadest international body and it had 100 member states. The patent applications filed under PCT are considered as 'international' because an international body (WIPO) initially processes them in an international phase before being formally introduced into designated national system. The international phase is concerned with the formal preliminaries, a prior art search and publication of the application.

European Patent Convention (EPC). This was held in 1973 and came into force in 1978. It established the European Patent Organization along with European Patent Office (EPO) and Administrative Council as its supportive organs. Each member country has its own regional Patent Office. A single patent application can be filed before EPO, which will be considered for its member states. The application may be made in any of the regional offices or the national Patent Offices, but will be examined by EPO in due course. When the patent is granted it will not be single patent as European patent, but will be a bundle of national patents individually for each member states. For example, European Patent (UK), European Patent (Germany), European Patent (France) etc. Thus a single patent application filed through EPO will develop into a single unitary indivisible object of property covering the whole of European Economic Community.

Budapest Treaty. This treaty known as "The Budapest Treaty on the International Recognition of the Deposit of Microorganism for the purposes of Patent Procedure" was signed in 1977 and came into force at the end of 1980. This treaty gives recognition for the microbial culture collections of the International Depository Authorities. A new strain of microorganism can be deposited with any of the International Depositories for the purpose of a patent application in any member state.

The main factions of a microbial culture collection are to accept, maintain and provide microbial cultures as a service to scientific community. They accept new strains of microorganisms isolated by scientist and its identity will be verified and properly classified. Then it is deposited into their culture collections and is maintained as a new strain. These depositories act as a source of microorganisms for use in scientific and industrial purposes. The new organism which is deposited with the culture collection are allotted an accession number and date by which the strain is identified and referred in future and in patent applications. Therefore, now it is a prerequisite that the microorganism should be deposited with the international culture collection and get an accession number and date for filing a patent application.

National Patent Laws

The following are the important developments in Indian scenario regarding the patent systems :

1. India joined the Paris Convention and Patent Cooperation Treaty (PCT) in December 7th, 1980.
2. The government of India has notified 130 countries including 20 non-Paris convention countries as convention countries for the purpose of claiming priority on 28th September 1998.

3. Similarly, Government of India notified 150 Paris convention countries as convention countries for the purpose of claiming priority in respect of patents and trademarks on 30th November 1998 and 3rd December 1998.

There are a number of advantages for India by joining the convention and the treaty advantages of joining the Paris Convention are :

- Reciprocity with 150 countries for patent and trademarks.
- It is a prerequisite for joining other treaties and conventions.
- India can attract more foreign investments, collaborations and joint ventures, and facilitation of technology transfer.

Advantages of joining Patent Cooperation Treaty (PCT) are :

- A single application for international patent in designated member countries.
- A centralized preliminary examination and international search.
- Time saving in getting international patents at least in the member countries.
- Saving in expenditure by dropping the member countries while filing a domestic patent application.
- There is a reduction 75% in fees for individual inventors of developing countries.

There are some points to remember regarding Indian patent system :

1. Living organisms are not patentable in India.
2. India is not granting any patents for software.

   But Indian can file PCT application for biotechnological inventions including living organism and software.
3. If an inventor is granted a patent for his invention in India it is not true that he will be protected in PCT and Paris convention.

9.5 PATENTING IN BIOTECHNOLOGY

Biotechnology in one form or another has been part of human development since the dawn of agriculture. Human ingenuity has led to increased production and greater diversity and quality of livestock and varieties of crops. Today's food crops and domestic animals embody the benefits of many generations of selection and breeding.

Biotechnology continues to offer considerable potential for enhancing human health and well being. Modern biotechnology, including gene technology, is finding increasing application in healthcare and in a host of industrial and agricultural industries. Effectively applied, modern biotechnology may contribute to economic growth, technological development and human welfare. Yet it has also raised concerns about ethical and moral issues, equitable sharing of the benefits of biotechnology, environmental impact, the accelerated pace of change and the regulatory challenges. Intellectual Property (IP) rights are not new in the biotechnology domain, but some of the concerns about modern biotechnology have focused on the nature, impact and legitimacy of IP rights as they are applied to gene technology and to inventions that draw on genetic resources and associated traditional knowledge.

Biotechnology patent law operates under the same general legal principles as other areas of patent law. However, the nature of biotechnology raises many unique patent law issues. Biotechnology law is becoming more complex in all industrialized nations of the world. Long lasting, broad, international patent protection is vital to ensure a financial return after an invention travels the long pipeline from lab to marketplace. The complexity in biotechnology patenting is that it involves patenting of living organisms and natural products, which are not patentable normally, in addition to the products and processes.

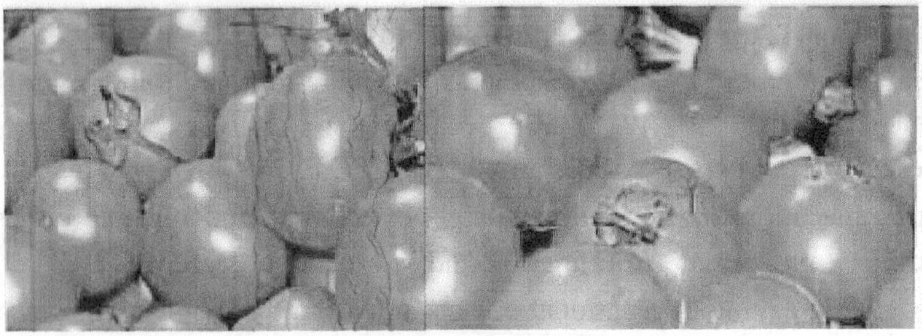

Figure 9.2 GM tomatoes showing prolonged shelf life.

For example, a gene that causes a fatal neurological disease and another gene that dramatically improves crop yields. Even though genes exist naturally and in that sense it is not patentable. It is not occurring in isolated and pure form. In this sense genes and other naturally occurring materials are patentable and commercialize these inventions. The recent awareness about patent and its elaborated studies is an outcome of biotechnological inventions.

Examples of Inventions that may be Patented

1. **Products**
    - Genes—including modified genes, expression vectors and probes;
    - Proteins—including modified proteins, monoclonal antibodies and receptors;
    - Other chemical compounds such as small organic molecules;
    - Cells and new strains or genetically modified organisms.
2. **Methods/Processes**
    - A new and non-obvious method of using a known compound;
    - Methods of medical treatment are not directly patentable in many countries, but they are patentable in the United States. However, methods of medical treatment can be protected indirectly.
3. **Compositions**
    - Formulation of products for specific uses.
4. **New Uses**
    - New uses for an previously known compound
    - Observing new abilities of known microorganisms for the synthesis of new compounds or known compounds like antibiotics.

5. **New methods of treatment or diagnosis**
   - New treatment or diagnosis procedures or methods for instruments, animals, plants and human.

Complex life forms such as genetically engineered plants and animals are not patentable in India. Plant varieties produced through traditional cross-breeding techniques may be protected under the laws of Plant Breeder's Rights.

How do you apply ?

The application process is the same as that for other patent applications. Biotechnology patenting includes a few unique twists. For example, applicants may have to make a cell or seed deposit with a culture collection before filing an application.

What protection can you get ?

The *Patent Act* gives biotechnology patents a maximum term of 20 years from the filing date (patent applications filed before October 1, 1989 have a term of 17 years from the issue date). Patent term extensions are not available in India.

How can the lawyers help in biotechnology patenting ?

The lawyers should have expertise in :

- Patenting strategy.
- Contracts: licenses, biological material transfer contracts, confidentiality agreements and sponsored research agreements.
- Litigation and dispute resolution.
- Designing around others' patents.
- Regulatory approvals of new drugs (including biologics) and foods.

9.6 VARIETAL PROTECTION

In conventional plant breeding experiments for the production of desirable hybrids, there is the plant breeders right and farmer's right. The plant breeder's right implies the protection of the new variety that has been created by the method of breeding procedures and selection. Farmer's right means that farmers have the right to produce the seeds from a hybrid plant and raise the seedling for their agricultural purpose.

In the year 2001, there were two major steps taken to protect the farmer's right in relation to the breeding of new varieties of crops. They are :

1. In the FAO's international treaty on plant genetic resources for food and agriculture there is a specific clause concerning the operation methods of farmer's right.
2. The Protection of Plant Varieties and Farmers Act, 2001 gives concurrent attention to the rights of farmers, plant breeders, and researchers and also the protection of public interests. This include :

   (*a*) Compulsory licensing of rights

(*b*) Prevention of import of varieties incorporated with Genetic Use Restriction Technology (GURT) like the seeds with terminator gene (the terminator technology of Monsanto).

In this technology, the plant are genetically engineered with a gene known as terminator gene, that make the seeds non viable and will not germinate. Therefore, at each time of cultivation the farmers have to purchase the seeds from the companies, which produce the seedlings by tissue culture technology.

9.7 ETHICAL ISSUES IN BIOTECHNOLOGY—AGRICULTURE AND HEALTH CARE

Government's promotional activities surrounding biotechnology have inspired debate. The use of public funds to further biotechnology research supports the use of a method which has social implications and around which there exists significant social disagreement. Government's decision to allocate funds to certain niche areas determines the applications, which are most likely to be commercialized. This history has established a relationship between government and the public. It has created trust and support among some, and distrust and opposition among others because of the socio-ethical position implied in the policy.

Using public resources for biotechnology's development raises objections from stakeholders who do not believe this technology will benefit society, or who believe that its current course of development poses socio-ethical risks. The use of public resources for the development of biotechnology is supported by biotech industry who believe that this technology offers significant health, environmental and economic benefits to public, and who believe that facilitating the development of this industry for those benefits is a government responsibility.

The key questions that arise when ethical issues are raised are :

1. Should government decide the ethics ? And
2. What should those ethics be ?

The current debate on ethical issues related to biotechnology in agriculture and health care has raised certain debating points in the following fields.

1. Release of genetically modified plants and animals to the environment could cause disturbance in the existing ecological balance in an unpredictable manner.
2. The release or use of genetically engineered microorganisms for industrial purposes can leads to the generation of new infectious types of organisms.
3. Introduction of artificial agents (such as genetically engineered bacteria, fish etc.) as biological control agents can cause serious imbalance in the environment and can lead to disastrous consequences.
4. Genetic engineering of plants for herbicide resistance and enhanced photosynthesis could result in more tolerant weeds, as a result of cross-pollination with related plant species.
5. Cloning of animals and human created a lot debate on ethical issues. There are also the social values that will be affected by biotechnology. For example, the social and ethical problems associated with new Reproductive Technologies, which includes the *in vitro* fertilization and human cloning. There are also technological risks associated with human and animal cloning. There is uncertainty with the current state of technology.

6. The new reproductive technologies may harmful for the individual autonomy, equality, protection of the vulnerable, accountability, respect for human life and dignity, non-commercialization of reproduction, appropriate use of resources, and balancing individual and collective interests.

Proponents of biotechnology argue, correctly, that biotechnology is being held to standards that are not demanded of other advanced technologies such as computers or the new information highway. There is considerable debate over whether biotechnology should be held to distinct standards because it is a power that raises special social questions. What is unquestionable is that biotechnology requires considerably more public trust than almost any other technology because of its potential power to transcend, quickly and intentionally, any God-given or natural limits to human activity.

• REVIEW QUESTIONS •

1. What is the public perception about biotechnology ?
2. What is meant by sustainable development ?
3. What is IP and IPR ?
4. What are the various types of IPR in India ?
5. What is a Patent ? What are its advantages in scientific studies and industries ?
6. What is the meaning of Farmer's Right and Plant breeder's Right ?
7. Give the general structure of a Patent. What are the criterions for a finding to be patented ?
8. What are the special problems in the patent laws of biotechnology ?
9. What is the justification for the patenting of natural materials like microbial strains, molecules like proteins, nucleic acids and genes ?
10. Expand the following :
    (*a*) PCT
    (*b*) GM Plants
    (*c*) GURT
    (*d*) WTO
    (*e*) EPC.

• SUGGESTED READING •

1. People's Perception of Biotechnology, Renato Schibeci, Ian Barns.
2. Biotechnology and Ethics: A Blueprint for the Future, Daniel Callahan President, Hastings Center, Center for Biotechnology, Northwestern University.
3. Social Issues in Science and Technology: An Encyclopedia, David E. Newton (ABC-CLIO, Santa Barbara), 1999.
4. Biotechnology: Issues, Ethics, and Regulations, Tina M. Prow, Communications Specialist, Office of Agricultural Communications and Education.

5. Biotechnologies and the Public: An International Study of Policy, Media Coverage and Public Attitudes from 1973 to 1996 (1995-1998), Helge Torqersen, Institute of Technology Assessment.
6. Patenting in Biotechnology—Part I, R. Stephen Crespi, Tibtech, Vol.9, 117-122, 1991.
7. A Passion for DNA: Genes, Genomes, and Society, James D. Watson (Cold Spring Harbor Laboratory Press, New York).
8. Plant Biotechnology: Facts and Public Perception, D. Boulter, Department of Biological Sciences, University of Durham, South Road, Durham DH1 3LE, U.K. *'Phytochemistry' (Vol. 40, No. 1, pp. 1-9, 1995).*
9. Public Attitudes to Genetically Engineered Products, Wendy Ross, Katy Marsh, Alexi Jackson, Jaqui Skoyles , (1998), John Innes Centre, Norwich, U.K.